Signals and Communication Technology

More information about this series at http://www.springer.com/series/4748

Frank Nielsen · Frank Critchley
Christopher T.J. Dodson
Editors

Computational Information Geometry

For Image and Signal Processing

Springer

Editors
Frank Nielsen
Laboratoire d'Informatique (LIX)
Ecole Polytechnique
Palaiseau
France

and

Sony Computer Science Laboratories, Inc.
Tokyo
Japan

Frank Critchley
School of Mathematics and Statistics
The Open University
Milton Keynes
UK

Christopher T.J. Dodson
Department of Mathematics
University of Manchester
Manchester
UK

ISSN 1860-4862 ISSN 1860-4870 (electronic)
Signals and Communication Technology
ISBN 978-3-319-47056-6 ISBN 978-3-319-47058-0 (eBook)
DOI 10.1007/978-3-319-47058-0

Library of Congress Control Number: 2016952891

Printed on acid-free paper

This Springer imprint is published by Springer Nature
The registered company is Springer International Publishing AG
The registered company address is: Gewerbestrasse 11, 6330 Cham, Switzerland

The original version of the book was revised:
For detailed information please see erratum.
The erratum to this book is available at
DOI 10.1007/978-3-319-47058-0_12

Preface

This book is the outcome of a workshop entitled "*Computational information geometry for image and signal processing*" (http://www.icms.org.uk/workshops/infogeom) that was hosted by the International Centre for Mathematical Sciences (ICMS, Edinburgh, UK), with funding from the London Mathematical Society and the British Engineering and Physical Sciences Research Council. The workshop took place during September 21–25, 2015. Participants at this workshop (see group photo of Fig. 1) were kindly asked to submit for peer-review a chapter summarizing some of their recent research achievements in this area.

First, let us give some background on the workshop. The workshop began with lectures providing overviews of information geometry and its applications, computational aspects, and in particular applications in machine learning, cognition, medical imaging of the brain and radar processing. The common ground is a scientific context in which the experimenter has a model with, possibly intricately connected, statistical properties. Then, the desired identification and extraction of features of interest depends on an optimisation process that exploits the existence of a metric structure on smooth families of probability density functions, or approximations thereto from observed frequency data for example, typically using relative information entropy. This was followed by a session on cases when the statistical aspects could be represented *a priori* by mixtures of given probability density functions notably multivariate Gaussians where, typically, the analytic information metric is not known and approximations need to be made. One of the challenges in computational anatomy is the identification of anomalous shape features. In this context deformed exponential families of probability density functions, originally introduced in the study of statistical physics of strongly correlated systems, can be used to select a model for statistical shape via a segmentation of the target organ region and suitable training data. A number of presentations were concerned with various aspects of nonlinear filters where a signal process is progressively estimated from the history of a related process of observations. Information theoretic properties of filters have relevance in non-equilibrium statistical mechanics. The nonlinear filtering situation is typically infinite dimensional and the role of finite-dimensional approximations is an important research area. For example, the

Fig. 1 Group picture (photo courtesy of ICMS, UK)

manifold projection method for stochastic nonlinear filtering gives the projection filter. Using a Hilbert space structure on the space of probability densities, the infinite-dimensional stochastic partial differential equation for the optimal filter can be projected onto a finite dimensional exponential or mixture family, respectively, with two different metrics: the Hellinger distance and the L_2 direct metric. Such projection filters can match the performance of numerical methods based on much higher numbers of parameters, thereby providing both estimation and computational efficiency. Hamilton Monte Carlo methods have an important role in generic Bayesian inference problems and presentations emphasised its geometric foundation, and how they can be used to understand properties like ergodicity and guide the choice of tuning parameters in applications. The formal foundations of the Hamilton Monte Carlo algorithm were examined through the construction of measures on smooth manifolds to demonstrate an efficient and robust implementation.

The balance of presentations and a novel departure including substantial discussion sessions with lively exchanges seems to have proved correct to judge from the very positive feedback. Evidently this balance achieved some of the aims of the workshop since the feedback reported the transfer of new skills and new ideas generating new approaches and applications. Moreover, new and renewed contacts were reported and expected to lead to prospective collaborations.

The workshop gave us an opportunity to discuss further research that we overview concisely below. The final session drew together themes from the lengthy, thought-provoking and, ultimately, very fruitful discussion sessions that were held throughout the workshop. A number of particularly promising areas for future work were highlighted:

- Dimension reduction: generically important, a strategic emphasis for future work is the infinite to finite case; this builds upon and develops recent successes in nonlinear filtering.

- Simplex approach: a variety of contributors, from a variety of perspectives, pointed to the great challenges and, correspondingly, large potential benefits associated with a move from traditional manifold-based methodologies to their simplicial counterparts; these being suitable closures, there are strong connections to ongoing parallel work on the closure of exponential families.
- Geometry of model—dually, data—space: this was a recurring theme throughout the Workshop and, especially, its Discussion sessions; among other novel possibilities, the potential benefit of a twisted product of Wasserstein and Fisher–Rao geometries was noted.
- Computational Information Geometry: its continuing evolution in three streams, and their potential fruitful interactions, were evident throughout the Workshop; in brief:

  - Statistical perspective:

    Offers an operational universal model space modulo, where appropriate, (coarse) binning;
    delivers novel, unique contributions to ubiquitous, challenging inference problems including: handling model uncertainty, estimating mixture models and big discrete data;
    overall motivation here is the long-term aim of providing experimenters with new operational tools for handling their data.

  - Engineering perspective: The workshop provided a unique venue for discussing fruitfully more precisely what is meant by "Computational Information Geometry", especially with respect to the established field of "Information Geometry". Several strong points have been acknowledged and promising research avenues sketched. For example:

    Since we deal with large but finite data sets, instead of considering parametric probability families (IG), consider dealing with high-dimensional data simplices with potentially many empty bins (therefore technically different from the regular probability simplex of IG). This raises the theoretical issue of discrepancy between dealing with continuous versus discrete models, etc.
    Some problems change fundamentally from the computational point of view if we consider the standard reduction by sufficient statistics (IG): See for example, Montanari, A. Computational implications of reducing data to sufficient statistics. Technical Report 2014-12, Stanford University, 2014.
    The number of local extrema in deep learning networks and how to analyse computationally the set of singularities. The workshop had about one third of papers dealing with implementations for applications ranging from signal processing, to medical imaging, to clustering, and to statistical mixture estimation, among others.

  - Artificial Intelligence and Machine Learning: Papers included several aspects of machine learning and optimisation thereof. General parametrised geometric objects allow design of efficient learning systems by imposing natural geometric constraints. We learnt of promising possibilities for spontaneous data learning introducing a novel explanatory paradigm beyond the discussion for misspecification of a parametric model. A recent breakthrough in the computation of optimal transport barycenters has the potential to impact deeply the machine learning and imaging communities.

- Singularities and unboundedness: once again, this new theme emerged during the Workshop, differing emphases being to the fore in different presentations; unifying these, and of central importance, is the fundamental move from open to closed structures, and consequences thereof.
- Divergences: the fundamental links between these naturally asymmetric objects and information geometry and, especially, their strategic actual or potential benefits in application were underscored by vital new contributions made during the meeting.
- Optimisation: a variety of opportunities for new or improved optimisation methodologies were opened up by advances reported during the Workshop notably, those based on natural gradient methods and still others arising in the engineering strand of computational information geometry.
- Markov Chain Monte Carlo: Hamiltonian geometry: distinctive results and methodologies, summarised above, reported during the Workshop open up a wide range of new potential applications, especially in Bayesian statistics; intriguingly, it will be of great interest to see how far links can be established between typical sets, support sets and structural zeroes.

We thank all the invited participants at this workshop:

Amari, Shun-ichi	RIKEN Brain Science Institute
Anaya-Izquierdo, Karim	University of Bath
Armstrong, John	King's College London
Ay, Nihat	Max Planck Institute for Mathematics in the Sciences
Barbaresco, Fréderic	Thales Land and Air Systems
Belavkin, Roman	Middlesex University
Betancourt, Michael	University of Warwick
Brigo, Damiano	Imperial College London
Byrne, Simon	University College London
Critchley, Frank	the Open University
Dodson, Kit	University of Manchester
Eguchi, Shinto	Institute of Statistical Mathematics
Galanis, George	Hellenic Naval Academy
Goh, Alvina	National University of Singapore
Jupp, Peter Edmund	University of St Andrews
Komori, Osamu	the Institute of Statistical Mathematics
Marti, Gautier	École Polytechnique & Hellebore Capital Management
Matsuzoe, Hiroshi	Nagoya Institute of Technology
Matúš, František	Institute of Information Theory and Automation
Newton, Nigel	University of Essex

Nielsen, Frank	École Polytechnique & Sony Computer Science Laboratories
Nock, Richard	NICTA & the Australian National University
Ohara, Atsumi	University of Fukui
Perrone, Paolo	Max Planck Institute for Mathematics in the Sciences
Peter, Adrian M.	Florida Institute of Technology
Peyré, Gabriel	Université Paris-Dauphine
Pistone, Giovanni	Collegio Carlo Alberto
Sabolova, Radka	the Open University
Saint-Jean, Christophe	Université de La Rochelle—UFR Sciences
Sampson, W.	University of Manchester
Schwander, Olivier	LIP6, UPMC
Szkola, Arleta	Max Planck Institute for Mathematics in the Sciences
Takatsu, Asuka	Tokyo Metropolitan University
Van Bever	Germain, the Open University
Vos, Paul	East Carolina University
Zhang, Jun	the University of Michigan at Ann Arbor

This book is a collection of eleven chapters that span both the theoretical side and practical applications of computational information geometry for image and signal processing:

1. Information Geometry and Its Applications: *An* Overview by Frank Critchley and Paul Marriott
2. The Geometry of Model Sensitivity: An Illustration by Karim Anaya-Izquierdo, Frank Critchley, Paul Marriott, and Paul Vos
3. On the Geometric Interplay Between Goodness-of-Fit and Estimation: Illustrative Examples by Karim Anaya-Izquierdo, Frank Critchley, Paul Marriott and Paul Vos
4. Spontaneous Learning for Data Distributions via Minimum Divergence by Shinto Eguchi, Akifumi Notsu, and Osamu Komori
5. Extrinsic Projection of Itô SDEs on Submanifolds with Applications to Non-linear Filtering by John Armstrong and Damiano Brigo
6. Fast $(1 + \epsilon)$ -Approximation of the Löwner Extremal Matrices of High-Dimensional Symmetric Matrices by Frank Nielsen and Richard Nock
7. Dimensionality Reduction for Information Geometric Characterization of Surface Topographies by C.T.J. Dodson, M. Mettaneny and W.W. Sampson
8. On Clustering Financial Time Series: A Need for Distances Between Dependent Random Variables by Gautier Marti, Frank Nielsen, Philippe Donnat, and Sébastien Andler
9. The Geometry of Orthogonal-Series, Square-Root Density Estimators: Applications in Computer Vision and Model Selection by Adrian M. Peter, Anand Rangarajan and Mark Moyou
10. Dimensionality Reduction for Measure Valued Evolution Equations in Statistical Manifolds by Damiano Brigo and Giovanni Pistone

11. Batch and Online Mixture Learning: A Review with Extensions by Christophe Saint-Jean and Frank Nielsen

We express our gratitude to the peer reviewers for their careful feedback, which led to the polished final form of these chapters.

Throughout the planning period, and during the running of the Workshop, the ICMS staff provided excellent support to the organisers. The facilities and catering were excellent. We mention in particular Moira Spencer for her tireless attention to detail and friendly unflappable handling of inevitable complications arising from a large number of international participants.

There is an exciting time ahead for computational information geometry, studying further the fundamental concepts and relationships of information, geometry and computation, and we envision many more applications in signal and image processing.

Palaiseau, France
Milton Keynes, UK
Manchester, UK
June 2016

Frank Nielsen
Frank Critchley
Christopher T.J. Dodson

Contents

Information Geometry and Its Applications: *An* Overview

Frank Critchley and Paul Marriott

Introduction

This paper is *an* overview of information geometry (IG) and it is important to emphasize that ours is one of many possible approaches that could have been taken. It is, necessarily, a somewhat personal view, with a focus on the authors' own expertise. We, the authors, both have our key interest in statistical theory and practice, and were both strongly influenced, just after its publication, by Professor Amari's monograph, Amari (1985). Recently we, and co-workers, have focused our attention on what we call *computational information geometry* (CIG). This, in all its forms – see, for example, Liu et al. (2012), Nielsen and Nock (2014a, b), Anaya-Izquierdo et al. (2013a), and Critchley and Marriott (2014a) – has been a significant recent development, and this paper includes further contributions to it. In our conception, CIG provides novel approaches to outstanding, major problems in analysing statistical data. In particular, its (uniquely) operational universal space enables new, computable ways to handle model uncertainty and estimate mixture distributions. For reasons of space, we will be forced to make limited reference to a number of exciting areas in, and related to, IG. In particular: Section 1.1 quantum information geometry, where the interested reader could look at Nielsen and Barbaresco (2014) and references therein, Sect. 1.2 Hessian geometries, Shima (2007), and Sect. 1.3 what might be called *sample space information geometry*, including manifold learning, Lee and Verleysen (2007) and statistics on manifolds, Bhattacharya (2008).

Frank Critchley: This work has been partly funded by EPSRC grant EP/L010429/1
Paul Marriott: This work has been partly funded by NSERC discovery grant 'Computational Information Geometry and Model Uncertainty'

F. Critchley
The Open University, Walton Hall, Milton Keynes, Buckinghamshire MK7 6AA, UK
e-mail: f.critchley@open.ac.uk

P. Marriott (✉)
University of Waterloo, 200 University Avenue West, Waterloo, ON, Canada
e-mail: pmarriot@uwaterloo.ca

F. Nielsen et al. (eds.), *Computational Information Geometry*,
Signals and Communication Technology, DOI 10.1007/978-3-319-47058-0_1

This paper is not intended to be an introduction to the area for the complete novice, rather it was written as a keynote address for the workshop 'Computational information geometry for image and signal processing' (*ICMS*, Edinburgh, September 2015), where the audience included many experts in IG with different perspectives. It has always been a problem for us when asked: 'what is the best book to read as an introduction to IG?'. The answer depends very much on what the questioner already knows, of course. For example we, the authors, represented two extremes when we started working together: one with no statistical background and one with no differential geometry. One aim of the paper is to point to what we, at least, regard as key references in each of the subject areas. We note, to start with, that there are now a number of volumes in the area of IG, for example the early work in Chentsov (1972) that developed the concept of a statistical manifold, Barndorff-Nielsen (1978), Amari et al. (1987), Dodson (1987), Murray and Rice (1993), Marriott and Salmon (2000), Amari and Nagaoka (2007), Arwini and Dodson (2008), Kass and Vos (2011), Nielsen and Bhatia (2013) and Nielsen (2014).

In this paper, we deliberately do not try to give a formal definition of exactly what information geometry is. Rather, we treat it as an evolutionary term. While IG started as the application of differential geometry to statistical theory, it has – and continues to develop – both with the types of geometry used and in its application areas. Early work was based on, what Amari and Nagaoka (2007) call *dualistic differential geometry* but more recently, wider classes of geometry have been used in IG. For example, links between convex geometry and exponential families are well known, Barndorff-Nielsen (1978), Brown (1986), and their geometric closures have been recognised in IG, see Csiszár and Matus (2005). The importance of affine geometry is explored in this paper in Sect. 1. We will not have space to explore the exciting links with algebraic geometry but point the interested reader to Pistone et al. (2000), Watanabe (2009) and Gibilisco et al. (2010). Symplectic geometry also plays an important role, Barndorff-Nielsen and Jupp (1997) and recent advances in Markov chain Monte Carlo theory, arising from the seminal paper Girolami and Calderhead (2011), has led to the development of applications of Hamiltonian geometry, see Betancourt (2013) and Betancourt et al. (2014). Of recent interest has been Wasserstein geometry and its links with IG, Takatsu (2013). The geometry of functional analysis also has important applications in non-parametric statistics, for an excellent review see Pistone (2013). In this paper we emphasize how the key geometric objects are not always smooth Riemannian manifolds, but that boundaries, changes in dimension, singularity and unboundedness in tensor fields will all play important roles. We also follow a non-traditional route for defining IG structures; starting with embedding spaces in Sect. 1, rather than directly with manifolds. See Sect. 6 for a discussion of this approach.

The conference that motivated writing this paper focused on the applications of IG to image and signal processing, giving examples of applications areas moving away from just statistical theory. Other areas where IG has made an impact include quantum systems, neuronal networks (both biological and artificial), image analysis, and optimization problems.

Throughout this paper we always start each section with a simple – potentially 'toy' – motivating example (Sect. 1.1), which we try and make as visual as possible, returning to this example repeatedly as a concrete illustration. One of the appeals, at least to us, of geometry is its visual aspect and we feel that this can often be lost when ideas become formalised. We follow up this motivating example with a discussion of general theory and point to key references for details and proofs (Sect. 1.2). Each section ends with important examples of the application of the theory (Sect. 1.3).

1 Dual Affine Families

1.1 Illustrative Example

Example For fixed integers m_1, m_2, consider the set of $m_1 \times m_2$ arrays of binary valued pixels. Figure 1 illustrates elements of this state space with a realisation for $m_1 = m_2 = 10$ in Panel (a), while Panel (b) shows the complete state space for the $m_1 = 1, m_2 = 2$ case. Let $(\pi_0, \ldots, \pi_k)$ be a probability vector on such a state space, where $k = 2^{m_1 m_2} - 1$. Here, and throughout the paper, we use the weak inequality $\pi_i \geq 0$. The set of all possible probability models is geometrically a closed k-dimensional simplex:

$$\Delta^k := \left\{ (\pi_0, \ldots, \pi_k) \, : \, \pi_i \geq 0 \, , \, \sum_{i=0}^{k} \pi_i = 1 \right\}. \tag{1}$$

Statistically (1) is an extended multinomial family, Critchley and Marriott (2014a), which is an example of the closure of an exponential family, studied by Barndorff-Nielsen (1978), Brown (1986), Lauritzen (1996) and Csiszár and Matus (2005).

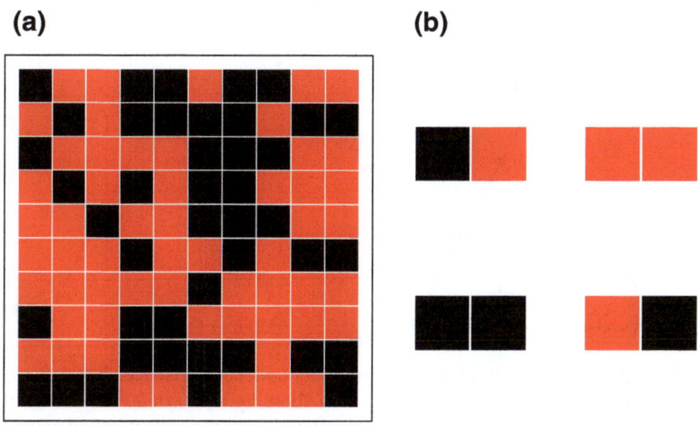

Fig. 1 **a** Realisation for 100 binary pixels. **b** Sample space for 2 binary pixels

The sample space for n independent realisations from an extended multinomial distribution is represented by the set of counts $(n_0, \ldots, n_k)$ where $n_i \geq 0$ and $n = \sum_{i=0}^{k} n_i$, and there is the natural correspondence between the sample and model spaces given by the maximum likelihood estimate

$$(\widehat{\pi}_0, \ldots, \widehat{\pi}_k) := \left(\frac{n_0}{n}, \ldots, \frac{n_k}{n}\right). \tag{2}$$

Why do we insist here on allowing probabilities to be zero? – after all this prevents the geometric objects being manifolds and contradicts the first regularity condition of Amari (1985, p. 16) of distributions having common support. One of the key ideas behind IG is to exploit the link between sample and model spaces – a duality which gives IG its own special flavour – and we want this relationship to be as clean as possible. Since counts in the identification Eq. (2) can be zero we also want to allow probabilities to have that value. We will also see, later in this paper, how the geometry of the boundary dominates the global IG in the relative interior. Hence explicitly including the boundary makes for a much cleaner analysis.

Example (1.1 revisited) For the $m_1 = 1, m_2 = 2$ example both the sample space and the model space can be represented in terms of the 3-simplex, see Fig. 2. The left panel shows the sample space for $n = 3$ with dots representing attainable values. The right panel shows the corresponding parameter space. The red surface in this panel is the set of models where the colour values of the pixels are independent of each other.

The relative interior of the simplex, $r.i.(\Delta^k)$, is commonly parametrized by $(\pi_1, \ldots, \pi_k)$ – which are (-1)-affine, or expectation, parameters in the terminology of Amari (1985) – or by

$$(\theta_1, \ldots, \theta_k) := \left(\log\left(\frac{\pi_1}{\pi_0}\right), \ldots, \log\left(\frac{\pi_k}{\pi_0}\right)\right),$$

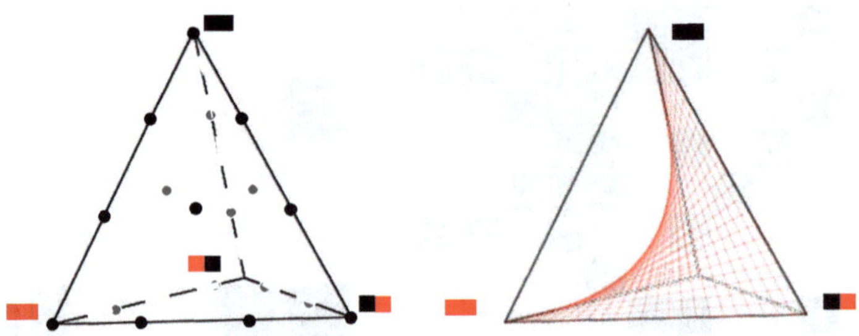

Fig. 2 The 3-simplex: sample, model space and independence subspace. The *left plot* shows the sample space embedded in the simplex for $n = 3$ by showing with *circles* the subset of achievable points. The *right plot* is the model space – the simplex – with the subset of independence models which is a ruled surface. In each plot, the element of the sample space shown in Fig. 1b is shown by the corresponding pair of pixels

the natural, canonical or (+1)-affine parameters.

Key Issue 1 *(Fisher information as change of basis) The matrix of partial derivatives between these smooth parameterisations, of the relative interior, is*

$$\left(\frac{\partial \theta_j}{\partial \pi_i}\right) = \left(\frac{\delta_{i,j}}{\pi_i} + \frac{1}{\pi_0}\right), \tag{3}$$

where $\delta_{i,j} = 1$ if $i = j$, and 0 otherwise. This matrix will be a key tool for moving between representations of geometric objects in the two parameterisations, and we note that it is the Fisher information. Its inverse matrix gives the corresponding inverse transformation.

A parametric statistical model of the set of images can be thought of as a subset of Δ^k, typically selected to have 'nice' mathematical properties. Examples might be that the family is a low dimensional affine subset with respect to the (+1) or (−1)-parameters. For example, the red surface shown in Fig. 2 is the set of independence models, which is an affine subset of the (+1)-parameters.

1.2 Dual Affine Parameters

The two types of parameters illustrated above are familiar from the theory of exponential families of the form

$$f(x;\theta) := \nu(x)\exp\left(\langle\theta, S(x)\rangle - \psi(\theta)\right), \tag{4}$$

where $\nu(x)$ is a positive measure, $\theta := (\theta_1, \ldots, \theta_p)^T$ are the natural (+1) parameters, $S(x) := (S_1(x), \ldots, S_p(x))^T$ are the sufficient statistics and $\mu := (E_\theta(S_1), \ldots, E_\theta(S_p))^T$ are the expectation (−1) parameters and $\psi(\theta)$ is the normalising term. The natural parameter space requires definition and is the set

$$\Theta := \{\theta \mid \psi(\theta) < \infty\}.$$

The boundary behaviour of ψ on this set will play an important role in what follows.

These affine structures are, in fact, much more general than their role in finite dimensional exponential families might suggest.

Key Issue 2 *(Existence of affine structures) There is a natural (+1)-affine structure on the space of positive measures and a (−1)-affine structure on the space of unit measures on a given set. The set of probability measures inherits both structures.*

Murray and Rice (1993) first described the (+1)-affine structure in Key Issue 2, while Marriott (2002) shows the existence of a (−1)-affine structure in unit measure space. The intersection of positive and unit measures is, of course, the set of probability measures, thus this space inherits both affine structures. However, we note that the ± 1-boundaries, where either positivity (-1) or finiteness $(+1)$ fails, will be important in understanding the underlying geometry of 'distribution space'.

The affine structures defined in Key Issue 2 are particularly important when we look at finite dimensional subsets. For example, Murray and Rice (1993, Sect. 1.5.1) show that being a finite dimensional affine subspace of the $(+1)$-affine structure characterises exponential families, while Anaya-Izquierdo and Marriott (2007) show how understanding finite dimensional affine subsets of the (-1)-affine structure explains important identification issues in mixture modelling. An example of a finite dimension subset of $(+1)$-affine space is the independence space plotted in Fig. 2. In the plot it looks 'curved' since the (-1)-affine geometry is used for the illustration.

Key Issue 3 *(Inner product form) Perhaps the crux of understanding duality ideas in IG is the geometric interpretation of the term*

$$\langle \theta, S(x) \rangle := \sum_{i=1}^{p} \theta_i S_i(x), \tag{5}$$

which appears in (4). We have intentionally chosen a suggestive notation which looks like an inner product but, while it is bilinear, the arguments of $\langle \cdot, \cdot \rangle$ lie in different spaces. The first argument lies in the parameter, or model, space and the second lies in the sample space. Of course, as we have seen these spaces can be closely connected. The $(+1)$-affine structure is most 'natural' for the first of these, while the (-1)-affine is most 'natural' for the second.

As illustrated by Example 1.1, these spaces are typically only convex subsets of affine spaces, not affine spaces themselves. However, as also illustrated by Example 1.1, these two spaces have strong links and this gives rise to the principal duality of IG.

There is one instance where all these spaces agree and $\langle \cdot, \cdot \rangle$ is indeed an inner product. This is the statistically very important case of normal linear regression. We can view the structure of classical information geometry as a way of extending the geometric foundation of regression to much more general contexts, see Vos and Marriott (2010).

To give Expression (5) an inner product interpretation we need to make some changes of perspective. Firstly, since we need affine spaces, we work with best linear approximations – tangent spaces – giving each the affine structure described in Key Issue 2. Secondly, we need to be able to map between the $(+1)$-representation of the tangent space and the (-1)-representation. This is the classical change of basis formula from differential geometry, instanced by Eq. (3) in the multinomial case. In general the change of basis between $(+1)$ and (-1)-coordinates, for exponential families, is the Fisher information matrix, see Sect. 4. Thus by searching for an inner product interpretation of $\langle \cdot, \cdot \rangle$, the Fisher metric structure has naturally arisen. We denote the Fisher information based inner product at a tangent space by $\langle \cdot, \cdot \rangle_F$.

We have therefore, at least where the underlying models are smooth manifolds, arrived at the classical IG structure described in Amari (1985). We have sets of distributions with enough smooth structure to be manifolds, different but related affine structures, and a change of basis formula which has the properties of being a metric tensor.

Before we briefly review the elegant mathematical structures associated with this structure, we make some observations. Historically an important paper was Lauritzen (1987), which described the structure (M, g, ∇^{α}) of manifold, metric and family of connections which characterise the affine structures. This united the 'expected' IG of Amari and the 'observed' IG as described in Barndorff-Nielsen (1987). These differ in the choice of metric associated with using unconditional or conditional sample spaces.

Secondly, while we always use the term manifold, much of IG only uses the local geometric structures – that is the tangent space. At least in our experience in statistics, most parameterisations are global and the powerful geometric structure associated with the term manifold – non-trivial topology, local charts, atlas etc. – are rarely used. This has been a drawback for practitioners since it appears that there is a bigger overhead of mathematical structure required than is really needed.

Thirdly, there are very simple, but practically important, models in statistics – two component mixtures of exponential distributions for example, Li et al. (2009) – where the Fisher information does not exist and yet there is still a very interesting geometry structure, see Sect. 4.

Finally, as we saw in Example 1.1 – but also in the important classes of mixture, graphical and conditional independence models – boundaries and singularities play a critical role and so these models are not manifolds but do, again, have very interesting geometry.

Key Issue 4 *(The pillars of IG) We can now review the keys pillars of IG. First, we note that we use Fisher information to define a Riemannian structure on the statistical manifold. The affine structures can be characterized by the differential geometric tool of an affine connection* ∇*, Amari and Nagaoka (2007, p. 17). There is a one dimensional family of such connections defined by*

$$\nabla^{(\alpha)} = \frac{1+\alpha}{2}\nabla^{(+1)} + \frac{1-\alpha}{2}\nabla^{(-1)} \tag{6}$$

Amari and Nagaoka (2007, p. 33) for $\alpha \in \mathbb{R}$*. Here the* $\alpha = \pm 1$ *connections agree with the affine structures defined in Key Issue 2. The* $\alpha = 0$ *connection is also of interest since it is the Levi–Civita connection Murray and Rice (1993, p. 115) associated with the Fisher information, see Sect. 4. The relationship between dual connections and the metric is encoded in the duality relationship*

$$X\langle Y, Z\rangle_F = \langle \nabla^{(\alpha)}_X Y, Z\rangle_F + \langle Y, \nabla^{(-\alpha)}_X Z\rangle_F, \tag{7}$$

where X*,* Y *and* Z *are smooth vector fields, Amari and Nagaoka (2007, p. 51). From this relationship we have two fundamental results: the dual flatness theorem, Amari*

(1985, Theorem 3.2, p. 72), and the Pythagoras theorem, Amari (1985, Theorem 3.9, p. 91).

The first of these fundamental results says that if a statistical manifold is α-flat (i.e. there exists a parameterisation in which α-geodesics are defined by affine functions of the parameters) then it is also $-\alpha$-flat. The classic example is the exponential family defined in Eq. (4), which has θ as $(+1)$-affine parameters and μ as (-1)-affine parameters. This result is very powerful since affine parameters are typically hard to find but very useful; they reduce much of the geometry to that of a Euclidean space. To get a 'free' set of affine parameters is thus excellent news. The dual nature of these affine parameters and the relationship with the metric is also exploited in Sect. 5. The second result is the Pythagoras theorem and this is discussed in Sect. 3 once we have introduced the concept of a divergence function.

1.3 Application Areas

Application Area 1 (Exponential families in Statistics) The primary application of finite dimensional dual affine structures in statistics is, of course, the full exponential family, Brown (1986), Barndorff-Nielsen (1978). The finite dimensional $(+1)$-structure induced by (4) has the property that under i.i.d. sampling the dimension of the sufficient statistic does not change as the sample size increases, meaning that information about the parameters of the model can accumulate with increasing sample size. Closely related are exponential dispersion models Jorgensen (1987) which form the probabilistic backbone of generalised linear models, McCullagh and Nelder (1989). These are the workhorses of much applied statistical modelling. The generalisation from the standard normal linear model – where $(+1)$ and (-1) structures are indistinguishable – is through the separation of the ± 1-affine structures of exponential dispersion models, Vos and Marriott (2010).

Application Area 2 (Maximum entropy models) Exponential families are also naturally generated through the maximum entropy principle, Jaynes (1978, 1982), Skilling (1989), Buck and Macaulay (1991). The principle of maximum entropy here has strong links with the material on divergences in Sect. 3 of this paper, and was motivated by notions of entropy as a measure of uncertainty in both statistical physics and information theory.

Application Area 3 (Curved exponential families) One of the most influential papers in the development of IG was Efron (1975) which first demonstrated that notations of curvature have application in statistical theory. The immediate applications in that paper were to information loss and asymptotic efficiency in inference for a curved exponential family – a submanifold in an exponential family. This class of curved models has important applications in applied statistics including, among many others, Poisson regression, auto-regressive models in time series analysis and common factor models in Econometrics, Marriott and Salmon (2000).

Application Area 4 (Graphical models and exponential families) In signal, image and speech processing, one area where the dual affine structure of exponential families has found many applications is through their representation of graphical models. We highlight the paper Wainwright and Jordan (2008) and references in Jordan et al. (2010). Models in these areas can be very high dimensional and direct computation of the normalising constant in Expression (4) – which encodes the full IG structure of such families – can be intractable. The paper points to variational methods in this context, see also Zhao and Marriott (2014) for links with IG.

Application Area 5 (Models in neuroscience) Exponential random graph models (ERGMs) have found important applications in connectivity research in neuroscience, Simpson et al. (2011). The geometry of such models is explored in Rinaldo et al. (2009). Related ideas in belief propagation – a universal method of stochastic reasoning – can be found in Ikeda et al. (2004), while Amari (2015) reviews the IG of, so-called, neural spike data. For related models in neuroscience see Tatsuno and Okada (2003); Tatsuno et al. (2009).

2 Boundaries in Information Geometry

2.1 Illustrative Example

Example In the example of modelling sets of binary pixels, consider again the set of independence models, illustrated in Fig. 2. In the case $k = 2$ we can show this space in both its (-1)-affine (Fig. 3a) and $(+1)$-affine (Fig. 3b) parameters. For the independence model, the expectation parameters are the marginal probabilities of

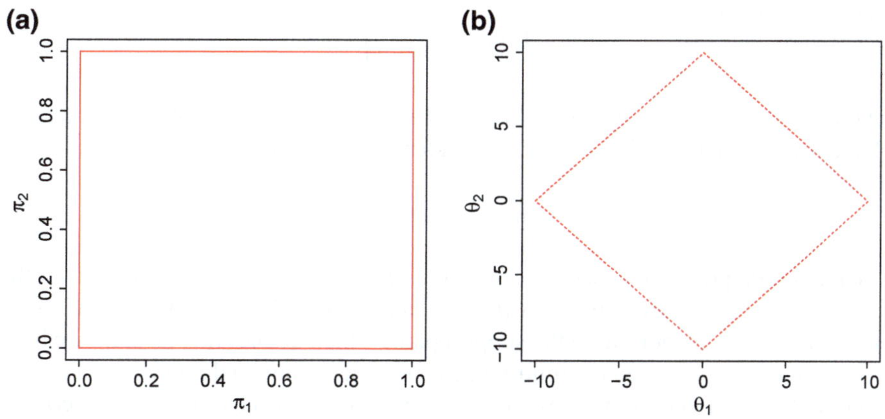

Fig. 3 **a** Expectation parameters. **b** Natural parameters

being a colour, π^M. The boundaries for this space are shown in Panel (a) with solid lines.

The relative interior of this space, which is an exponential family, can be parameterised by its natural parameters – the marginal log-odds. We can ask the question of how to represent the boundary in the natural parameters. In Panel (b) we represent this with the red dashed lines 'at infinity'. They represent the 'directions of recession' for this model, Geyer (2009). There is a duality between the two forms of the boundary, with vertices in one representation corresponding to edges in the other, and *vice versa*. To formalize the correspondence between the two we need to understand the closure of the exponential family, Barndorff-Nielsen (1978). That is, what happens to $\theta(\pi^M)$ as at least one component of π^M tends to zero?

In our running example, from Sect. 1.1, a boundary point in model space corresponds to a degenerate distribution. So in the independence model, shown in Fig. 2, boundary points correspond to particular pixels being always the same colour.

2.2 Boundaries and Polar Duals

Key Issue 5 *(Polar duals) We can understand boundary behaviour in extended exponential families by considering the polar dual Critchley and Marriott (2014b) or, alternatively, the directions of recession, Geyer (2009), Rinaldo et al. (2009).*

For simplicity we consider discrete p-dimensional exponential families, given by (4), which are subsets of Δ^k described by Eq. (1). For more general results on closures of exponential families see Barndorff-Nielsen (1978), Brown (1986), Lauritzen (1996) and Csiszár and Matus (2005).

We want to consider the limit points of the p-dimensional exponential family, so we consider the limiting behaviour of the path $\theta(\lambda) := \lambda q$ as $\lambda \to \infty$, where $q \in \mathbb{R}^p$ and $\|q\| = 1$. The support of the limiting distribution is determined by the maximal elements of the set

$$\{s_0^T q, \ldots, s_k^T q\}$$

where $s_i := \left(S_0(i), \ldots, S_p(i)\right)^T$. Let $\mathcal{F}_q$ be the set of indices of these maximal elements, so that $1 \le |\mathcal{F}_q| \le k+1$. Consider the convex hull, C, of the set

$$\{s_0, \ldots, s_k\} \subset \mathbb{R}^p.$$

The maximum principle for convex functions tells us that $s^T q$ is maximised over the face of C defined by the vertices $\{s_i | i \in \mathcal{F}_q\}$ and, as Critchley and Marriott (2014b) easily show, q is the normal to the support plane which defines this face. So we have a correspondence between the limiting behaviour of exponential families in a certain direction – the direction of recession – and the set of normals to faces of a convex polygon. The set of outward pointing normals to a polygon is called its polar dual, Tuy (1998).

Example (2.1 revisited) In Fig. 3 the polygons in Panels (a) and (b) are polar duals of one another. As the point approaches the boundary in Panel (a) its $(+1)$-parameters will go to infinity in the direction indicated by the corresponding point on its polar dual.

Often the computation of the boundary polytopes are completely straightforward and there are many cases where the key step, computing the convex hull of a finite number of points in $\mathbb{R}^p$, can be done with standard software. We note however, as the number of parameters and the sample size grows, complete enumeration of the boundary becomes computationally infeasible, see Fukuda (2004).

Key Issue 6 *(Convex geometry) We see here that the key to understanding the closures of exponential families is convex, rather than differential, geometry, and the important geometric objects are convex hulls rather than manifolds. We will also see the important role that convex geometry plays in Sect. 3.*

Another place where the dominant geometric tools come from convex geometry is in the analysis of mixture models. A major highlight is found in Lindsay (1995), where convex geometry is shown to give great insight into the fundamental problems of inference in these models and helps in the design of corresponding algorithms. Other differential geometric approaches for mixture models in image analysis can be found in Mio et al. (2005a). Explicit links between this literature and IG can be found in Anaya-Izquierdo et al. (2013b). The boundaries in this geometry are natural generalisations of the simplest mixture model,

$$\rho f(x) + (1 - \rho)g(x),$$

where $\rho \in [0, 1]$ with boundaries at $\rho = 0, 1$. Example 7 of Critchley and Marriott (2014a) gives an example of very different statistical behaviour at each boundary point when mixing is between a normal and a Cauchy distribution.

2.3 *Application Areas*

Application Area 6 (The finite moment problem) A classical topic in statistics is the moment problem; which distributions can be represented by a finite set of moments? Very early applications of convex geometry in statistical theory can be found in Karlin and Shapley (1953). This work uses convex sets and their conjugate duals to show how moment spaces – sets of achievable moments – are convex bodies whose extreme points can be characterized, often by algebraic means.

Application Area 7 (Boundaries in ERGMs) We have already discussed applications of exponential family random graph models in Application Area 5. The geometry of ERGMs has a number of very interesting features. As pointed out in Geyer

(2009) the existence of the maximum likelihood estimate, and corresponding inferences, depends on the boundary behaviour of the closures of the corresponding exponential families. This boundary geometry also dominates the shape of the likelihood and hence also is important in Bayesian inference. Key references here include Rinaldo et al. (2009) and the recent Critchley and Marriott (2014a).

Application Area 8 (Logistic regression) The classical workhorse of statistical modelling with binary data, logistic regression, relies on standard first order asymptotic inference methods using the likelihood. The paper Anaya-Izquierdo et al. (2014) looks at the way that analysing the boundary behaviour of these models generates a simple diagnostic which gives a necessary condition that these first order methods are justified.

Application Area 9 (Marginal polytopes) Connected with these ideas of convex boundaries of exponential families is the idea of a marginal polytope. These are geometric objects associated with any undirected graphical model. They are defined as the set of all marginal probabilities that are realizable under the dependency structure defined by the graphical model. Applications of these geometric ideas can be found in the analysis of Markov Random Fields, which are important in image analysis and many other places. References for this topic include Wainwright and Jordan (2003), Sontag and Jaakkola (2007), and Kahle et al. (2010).

3 Divergences

3.1 Illustrative Example

The two previous sections looked at basic geometric issues of affineness (i.e. what is a straight line?), convexity, and what happens at boundaries. Section 4 will look at how to measure angles and orthogonality. One major geometric issue not so far mentioned concerns measuring 'distance' in IG and then how to minimize such 'distances'. These questions have been a major driving force in the development of IG, with the following as a key example.

Example If $f(x;\xi_1)$ and $f(x;\xi_2)$ are two density functions in a parametric model, then we define the Kullback-Leibler divergence, from $f(x;\xi_1)$ to $f(x;\xi_2)$, as

$$K(\xi_1;\xi_2) := E_{f(x;\xi_1)}\left[\log\left(\frac{f(X;\xi_1)}{f(X;\xi_2)}\right)\right], \tag{8}$$

when the expectation exists. Of course, this is not a metric distance, as there is no corresponding triangle inequality and symmetry also fails, Kass and Vos (2011, p. 51). It does, however, have the distance like properties of being greater than, or equal, to zero, with equality if and only if $\xi_1 = \xi_2$.

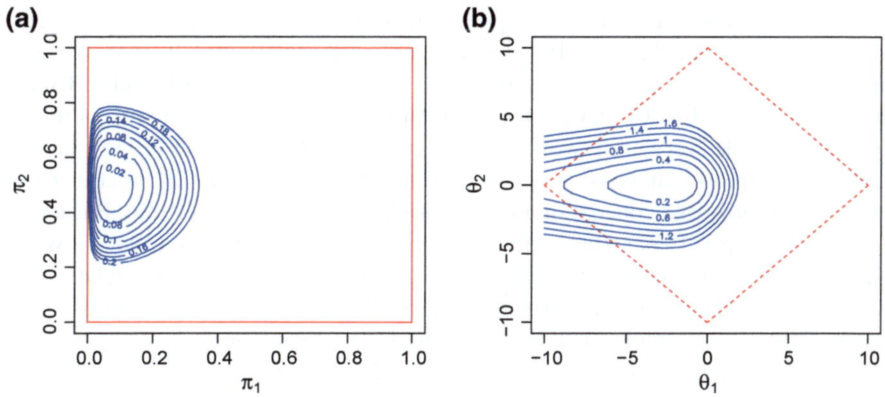

Fig. 4 **a** KL divergence expectation parameters. **b** KL divergence natural parameters

Figure 4 shows concentric KL-spheres in the independence model from the running pixel-based Example 1.1. The level sets are measuring the divergence between two models for the distribution of the pixels in the array. When one of the distributions is degenerate then this distance can be unbounded. As would be expected, from general principles, divergence locally behave qualitatively like the Fisher information spheres of Example 4.1. This is expected since, locally, this divergence is well approximated by a quadratic form based on the Fisher information.

Further, we see how the boundaries in each model determine the global behaviour of the spheres. In Panel (b) the K-L sphere are stretched 'to infinity' in the direction of recession determined by a vertex of the boundary. This vertex is dually equivalent to the edge in Panel (a) which are 'distorting' the shape of the spheres.

Key Issue 7 *(Convexity) If a function is going to have distance-like properties then how to minimize it over subsets is a natural question. It is therefore very convenient if the function has nice convexity properties, but since convexity is not invariant to all reparameterisations the link between choice of divergence and the parametrisation used is critical.*

3.2 Divergences in IG

While the KL-divergence is very popular, for a number of reasons, it is far from the only possibility. In fact the opposite is true, there is a bewildering number of possible choices which could have been made, depending on what conditions are needed. To help the novice, a useful reference is the annotated bibliography, Basseville (2013) while other important reviews include Kass and Vos (2011, Chap. 9), Cichocki et al. (2009, Chap. 2) and references therein.

One of the most influential developments in basing IG around distance/divergence ideas came from Eguchi et al. (1985), which looked at constructing IG from the point of view of a contrast (divergence) function. Related work can be found in Eguchi et al. (1992), Eguchi (2009), Eguchi et al. (2014). Other important streams of related concepts include: very early work by Csiszár et al. (1967), Csiszár (1975), Csiszár (1995); asymptotic analysis of related estimators, Pfanzagl (1973); metric based ideas, Rao (1987); the concept of a yoke, Barndorff-Nielsen et al. (1989), Barndorff-Nielsen and Jupp (1997), Blaesild (1991), Barndorff-Nielsen et al. (1994) – which has similar structure to a divergence and also generates IG structures; the relationship with preferred point geometry, Critchley et al. (1994), Critchley et al. (1996); and also Zhang (2004), which looks at convexity properties of divergences, f-divergences for affine exponential families Nielsen and Nock (2014b), and Belavkin (2013) which looks at optimization problem for measures. A stream of related ideas which was developed rather independently of IG can be found in Cressie and Read (1984), Read and Cressie (2012).

In this paper, for reasons purely of space, we will focus on only one part of this development. In the definition of Bregman (1967), a (Bregman) divergence is a function $D : S \times S \to \mathbb{R}$ where S is a convex set in a linear topological space satisfying certain positivity, projection, convexity and smoothness conditions while, to be precise, the second argument of D should belong to the relative interior of S. Under the conditions of the paper, the function can be expressed using a strictly convex smooth function τ as

$$D_\tau(\xi_1; \xi_2) = \tau(\xi_1) - \tau(\xi_2) - \langle \tau'(\xi_2), \xi_1 - \xi_2 \rangle. \tag{9}$$

for $\xi_i \in S$. Under certain conditions, this can be expressed as

$$D_\tau(\xi_1; \xi_2) = \tau(\xi_1) + \tau^*(\xi_2) - \xi_1^T \xi_2^* \tag{10}$$

where a dual parameter system is defined to ξ by $\xi^*(\xi) := \tau'(\xi)$ and $\tau^*(\xi) := \xi^T \xi^* - \tau(\xi)$ is the Legendre transform when it exists, Rockafellar (1997). We note here that, appropriately interpreted, (10) is a 'dualistic form' of the cosine law. Further, again appropriately dualistically interpreted, (11) below shows that divergence behave like half a squared distance.

Example (3.1 **revisited**) We note that the expression of a divergence in form (10) requires a parameter system ξ and a function which is strictly convex in this parameter system. Since convexity is not invariant to non-linear reparametrisations, each Bregman divergence is associated with particular classes of parameters, called by Kass and Vos (2011, p. 242) the divergence parameter. For the KL divergence in Example 3.1 in an exponential family, (4), the expectation parameter is the divergence parameter, since we have

$$K(\mu_1; \mu_2) = \tau(\mu_1) - \tau(\mu_2) - \langle \tau'(\mu_2), (\mu_1 - \mu_2) \rangle,$$

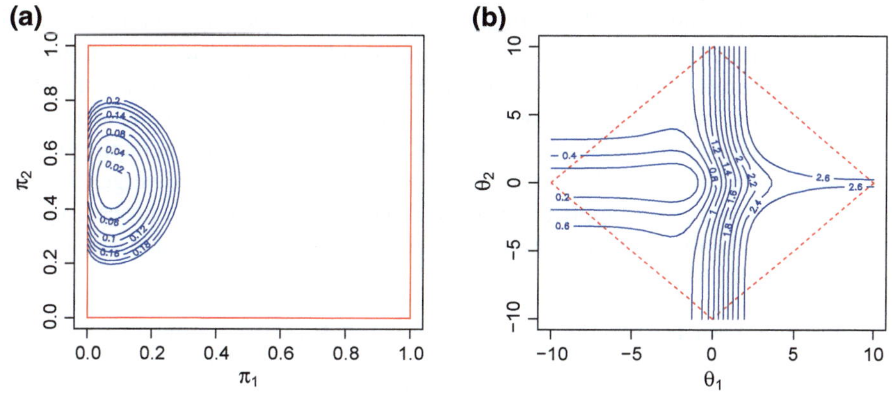

Fig. 5 **a** Dual KL divergence expectation parameters. **b** Dual KL divergence natural parameters

where $\tau(\mu) := \theta(\mu)^T\mu - \psi(\theta(\mu))$. The 'reverse' KL-divergence, $K^*(\xi_1; \xi_2) := K(\xi_2; \xi_1)$, can be written as

$$K^*(\theta_1; \theta_2) = \tau^*(\theta_1) - \tau^*(\theta_2) - \langle \tau^{*\prime}(\theta_2), (\theta_1 - \theta_2) \rangle$$

where $\tau^*(\theta) = \psi(\theta)$. So, cf. (9), we see the dual affine parameters have corresponding dual divergences.

The divergence spheres for these dual divergences are shown in Fig. 5. The boundaries in each panel are determining the shape of the contours. The vertices in (b) – which correspond to the edges in (a) – are controlling the global shapes associated with the level sets. We also note the lack of convexity in Panel (b) since here the level sets are not being plotted in the affine parameters associated with the Bregmann divergence, Kass and Vos (2011).

For a Bregman divergence in its corresponding affine parametrisation we have the formula

$$D_\tau(\xi_1; \xi_2) + D^*_\tau(\xi_1; \xi_2) = (\xi_1 - \xi_2)^T \left(\xi_1^* - \xi_2^*\right). \tag{11}$$

We note the 'doubly dualistic' structure of Eq. 11 where, on the right, we have the dual version of the 'inner product' – it might be helpful to refer again to our key Eq. (5) – and we also have the pair of dual divergences on the left. We can, in fact, build the IG structure of Sect. 1 by starting with a pair of dual Bregman divergences and their corresponding dual divergence parameters, see Kass and Vos (2011, Sect. 9.3).

3.3 Application Areas

Application Area 10 (Statistical pattern recognition) The paper Eguchi (2006) looks at ways to apply IG, through a divergence function representation, to statistical pattern recognition. In particular, it looks at boosting algorithms. Boosting is a way of combining the results from simple models, so called weak learners, into a combined result which is much stronger. The paper uses divergences, in this case U-divergences, and their projection properties to construct new boosting algorithms and to give insight into the popular AdaBoost algorithm Freund and Schapire (1995). See also Collins et al. (2002) for more links between boosting and divergence functions. Other, more recent, applications to machine learning and signal processing can be found in Takenouchi et al. (2008), Kawakita and Eguchi (2008), and Takenouchi et al. (2012, 2015).

Application Area 11 (Audio stream processing) The paper Cont et al. (2011) applies IG methods, in particular, using Bregman divergences, to build a framework for the analysis of audio signals and shows concrete applications for online audio structure discovery and audio matching.

Application Area 12 (Non-negative matrix factorisation) The book Cichocki et al. (2009) looks at the area of non-negative matrix and tensor factorisation. This is a technique with applications in computer vision, signal processing and many other areas. The mathematical problem is to factorize a 'large' (non-negative) matrix into the product of two 'smaller' (non-negative) matrices. This is often not always possible exactly and so approximation methods are used and measures are needed to measure the size of the error. The geometry found in Cichocki et al. (2009) uses gradient algorithms, often based on different types of divergence to measure the quality of approximation.

We note that the divergences are here not defined on probability spaces, but, rather, on positive measure spaces. This is a good example of how IG has moved beyond the area of probability and statistics.

Application Area 13 (Tsallis entropy) The link between divergence functions and entropy is clear in Example 3.1. The concept of entropy itself has one of its roots in equilibrium statistical mechanics, another being in information theory. Tsallis entropy is a non-additive entropy, which differs from the classical Boltzmann-Gibbs entropy, and has applications in non-extensive statistical mechanics, see Tsallis (1988, 2009) and with a focus on IG issues, Amari and Cichocki (2010) and Amari and Ohara (2011).

4 Tangent Spaces and Tensors

4.1 Illustrative Example

Example The most familiar object which is a tensor in IG is the Fisher information matrix, already discussed in Sect. 1. In that section we highlighted its role defining changes of coordinates on tangent spaces as we change parameterisations. It, of

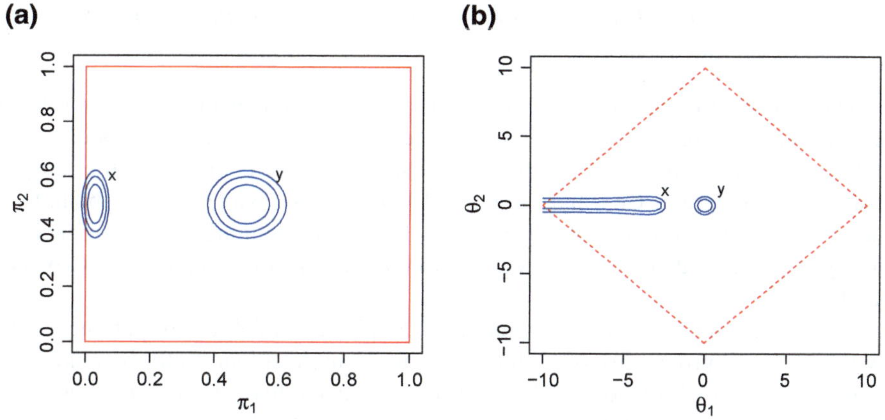

Fig. 6 **a** Expectation parameters. **b** Natural parameters

course, has an alternative statistical role. If $\ell(\eta; D)$ is the log-likelihood function in some arbitrary parameterisation, when D is the observed data, then the Fisher information matrix for η is

$$Cov_\eta\left[\frac{\partial \ell}{\partial \eta_i}(\eta; D), \frac{\partial \ell}{\partial \eta_j}(\eta; D)\right] \equiv -E_\eta\left[\frac{\partial^2 \ell}{\partial \eta_i, \partial \eta_j}(\eta; D)\right], \tag{12}$$

where Cov denotes the covariance operator. The form of the matrix obviously depends on the choice of parameters, and it is convenient that it has a tensorial transformation rule. We say 'convenient' because it makes it easy to check when objects constructed using tensors have invariant meanings.

In statistics the Fisher Information is familiar since its inverse determines the variance-covariance matrix for the first order asymptotic distribution of the maximum likelihood estimate, Cox and Hinkley (1979). Figure 6 shows for our running example the $p = 2$ dimensional extended exponential family in its expectation and natural parameters. The red line in (a) is the boundary, a polygon, and the corresponding line in (b) is its polar dual. The blue ellipses represent the variability of the maximum likelihood estimates for different data generation distributions across the model. The different 'shapes' and 'scales' in the different parameterisations are given by the tensorial rules of transformation.

Again we note the way that the dual boundaries determine the global behaviour of the shapes of these contours. In Panel (b) the direction of recession is pulling the boundary to infinity, and this vertex corresponds to the edge in Panel (a) which the contours are cutting.

Key Issue 8 *(Two roles of Fisher information) We have seen that the Fisher information has two distinct roles in IG: first, as the key change of basis matrix between* *expectation and natural parameters and, second in its role in the Cramér-Rao theorem and asymptotic theory.*

The Fisher information was recognised to be a Riemannian metric by Rao (1945) and in Sect. 4.3 we will discuss some aspects to its corresponding geodesics. In statistical theory, outside asymptotic analysis, its key role comes from the famous Cramer-Rao theorem, Cox and Hinkley (1979, p. 254), which gives a bound on the accuracy of estimation of a parameter. Its role in defining the importance of orthogonality in statistical theory was explored in a very influential paper, Cox and Reid (1987).

In the independence case of our running example, all models can be parameterised by the marginal probability of each pixel being a single colour. In this example the Fisher information is diagonal. We could also parameterize by the marginal log-odds of each pixel's colour, and the Fisher information would change by an appropriate tensorial transformation.

Also of interest is the behaviour of the Fisher information near the boundary. This is explored in Anaya-Izquierdo et al. (2014) which shows how first order asymptotic analysis can break down when the boundary is 'close' as measured by the Fisher information. Furthermore, in Critchley and Marriott (2014a), the limiting behaviour of the Fisher information, as it approaches a boundary, is studied by analysing its spectrum.

4.2 Tensorial Objects

Key Issue 9 *(Invariance) In differential geometry a great deal of attention is paid to understanding the problem of invariance to reparameterisation. The idea here is simply that, at least as far as a geometer is concerned, parameters are just constructs, the manifold is the object of interest, and no results on the manifolds should depend on arbitrary choices. We feel that it is not completely clear that these ideas should be taken without some thought directly into IG in all cases. In statistics it is common that a parameter, such as a mean or probability, has real world meaning in its own right. Indeed this meaning can exist independently of the model selected. In this case we have – what might seem a paradoxical situation to a geometer's eyes – that the parameter is the object of interest while the manifold (model) is the arbitrary construct.*

Nevertheless, the study of invariance has played an important role in the development of IG. Example 4.1 has two aspects which are key. Firstly the tensorial nature of the Fisher information and secondly its role in sample size asymptotic expansions.

Good references for the general structure of tensors are Dodson and Poston (2013), which focuses on the geometric aspects of tensorial analysis, and McCullagh (1987), which emphasises their statistical importance. In particular, for a reference to the tensorial properties associated to cumulants, see McCullagh (1987, pp. 57–62).

The introduction to McCullagh's book is also a good way of learning about the algebraic structure of tensor spaces.

To study the role of asymptotic analysis, in particular its geometrical aspects, good references are Barndorff-Nielsen et al. (1986), Barndorff-Nielsen and Cox (1989), Cox and Barndorff-Nielsen (1994), Barndorff-Nielsen et al. (1994) as well as McCullagh (1987), Murray and Rice (1993, Chap. 9) and Kass and Vos (2011, Chap. 3). This last reference also has material on asymptotic expansions in Bayesian theory.

4.3 Application Areas

Application Area 14 (Asymptotic expansions) The classical application of IG in statistics is, of course, the asymptotic analysis found in Amari (1985). A representative example is the bias correction of a first-order efficient estimator $\hat{\beta}$ which is defined by

$$b^a(\beta) = -\frac{1}{2n} g^{aa'} \left\{ g^{bc} \Gamma^{(-1)}_{a'bc} + g^{\kappa\lambda} h^{(-1)}_{\kappa\lambda a'} \right\},$$

and has the property that if $\hat{\beta}^* := \hat{\beta} - b(\beta)$ then

$$E_\beta(\hat{\beta}^* - \beta) = O(n^{-3/2}).$$

All terms in this expansion have a direct IG interpretation, see Amari (1985), and their dependence on the choice of parametrisation is made clear. Other important work on the geometry of asymptotic expansions includes Kass (1989), and the books Barndorff-Nielsen and Cox (1989), Cox and Barndorff-Nielsen (1994), and Kass and Vos (2011).

Application Area 15 (Laplace expansions) A related set of work concerns the geometry of the Laplace expansion, which has important applications in Bayesian analysis, Kass et al. (1988), Tierney et al. (1989), Kass et al. (1991), and Wong and Li (1992). Other related work exploiting information geometric properties of the Laplace expansion in mixture models includes Marriott (2002) and Anaya-Izquierdo and Marriott (2007). Other work looking at the local geometry of the likelihood includes Eguchi and Copas (1998).

Application Area 16 (Image analysis) The, so-called, Fisher-Rao geometry which is based on the 0-geodesics of the Fisher metric, has found application in image analysis. We point, in particular to Mio et al. (2005b), Mio and Liu (2006), Lenglet et al. (2006) and Peter and Rangarajan (2006).

Application Area 17 (Model uncertainty) Model uncertainty is a critical problem in applied statistics. The paper Copas and Eguchi (2005) provides an intriguing solution by proposing the 'double the variance' method for addressing the possibility of undetectably small departures from the model. The paper builds local neighbourhoods, using essentially metric based first-order geometric methods, of observationally equivalent models and then studies the inferential effects of working inside this set, which is geometrically a tubular neighbourhood. Much more detail on this area can be found in Anaya-Izquierdo et al. (2016).

Application Area 18 (Infinite Fisher Information) The tensorial structure of IG outside the familiar exponential family can have surprises. The paper Li et al. (2009) shows very simple examples of mixture models – such as a two component mixtures of Poisson or exponential distributions – where the Fisher information does not exist. This means that a great deal of standard statistical methodology does not hold. Nevertheless geometry has a great deal to say about these problems, see for example Morozova and Chentsov (1991), Lindsay (1995) or Anaya-Izquierdo et al. (2013b).

5 Dimensionality and Dual Parameters

5.1 *Illustrative Example*

Example (1.1 revisited) Let us return to our running example. We might want to model a binary array of pixels with an independence model, but we may have other modelling assumptions which further reduce the dimension. Accordingly, in Fig. 7 (left hand panel) we illustrate this with a one dimensional exponential family lying in the independence space. As an aside we note the way that such a family, typically, starts and ends at a vertex. Suppose we are interested in a more general model and in the spirit of random effects modelling allow mixing over the one-dimensional family. We show the resulting (-1)-convex hull in the right hand panel. This convex

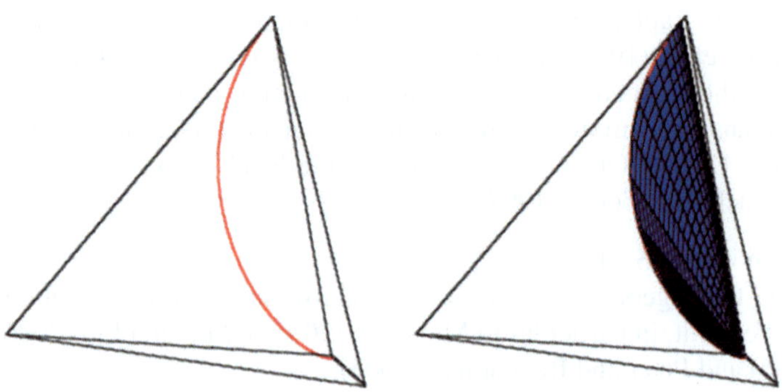

Fig. 7 The 3-simplex: (*left*) one dimensional exponential family in simplex (*right*) (-1)-convex hull which represents mixtures over the $(+1)$-family. The convex hull here is a three dimensional subset of the simplex

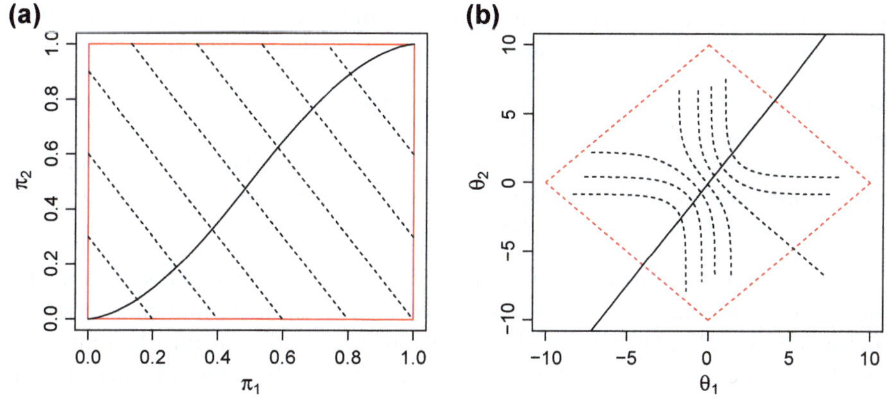

Fig. 8 Mixed parameters in the independence model: **a** expectation parameters and **b** natural parameters

hull is, generically, of full dimension, Critchley and Marriott (2014a). Thus, we have here an example where very low dimensional $(+1)$-objects have very large, indeed maximal, dimensional (-1)-convex hull.

Example (1.1 revisited) We can also consider the one dimensional family from Example 5.1 in another way. Figure 8 shows the one dimensional family considered above, in the two affine parameterisations. In both panels the family is shown by the solid line. The fact that it is an exponential family in its own right from its linearity is clear in from Panel (b). The duality relation, given by equation (7), allows us to define a set of (-1)-flat families which cut the model (Fisher) orthogonally. These are plotted in both plots by the dashed lines.

In the figure again we note the way that the dual boundaries are determining the global structure of the IG. In Panel (b), the (-1)-parallel set of (-1)-geodesics are pulled in the (recession) directions determined by the vertices of the boundary – which are dually equivalent to the edges in Panel (a). The one (-1)-geodesic which passes through a vertex in (a) corresponds to the one cutting an edge in (b).

In terms of our running example, a one-dimensional family of the form shown in Fig. 7 could come from a logistic regression model. This would be a low dimensional $(+1)$-affine subset of the independence space. Mixtures of such families can be derived from random effects models over such logistic regression models, Agresti (2013).

5.2 *Dual Dimensionality*

As shown in Critchley and Marriott (2014a), the results in Example 5.1 are general. From that paper we have that the (-1)-convex hull of an open subset of a generic one-dimensional exponential family in Δ^k is of full dimension, where generic here means that the one dimensional sufficient statistic for the model has no ties.

Key Issue 10 *(Dimensional duality) We can summarise this by saying that in general low dimensional* $(+1)$*-objects have maximal dimensional* (-1)*-convex hulls in the simplex of distributions. Results such as this follow from total positivity properties of exponential families, Karlin (1968). Such results, despite being classical, probably have not been sufficiently explored in IG.*

The mixed parameterisation of Example 5.1 is also very general, see Barndorff-Nielsen and Blaesild (1983), as is the related idea of an inferential cut, Barndorff-Nielsen and Koudou (1996), which gives geometric conditions on when inference on subparameters – often called interest parameters – can be achieved independently of the remaining 'nuisance' parameters. See also Pistone et al. (1999). In these constructions, we have a duality relationship between the ± 1-affine parts of the construction with the sum of the dimensions being constant. Thus, if one is 'small' the other will be 'big'.

The IG theory which is found in Amari (1985) is based on the differential geometry of finite dimensional manifolds. It is natural to ask if it can be extended to 'infinite dimensional' models, applications to non-parametric statistics being the stand-out motivation, Pistone (2013), see also Morozova and Chentsov (1991). We note that, at least in statistical applications, some thought is required as to what 'infinite dimensional' should mean. For example, in his elegant geometric theory, Lindsay (1995) defines a non-parametric maximum likelihood estimate (NPMLE) in a finite, but data dependent, geometric construction. In applied statistics, at least, the sample size is always finite despite useful tools coming from infinite dimensional ideas, Small and McLeish (2011). Accordingly, a potentially fruitful concept is to think of 'infinite dimensional' as being the case where the dimension is not fixed a priori, rather is a function of the data.

Nevertheless, we can still think about the truly unbounded dimensional case, but this needs care. For example, Amari notes the problem of finding an 'adequate topology', Amari (1985, p. 93). There has been work following up this topological challenge.

Key Issue 11 *(Infinite dimensional affine structures) We note that the affine structures defined in Issue 2 are naturally infinite dimensional. Of course, to link them in a standard IG way we need the Fisher information which does not always exist, see Li et al. (2009).*

To try and construct a more complete infinite dimensional IG, Pistone et al. (1999) use the geometry of a Banach manifold and Orlicz spaces – where local patches on the manifold are modelled by Banach spaces. This generates a form of infinite dimensional exponential family, with expectation, natural and mixed parameterisations. Interestingly, as pointed out in Fukumizu (2005), the likelihood function with finite samples is not continuous on the manifold with this Banach structure. He points out that a reproducing kernel Hilbert space structure has a stronger topology and can be usefully employed. Another approach to the infinite dimensional case can be found in Newton (2012). More discussion on infinite versions of the simplex geometry used here as a running example can be found in Critchley and Marriott (2014a); see also Zhang (2013).

5.3 Application Areas

Application Area 19 (Neural networks) The papers Amari (1995, 1998) look at the way divergences can be used to efficiently fit neural network models. It uses a dual geometric form of the EM algorithm to estimate hidden layers in a neural network. In particular, it exploits the idea of a mixed parameterisation and Fisher orthogonality. Applications in this paper include stochastic multilayer perceptron models, mixtures of experts, and the normal mixture model. Related applications in this area include Amari et al. (1992) and Amari (1997).

Application Area 20 (Image segmentation) Image segmentation is a key step in image analysis. The paper Fu et al. (2013) uses entropy methods in the class of Gaussian mixture models to undertake image segmentation. Related work can be found in Zhang et al. (2013).

Application Area 21 (Multi-scale analysis) The spike train analysis described in Application Area 5 can involve the estimation of intensity functions of point processes. The paper Ramezan et al. (2014) analysed the multi-scale properties of these intensity functions in the spike train context. Here, a critical aspect is the concept of an inferential cut, strongly associated with the IG structure of the mixed parametrisation and discussed above in Example 5.1. Inferential cuts are studied when we want to undertake inference on an interest parameter in the presence of nuisance parameters and, outside of the Bayesian inference approach, this is a difficult question. The work of Kolaczyk and Nowak (2004, 2005) gives the foundation for applying the idea of cuts to a multi-scale analysis of intensity functions of point processes, and in other areas.

Application Area 22 (Non- and semi-parametric modelling) Non-parametric and semi-parametric modelling are very popular approaches in statistical practice and they can be viewed from a geometric perspective. The Hilbert space methodology of Small and McLeish (2011) is closely related to the Hilbert bundle approach of Amari and Kumon (1988) and the geometry of the estimating function approach – often called a semi-parametric method – can also be seen in Amari (1997). We also note the work of Gibilisco and Pistone (1998) and Zhang (2013) in this area. A nice applied example of a Hilbert space approach to interest rate modelling can be found in Brody and Hughston (2001).

6 Closing Comments

In this paper we have seen a number of 'dual' objects and, albeit without a formal definition, this is a characteristic which enables us to recognise an IG object when we see it. In Sect. 1 we have the pair: sample and model (parameter) spaces, in Sect. 2 we have a polytope and its polar dual, in Sect. 3 we have a divergence and its 'dual' where arguments are reversed, in Sect. 4 we have tangent and cotangent spaces, and in Sect. 5 we have pairs of low dimension $(+1)$-affine spaces, and high dimensional (-1)-convex hulls. We also note the work of Zhang (2006, 2015) which looks at the closely related ideas of reference and representation duality in IG.

One point we would like to make is that to give these objects dual structures, which are truly symmetric, often requires stronger regularity conditions than the user might need, or be able to provide. For example, while the sample/parameter space pairing is attractive in some ways, these are very different objects. For any given sample size, n, the sample space, in the running example of this paper, is a lattice inside a convex set and not a convex set itself. The link becomes clear in the 'asymptotic limit', but the user might not have large enough n for this to be at all relevant. Another example is the duality in the divergence section. To have the cleanest links between $D(\cdot;\cdot)$ and $D^*(\cdot;\cdot)$ requires regularity conditions on the Legendre transform, Rockafellar (1997), which can fail in simple examples. Another example is the way that the Fisher information allows the duality seen in Sects. 1 and 4 but as Li et al. (2009) illustrate, there are very simple, and useful, statistical models where this object does not exist. What can we take away from these examples? We feel that it would be a mistake to aim for a very elegant mathematical theory *per se*, as attractive as that might be, requiring regularity conditions which contextual considerations indicate to be overly restrictive. Rather, we would like IG to be as inclusive as possible, while still remaining a coherent set of theories.

One issue, that has been a focus of this paper, is the importance of boundaries in IG. In this paper, we concentrated on sets where probabilities are allowed to be zero. In fact, there are other boundaries where normalising constants, or moments, fail to exist. There are important and interesting open questions as to the limiting properties of traditional information geometric objects at these boundaries. Some results in this direction already exist. In Critchley and Marriott (2014a), the behaviour of the Fisher information near a boundary is analysed, while in Critchley and Marriott (2014b) it is shown that the (0)-geodesic (i.e. the minimum path length geodesic) smoothly touches the boundary set.

We have deliberately taken a non-traditional approach to building information geometric structures. It is common, in the literature, to start with the manifold structure of statistical models, defining differential geometric structures, such as metrics and connections, on them. This follows a standard approach in differential geometry, where the geometry of a manifold is defined *implicitly* and independent of any embedding space. This has the advantage for the geometer that they would not have to check that any construction depends on the choice of embedding space. Rather, since there are natural (± 1)-affine embedding spaces, defined in Sect. 1, we deliberately exploit their simplicity, generality and natural duality. Furthermore, the boundaries which we regard as fundamental, occur completely naturally in this approach.

In this paper, we have taken a personal tour through the emerging subject that is Information Geometry. As we are statisticians, we have mostly focused on applications related to modern statistical practice but, as instanced in the introduction, we note that IG has become a broad church and that there are many other places where it has had an important impact. The general notions of geometric dualistic structures and ideas of divergence that we have seen here are, of course, very widely applicable.

To close, we would like to reiterate some of the key ideas that we have tried to emphasize above. First, we note that the fundamental geometric objects of interest are not always going to be smooth manifolds – boundaries and closures matter. Second, we started our tour with the existence of *very general* affine structures. This is not the only way to build the foundations of IG, of course, but we find it a very attractive one. Third, convexity and other ideas from convex geometry are key in understanding IG structures. This relates to our fourth point, that boundaries of convex sets, and in particular their polar duals, give a great deal of information about the global IG of a problem. Fifth, we note a very attractive duality in dimension inherent in IG that has perhaps not had the attention in the literature that it could have. Sixth, and finally, we note that singularities in tensor fields and boundary effects – which again would not be expected for a geometry based on smooth manifolds – do play an important part in understanding IG as a whole and, we feel, understanding them will be important in moving IG forward.

References

Agresti, A. (2013). *Categorical data analysis*. New Jersey: Wiley.

Amari, S.-I. (1985). *Differential-geometrical methods in statistics* (Vol. 28). Heidelberg: Springer-Verlag.

Amari, S.-I. (1995). Information geometry of the EM and em algorithms for neural networks. *Neural Networks*, *8*(9), 1379–1408.

Amari, S.-I. (1997). Information geometry of neural networks - an overview. *Mathematics of neural networks* (pp. 15–23). Heidelberg: Springer.

Amari, S.-I. (1998). Natural gradient works efficiently in learning. *Neural Computation*, *10*(2), 251–276.

Amari, S.-I. (2015). Information geometry as applied to neural spike data. *Encyclopedia of Computational Neuroscience*, 1431–1433.

Amari, S.-I., Barndorff-Nielsen, O. E., Kass, R., Lauritzen, S., & Rao, C. (1987). Differential geometry in statistical inference. *IMS Lecture Notes-Monograph Series*, 1–240.

Amari, S.-I., & Cichocki, A. (2010). Information geometry of divergence functions. *Bulletin of the Polish Academy of Sciences: Technical Sciences*, *58*(1), 183–195.

Amari, S.-I., & Kumon, M. (1988). Estimation in the presence of infinitely many nuisance parameters–geometry of estimating functions. *The Annals of Statistics*, 1044–1068.

Amari, S.-I., Kurata, K., & Nagaoka, H. (1992). Information geometry of Boltzmann machines. *IEEE Transactions on Neural Networks*, *3*(2), 260–271.

Amari, S.-I., & Nagaoka, H. (2007). *Methods of information geometry* (Vol. 191). Rhode Island: American Mathematical Society.

Amari, S.-I., & Ohara, A. (2011). Geometry of q-exponential family of probability distributions. *Entropy*, *13*(6), 1170–1185.

Anaya-Izquierdo, K., Critchley, F., & Marriott, P. (2014). When are first-order asymptotics adequate? a diagnostic. *Statistics*, *3*(1), 17–22.

Anaya-Izquierdo, K., Critchley, F., Marriott, P., & Vos, P. (2013a). Computational information geometry: foundations. *Geometric science of information* (pp. 311–318). Heidelberg: Springer.

Anaya-Izquierdo, K., Critchley, F., Marriott, P., & Vos, P. (2013b). Computational information geometry in statistics: Mixture modelling. *Geometric science of information* (pp. 319–326). Heidelberg: Springer.

Anaya-Izquierdo, K., Critchley, F., Marriott, P., & Vos, P. (2016). *The geometry of model sensitivity: An illustration. In Computational information geometry: For image and signal processing*. Heidelberg: Springer.

Anaya-Izquierdo, K., Marriott, P. (2007). Local mixture models of exponential families. *Bernoulli*, 623–640.

Arwini, K. A., & Dodson, C. T. J. (2008). *Information geometry: Near randomness and near independence*. Heidelberg: Springer.

Barndorff-Nielsen, O. (1978). *Information and exponential families in statistical theory*. New Jersey: Wiley.

Barndorff-Nielsen, O., & Blaesild, P. (1983). Exponential models with affine dual foliations. *The Annals of Statistics*, 753–769.

Barndorff-Nielsen, O., Cox, D., & Reid, N. (1986). The role of differential geometry in statistical theory. *International Statistical Review/Revue Internationale de Statistique*, 83–96.

Barndorff-Nielsen, O. E. (1987). Differential geometry and statistics: some mathematical aspects. *Indian Journal of Mathematics*, *29*(3), 335–350.

Barndorff-Nielsen, O. E., Blaesild, P., & Mora, M. (1989). Generalized higher-order differentiation. Acta Applicandae. *Mathematica*, *16*(3), 243–259.

Barndorff-Nielsen, O. E., & Cox, D. R. (1989). *Asymptotic techniques for use in statistics*. London: Chapman & Hall.

Barndorff-Nielsen, O. E., & Jupp, P. E. (1997). Statistics, yokes and symplectic geometry. *In Annales de la Faculté des sciences de Toulouse: Mathématiques*, *6*, 389–427.

Barndorff-Nielsen, O. E., Jupp, P. E., & Kendall, W. S. (1994). Stochastic calculus, statistical asymptotics, Taylor strings and phyla. *In Annales de la Faculté des sciences de Toulouse: Mathématiques*, *3*, 5–62.

Barndorff-Nielsen, O. E., & Koudou, A. E. (1996). Cuts in natural exponential families. *Theory of Probability & Its Applications*, *40*(2), 220–229.

Basseville, M. (2013). Divergence measures for statistical data processingan annotated bibliography. *Signal Processing*, *93*(4), 621–633.

Belavkin, R. V. (2013). Optimal measures and Markov transition kernels. *Journal of Global Optimization*, *55*(2), 387–416.

Betancourt, M. (2013). A general metric for Riemannian manifold Hamiltonian Monte Carlo. *Geometric science of information* (pp. 327–334). Heidelberg: Springer.

Betancourt, M., Byrne, S., Livingstone, S., & Girolami M. (2014). The geometric foundations of Hamiltonian Monte Carlo. arXiv:1410.5110

Bhattacharya, A. (2008). *Nonparametric statistics on manifolds with applications to shape spaces*. ProQuest.

Blaesild, P. (1991). Yokes and tensors derived from yokes. *Annals of the Institute of Statistical Mathematics*, *43*(1), 95–113.

Bregman, L. M. (1967). The relaxation method of finding the common point of convex sets and its application to the solution of problems in convex programming. *USSR Computational Mathematics and Mathematical Physics*, *7*(3), 200–217.

Brody, D. C., & Hughston, L. P. (2001). Interest rates and information geometry. In *Proceedings of the Royal Society of London A: Mathematical, Physical and Engineering Sciences* (Vol. 457, pp. 1343–1363). London: The Royal Society.

Brown, L. D. (1986). Fundamentals of statistical exponential families with applications in statistical decision theory. *IMS Lecture Notes-monograph Series*.

Buck, B., & Macaulay, V. A. (1991). *Maximum entropy in action: a collection of expository essays*. Oxford: Clarendon Press.

Chentsov, N. N. (1972). *Statistical decision rules and optimal inference* (Vol. 53). Rhode Island: American Mathematical Society.

Cichocki, A., Zdunek, R., Phan, A. H., & Amari, S.-I. (2009). *Nonnegative matrix and tensor factorizations: Applications to exploratory multi-way data analysis and blind source separation*. New Jersey: Wiley.

Collins, M., Schapire, R. E., & Singer, Y. (2002). Logistic regression. *Adaboost and Bregman Distances. Machine Learning*, *48*(1–3), 253–285.

Cont, A., Dubnov, S., & Assayag, G. (2011). On the information geometry of audio streams with applications to similarity computing. *IEEE Transactions on Audio, Speech, and Language Processing*, *19*(4), 837–846.

Copas, J., & Eguchi, S. (2005). Local model uncertainty and incomplete-data bias (with discussion). *Journal of the Royal Statistical Society: Series B (Statistical Methodology)*, *67*(4), 459–513.

Cox, D., & Barndorff-Nielsen, O. (1994). *Inference and asymptotics* (Vol. 52). Florida: CRC Press.

Cox, D. R., & Hinkley, D. V. (1979). *Theoretical statistics*. Florida: CRC Press.

Cox, D. R., & N. Reid (1987). Parameter orthogonality and approximate conditional inference. *Journal of the Royal Statistical Society. Series B (Methodological)*, 1–39.

Cressie, N., & Read, T. R. (1984). Multinomial goodness-of-fit tests. *Journal of the Royal Statistical Society. Series B (Methodological)*, 440–464.

Critchley, F., & Marriott, P. (2014a). Computational information geometry in statistics: theory and practice. *Entropy*, *16*, 2454–2471.

Critchley, F., & Marriott, P. (2014b). Computing with Fisher geodesics and extended exponential families. *Statistics and Computing*, 1–8.

Critchley, F., Marriott, P., & Salmon, M. (1994). Preferred point geometry and the local differential geometry of the Kullback-Leibler divergence. *The Annals of Statistics*, 1587–1602.

Critchley, F., Marriott, P., & Salmon, M. (1996). On the differential geometry of the Wald test with nonlinear restrictions. *Econometrica: Journal of the Econometric Society*, 1213–1222.

Csiszár, I. (1975). I-divergence geometry of probability distributions and minimization problems. *The Annals of Probability*, 146–158.

Csiszár, I. (1995). Generalized projections for non-negative functions. *Acta Mathematica Hungarica*, *68*(1–2), 161–186.

Csiszár, I., et al. (1967). On topological properties of f-divergences. *Studia Scientiarum Mathematicarum Hungarica*, *2*, 329–339.

Csiszár, I., & Matus, F. (2005). Closures of exponential families. *The Annals of Probability*, *33*(2), 582–600.

Dodson, C. T. (1987). *Geometrization of statistical theory*. In: Proceedings of the GST Workshop, University of Lancaster Department of Mathematics, 28–31 October 1987. ULDM Publications.

Dodson, C. T., & Poston, T. (2013). *Tensor geometry: the geometric viewpoint and its uses* (Vol. 130). Heidelberg: Springer Science & Business Media.

Efron, B. (1975). Defining the curvature of a statistical problem (with applications to second order efficiency). *The Annals of Statistics*, 1189–1242.

Eguchi, S. (2006). Information geometry and statistical pattern recognition. *Sugaku Expositions*, *19*(2), 197–216.

Eguchi, S. (2009). Information divergence geometry and the application to statistical machine learning. *Information theory and statistical learning* (pp. 309–332). Heidelberg: Springer.

Eguchi, S., et al. (1985). A differential geometric approach to statistical inference on the basis of contrast functionals. *Hiroshima Mathematical Journal*, *15*(2), 341–391.

Eguchi, S., et al. (1992). Geometry of minimum contrast. *Hiroshima Mathematical Journal*, *22*(3), 631–647.

Eguchi, S., & Copas, J. (1998). A class of local likelihood methods and near-parametric asymptotics. *Journal of the Royal Statistical Society: Series B (Statistical Methodology), 60*(4), 709–724.

Eguchi, S., Komori, O., & Ohara, A. (2014). Duality of maximum entropy and minimum divergence. *Entropy, 16*(7), 3552–3572.

Freund, Y., & Schapire, R. E. (1995). A decision-theoretic generalization of on-line learning and an application to boosting. *Computational learning theory* (pp. 23–37). Heidelberg: Springer.

Fu, W., Johnston, M., & Zhang, M. (2013). Gaussian mixture models and information entropy for image segmentation using particle swarm optimisation. *2013 28th International Conference of Image and Vision Computing New Zealand (IVCNZ)* (pp. 328–333). New Jersey: IEEE.

Fukuda, K. (2004). From the zonotope construction to the Minkowski addition of convex polytopes. *Journal of Symbolic Computation, 38*, 1261–1272.

Fukumizu, K. (2005). Infinite dimensional exponential families by reproducing kernel Hilbert spaces. In *Proceedings of the 2nd International Symposium on Information Geometry and its Applications* (pp. 324–333).

Geyer, C. J. (2009). Likelihood inference in exponential families and directions of recession. *Electronic Journal of Statistics, 3*, 259–289.

Gibilisco, P., & Pistone, G. (1998). Connections on non-parametric statistical manifolds by Orlicz space geometry. *Infinite Dimensional Analysis, Quantum Probability and Related Topics, 1*(02), 325–347.

Gibilisco, P., Riccomagno, E., Rogantin, M., & Wynn, H. (2010). *Algebraic and Geometric Methods in Statistics*. New York, NY: Cambridge University Press.

Girolami, M., & Calderhead, B. (2011). Riemann manifold Langevin and Hamiltonian Monte Carlo methods. *Journal of the Royal Statistical Society: Series B (Statistical Methodology), 73*(2), 123–214.

Ikeda, S., Tanaka, T., & Amari, S.-I. (2004). Stochastic reasoning, free energy, and information geometry. *Neural Computation, 16*(9), 1779–1810.

Jaynes, E. T. (1978). Where do we stand on maximum entropy. *The maximum entropy formalism*, 15–118.

Jaynes, E. T. (1982). On the rationale of maximum-entropy methods. *Proceedings of the IEEE, 70*(9), 939–952.

Jordan, M., Sudderth, E. B., Wainwright, M., Willsky, A. S., et al. (2010). Major advances and emerging developments of graphical models [from the guest editors]. *Signal Processing Magazine, IEEE, 27*(6), 17–138.

Jorgensen, B. (1987). Exponential dispersion models. *Journal of the Royal Statistical Society. Series B (Methodological)*, 127–162.

Kahle, T., et al. (2010). Neighborliness of marginal polytopes. *Contributions to Algebra and Geometry, 51*(1), 45–56.

Karlin, S. (1968). *Total positivity* (Vol. 1). California: Stanford University Press.

Karlin, S., & Shapley, L. S. (1953). Geometry of moment spaces. *Memoirs of the American Mathematical Society 12.*

Kass, R., Tierney, L., & Kadane, J. (1988). *Asymptotics in Bayesian computation. Bayesian statistics, 3*, 261–278.

Kass, R. E. (1989). The geometry of asymptotic inference. *Statistical Science*, 188–219.

Kass, R. E., Tierney, L., & Kadane, J. B. (1991). Laplace method in Bayesian analysis. *Contemporary Mathematics, 115*, 89–99.

Kass, R. E., & Vos, P. W. (2011). *Geometrical foundations of asymptotic inference* (Vol. 908). New Jersey: Wiley.

Kawakita, M., & Eguchi, S. (2008). Boosting method for local learning in statistical pattern recognition. *Neural computation, 20*(11), 2792–2838.

Kolaczyk, E. D., & Nowak, R. D. (2004). Multiscale likelihood analysis and complexity penalized estimation. *Annals of Statistics*, 500–527.

Kolaczyk, E. D., & Nowak, R. D. (2005). Multiscale generalised linear models for nonparametric function estimation. *Biometrika, 92*(1), 119–133.

Lauritzen, S. L. (1987). Statistical manifolds. *Differential geometry in Statistical Science* (pp. 163–216). CA: IMS Hayward.
Lauritzen, S. L. (1996). *Graphical models*. Oxford: Oxford University Press.
Lee, J. A., & Verleysen, M. (2007). *Nonlinear dimensionality reduction*. Heidelberg: Springer Science & Business Media.
Lenglet, C., Rousson, M., Deriche, R., & Faugeras, O. (2006). Statistics on the manifold of multivariate normal distributions: Theory and application to diffusion tensor MRI processing. *Journal of Mathematical Imaging and Vision*, *25*(3), 423–444.
Li, P., Chen, J., & Marriott, P. (2009). Non-finite fisher information and homogeneity: An em approach. *Biometrika*, *96*(2), 411–426.
Lindsay, B. (1995). *Mixture Models: Theory, Geometry, and Applications*. Institute of Mathematical Statistics.
Liu, M., Vemuri, B., Amari, S.-I., & Nielsen, F. (2012). Shape retrieval using heirarchical total Bregman soft clustering. *IEEE Transactions on pattern analysis and machine intelligence*, *34*, 2407–2419.
Marriott, P. (2002). On the local geometry of mixture models. *Biometrika*, *89*(1), 77–93.
Marriott, P., & Salmon, M. (2000). *Applications of differential geometry to econometrics*. Cambridge University Press.
McCullagh, P. (1987). *Tensor methods in statistics* (Vol. 161). London: Chapman and Hall.
McCullagh, P., & Nelder, J. A. (1989). *Generalized linear models* (Vol. 37). Florida: CRC Press.
Mio, W., Badlyans, D., & Liu, X. (2005a). A computational approach to fisher information geometry with applications to image analysis. *Proceedings of the EMMCVPR*, 18–33.
Mio, W., Badlyans, D., & Liu, X. (2005b). A computational approach to fisher information geometry with applications to image analysis. *Energy minimization methods in computer vision and pattern recognition* (pp. 18–33). Heidelberg: Springer.
Mio, W., & Liu, X. (2006). Landmark representation of shapes and Fisher-Rao geometry. *2006 IEEE International Conference on Image Processing* (pp. 2113–2116). New Jersey: IEEE.
Morozova, E. A., & Chentsov, N. N. (1991). Natural geometry of families of probability laws. Itogi Nauki i Tekhniki. Seriya "Sovremennye Problemy Matematiki. *Fundamental'nye Napravleniya"*, *83*, 133–265.
Murray, M. K., & Rice, J. W. (1993). *Differential geometry and statistics* (Vol. 48). Florida: CRC Press.
Newton, N. J. (2012). An infinite-dimensional statistical manifold modelled on Hilbert space. *Journal of Functional Analysis*, *263*(6), 1661–1681.
Nielsen, F. (2014). *Geometric Theory of Information*. Heidelberg: Springer.
Nielsen, F., & Barbaresco, F. (2014). *Proceedings of the 2nd International Symposium on Information Geometry and its Applications*. Heidelberg: Springer.
Nielsen, F., & Bhatia, R. (2013). *Matrix information geometry*. Heidelberg: Springer.
Nielsen, F., & Nock, N. (2014a). Optimal interval clustering: Application to Bregman clustering and statistical mixture learning. *IEEE Transactions on pattern analysis and machine intelligence*, *21*(10), 1289–1292.
Nielsen, F., & Nock, R. (2014b). On the chi square and higher-order chi distances for approximating f-divergences. *IEEE Signal Processing Letters*, *21*(1), 10–13.
Peter, A., & Rangarajan, A. (2006). Shape analysis using the Fisher-Rao Riemannian metric: Unifying shape representation and deformation. *3rd IEEE International Symposium on Biomedical Imaging: Nano to Macro, 2006* (pp. 1164–1167). New Jersey: IEEE.
Pfanzagl, J. (1973). Asymptotic expansions related to minimum contrast estimators. *The Annals of Statistics*, 993–1026.
Pistone, G. (2013). Nonparametric information geometry. In Geometric Science (Ed.), *of Information* (pp. 5–36). Heidelberg: Springer.
Pistone, G., Riccomagno, E., & Wynn, H. (2000). *Algebraic Statistics: Computational Commutative Algebra in Statistics*. London: Chapman and Hall.

Pistone, G., Rogantin, M. P., et al. (1999). The exponential statistical manifold: mean parameters, orthogonality and space transformations. *Bernoulli*, *5*(4), 721–760.

Ramezan, R., Marriott, P., & Chenouri, S. (2014). Multiscale analysis of neural spike trains. *Statistics in medicine*, *33*(2), 238–256.

Rao, C. (1945). Information and accuracy attainable in the estimation of statistical parameters. *Bulletin of the Calcutta Mathematical Society*, *37*(3), 81–91.

Rao, C. R. (1987). Differential metrics in probability spaces. *Differential geometry in statistical inference*, *10*, 217–240.

Read, T. R., & Cressie, N. (2012). *Goodness-of-fit statistics for discrete multivariate data*. Heidelberg: Springer Science & Business Media.

Rinaldo, A., Feinberg, S., & Zhou, Y. (2009). On the geometry of discrete exponential families with applications to exponential random graph models. *Electronic Journal of Statistics*, *3*, 446–484.

Rockafellar, R. T. (1997). *Convex analysis. Princeton landmarks in mathematics*. Princeton: Princeton University Press.

Shima, H. (2007). *The geometry of Hessian structures* (Vol. 1). Singapore: World Scientific.

Simpson, S. L., Hayasaka, S., & Laurienti, P. J. (2011). Exponential random graph modeling for complex brain networks. *PLoS One*, *6*(5), e20039.

Skilling, J. (1989). Classic maximum entropy. In *Maximum Entropy and Bayesian Methods* (pp. 45–52). Heidelberg: Springer.

Small, C. G., & McLeish, D. L. (2011). *Hilbert space methods in probability and statistical inference* (Vol. 920). New Jersey: Wiley.

Sontag, D., & Jaakkola, T. S. (2007). New outer bounds on the marginal polytope. *In Advances in Neural Information Processing Systems (NIPS)*, *20*, 1393–1400.

Takatsu, A. (2013). Behaviors of φ-exponential distributions in Wasserstein geometry and an evolution equation. *SIAM Journal on Mathematical Analysis*, *45*(4), 2546–2556.

Takenouchi, T., Eguchi, S., Murata, N., & Kanamori, T. (2008). Robust boosting algorithm against mislabeling in multiclass problems. *Neural computation*, *20*(6), 1596–1630.

Takenouchi, T., Komori, O., & Eguchi, S. (2012). An extension of the receiver operating characteristic curve and AUC-optimal classification. *Neural computation*, *24*(10), 2789–2824.

Takenouchi, T., Komori, O., & Eguchi, S. (2015). A novel boosting algorithm for multi-task learning based on the Itakuda-Saito divergence. In *Bayesian inference and Maximum Entropy methods in science and engineering (MAXENT 2014)* (Vol. 1641, pp. 230–237). Melville: AIP Publishing.

Tatsuno, M., Fellous, J.-M., & Amari, S.-I. (2009). Information-geometric measures as robust estimators of connection strengths and external inputs. *Neural computation*, *21*(8), 2309–2335.

Tatsuno, M., & Okada, M. (2003). How does the information-geometric measure depend on underlying neural mechanisms? *Neurocomputing*, *52*, 649–654.

Tierney, L., Kass, R. E., & Kadane, J. B. (1989). Fully exponential Laplace approximations to expectations and variances of nonpositive functions. *Journal of the American Statistical Association*, *84*(407), 710–716.

Tsallis, C. (1988). Possible generalization of Boltzmann-Gibbs statistics. *Journal of statistical physics*, *52*(1–2), 479–487.

Tsallis, C. (2009). *Introduction to nonextensive statistical mechanics*. Heidelberg: Springer.

Tuy, H. (1998). *Convex analysis and global optimization*. London: Klumer academic publishers.

Vos, P. W., & Marriott, P. (2010). Geometry in statistics. *Wiley Interdisciplinary Reviews: Computational Statistics*, *2*(6), 686–694.

Wainwright, M. J., & Jordan, M. I. (2003). Variational inference in graphical models: The view from the marginal polytope. In *Proceedings of the Annal Allerton Conference on Communication Control and Computing* (Vol. 41, pp. 961–971). Citeseer.

Wainwright, M. J., & Jordan, M. I. (2008). Graphical models, exponential families, and variational inference. *Foundations and Trends in Machine Learning*, *1*(1–2), 1–305.

Watanabe, S. (2009). *Algebraic geometry and statistical learning theory* (Vol. 25). Cambridge University Press.

Wong, W. H., & Li, B. (1992). Laplace expansion for posterior densities of nonlinear functions of parameters. *Biometrika, 79*(2), 393–398.

Zhang, H., Wu, Q., & Nguyen, T. M. (2013). Image segmentation by a robust modified gaussian mixture model. *2013 IEEE International Conference on Acoustics, Speech and Signal Processing (ICASSP)* (pp. 1478–1482). New Jersey: IEEE.

Zhang, J. (2004). Divergence function, duality, and convex analysis. *Neural Computation, 16*(1), 159–195.

Zhang, J. (2006). Referential duality and representational duality on statistical manifolds. In *Proceedings of the Second International Symposium on Information Geometry and Its Applications, Tokyo* (pp. 58–67).

Zhang, J. (2013). Nonparametric information geometry: From divergence function to referential-representational biduality on statistical manifolds. *Entropy, 15*(12), 5384–5418.

Zhang, J. (2015). Reference duality and representation duality in information geometry. In *Bayesian inference and Maximum Entropy methods in science and engineering (MAXENT 2014)* (Vol. 1641, pp. 130–146). Melville: AIP Publishing.

Zhao, H., & Marriott, P. (2014). Variational Bayes for regime-switching log-normal models. *Entropy, 16*(7), 3832–3847.

Towards the Geometry of Model Sensitivity: An Illustration

Karim Anaya-Izquierdo, Frank Critchley, Paul Marriott and Paul Vos

1 Introduction

This paper is an introduction to a new approach to ubiquitous problems of modelling - in particular model building, sensitivity and uncertainty. By exploring simple, but illustrative, examples we demonstrate that Computational Information Geometry (CIG) delivers both concrete and unexpected results. The key idea is to construct, in a geometrical way, universal operational spaces which allow perturbations of parametric models to be explored and also throws light on the relationship between parametric and non-parametric approaches to inference. We deliberately restrict attention to a particular class of models and type of associated inference problems, see Definition 1, and use them to illustrate a much wider theory which will be explored in related papers.

We positively agree with Box's view of science, put forward in his landmark paper 'Science and Statistics' Box (1976) and developed in Box (1980). In these

F. Critchley—This work has been partly funded by EPSRC grant EP/L010429/1.
P. Marriott—This work has been partly funded by NSERC discovery grant 'Computational Information Geometry and Model Uncertainty'.

K. Anaya-Izquierdo (✉)
Department of Mathematical Sciences, University of Bath, Bath BA2 7AY, UK
e-mail: kai21@bath.ac.uk

F. Critchley
The Open University, Walton Hall, Milton Keynes, Buckinghamshire MK7 6AA, UK
e-mail: f.critchley@open.ac.uk

P. Marriott
University of Waterloo, 200 University Avenue West, Waterloo, ON, Canada
e-mail: pmarriot@uwaterloo.ca

P. Vos
East Carolina University, Greenville, NC 27858-4353, USA
e-mail: VOSP@ecu.edu

F. Nielsen et al. (eds.), *Computational Information Geometry*,
Signals and Communication Technology, DOI 10.1007/978-3-319-47058-0_2

papers, scientific knowledge is seen as advancing by 'a motivated *iteration* between theory and practice' (his italics), 'efficient scientific iteration evidently requiring unhampered feedback', adding that: 'since all models are wrong the scientist must be alert to what is importantly wrong.' We are therefore developing operational tools to implement these powerful ideas.

Let us assume that we are starting with a well-defined question of interest which we are trying to answer using a set of data, for example wanting to learn about a population mean. We have a working problem formulation – our current statistical model – which has been constructed by using prior knowledge about the experiment and also diagnostic testing to evaluate the adequacy of the model. This model, of course is just one of many that could have been used and we construct an operational universal space (using the tools of high-dimensional extended sparse multinomial models, Anaya-Izquierdo et al. (2013)) which allows us to define the geometry of the 'space of all models'. Within this space we make extensive use of the tools of CIG (Critchley and Marriott 2014a) and the inferential ideas of orthogonal, mixed parameterizations, Barndorff-Nielsen and Blaesild (1983), Cox and Reid (1987), and the related idea of an (approximate) cut, Barndorff-Nielsen and Koudou (1995). All these ideas benefit from a direct computational implementation. In particular since the dimension of the operational space could be very large we need computational tools that are adapted thereto, such as linear programming. Within this operational space we iteratively construct a – as it turns out surprisingly simple – space of all important perturbations of the working model, where important is relative to changes in inference for the given question of interest. The iterative search first looks for the directions of most sensitivity. It also carefully distinguishes between possible modelling choices that are empirically answerable and those which must remain purely putative. For example, observed data may contain a great deal of information about a population mean, but almost none about a high quantile value. In this we follow the principle spelt out in Critchley and Marriott (2004) of 'learn what you can, explore what you can't. Aspects which must be putative exploration can then inform future scientific experiments.

1.1 A Cartoon of Modelling

In parametric statistics the question of model specification is a critical one about which there has been a great deal of research. Here we take a new, information geometric approach to the problem. To illustrate ideas we deliberately select simple models and focus on the independent, identically distributed (i.i.d.) case and one particular form of question of interest. Despite this apparent simplicity we show non-trivial geometric results. Other questions of interest, and the much more important for practitioners, case of models with covariates and/or dependent data, will be studied in following papers.

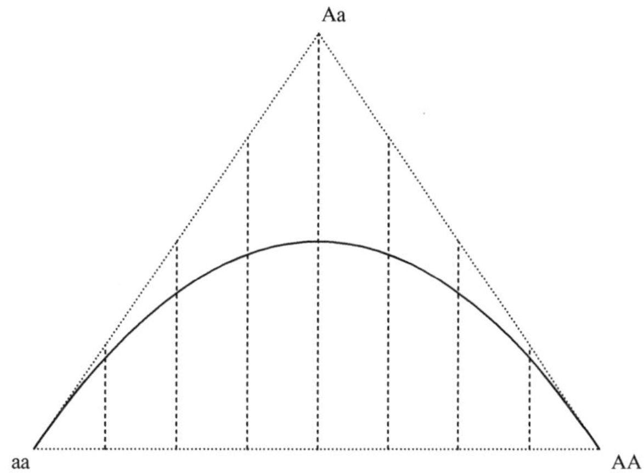

Fig. 1 The Hardy–Weinberg model embedded in the simplex

We look at the very general question of where do statistical models come from? To start the exploration Examples 1 and 2 describes two extremes. In the first, information about model specification comes from theoretical considerations.

Example 1 The Hardy–Weinberg model in genetics states that allele frequencies in a population in equilibrium follows a parametric model. The three classes, coded by aa, Aa, and AA, are assumed to follow a one-dimensional model

$$P(AA) = p^2, P(Aa) = 2pq, P(aa) = q^2,$$

where the marginal probabilities are $P(A) = p$, $P(a) = q$ and $p = 1 - q$. If we observed frequencies (n_0, n_1, n_2), with fixed $n := n_0 + n_1 + n_2$, the number of independent realisations, then we get a one dimensional exponential family $P((n_0, n_1, n_2) | p)$ with a sufficient statistic $n_0 - n_2$ and natural parameter $\phi := \log(p/(1-q))$. The mean of the sufficient statistic is $\mu = n(2p - 1)$, and we will consider the case where this, or equivalently p, is of inferential interest. We illustrate this in the space of trinomial models in Fig. 1.

Often, though, models are derived from purely empirical considerations without an underlying theoretical model, thus leaving the problem of finding exactly what the model has contributed to the inference problem.

Example 2 Here we have the question: what is the population mean? To answer this we consider the (simulated) dataset shown in Fig. 2, Panel (a). It is count data with a known support of [0, 40] and an analyst considered that it might be plausibly modelled by a binomial distribution. However, after fitting it is noted that there is some over-dispersion relative to the binomial.

Figure 2b shows the fitted distribution corresponding to a binomial assumption. Panel (c) shows one way of undertaking the inference problem, given we accept the model, using the asymptotic distribution of the deviance – i.e. twice the log-likelihood function. For this example the sample size and simplicity of the model

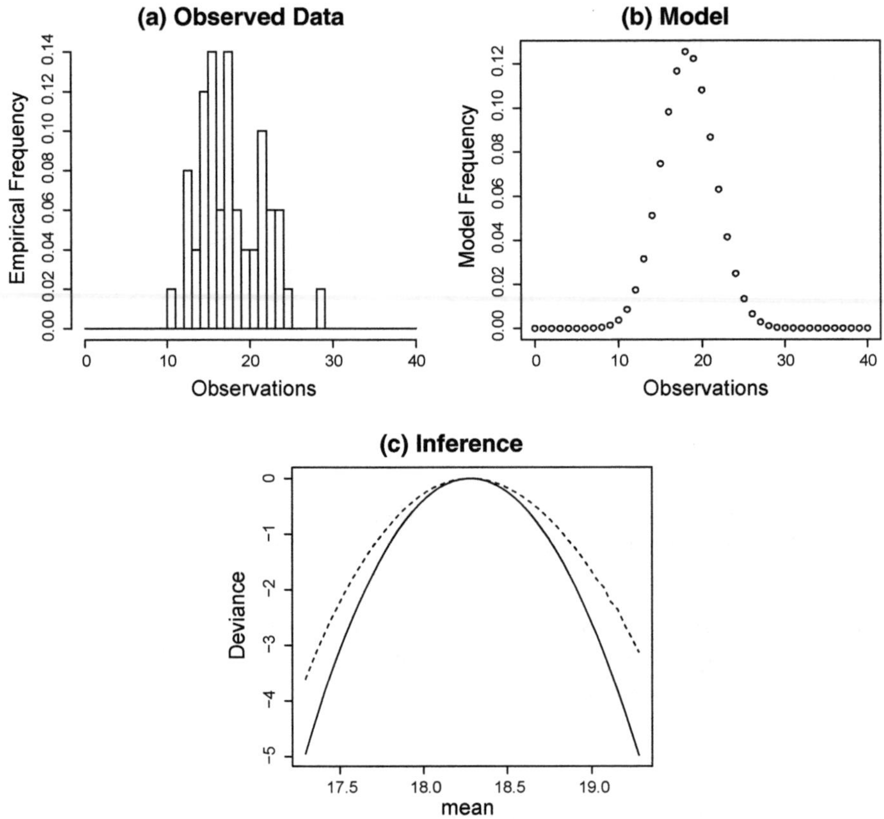

Fig. 2 The data for Example 2: Panel **a** shows the empirical distribution of observed data, **b** shows the fitted binomial model, **c** show the model based deviance (*solid line*) and the empirical deviance (*dashed line*)

mean first order asymptotic arguments are appropriate. In this panel we contrast the model based inference with a model free ones, such as the t-test, justified here on asymptotic grounds, or the empirical likelihood (Owen 1988), which is shown in the panel with the dashed line. Here the empirical likelihood is computed by profiling over the whole multinomial; see "Appendix 2: Empirical Likelihood for the Mean Parameter in a Multinomial Setting" for details. We can clearly see how much the choice of model is contributing to the inference question by comparing these two methods. We clarify that we are not, here, judging which one of these methods is "best", rather just noting that there are significant differences which the analyst needs to be aware of. In this toy example we see the model choice is affecting the uncertainty associated with the estimate by a moderate amount. It is one of the aims of this paper to show how the geometry of the model can be used to understand the effect of model choice.

Above we have discussed so-called 'model free methods' whose justification is through asymptotic analysis. We would like to make the point here that most asymptotic arguments are not uniform across the simplex, Anaya-Izquierdo et al. (2014). That is for a given sample size the quality of the asymptotic approximation depends on where we are in the simplex. Thus, strictly speaking, these methods are not truly 'model free', nevertheless they do provide a sensible and practical base line for comparison.

Suppose we have a working, putative model and we want to check that the model is concordant with the data. One general approach is to perform an appropriate goodness of fit test such as Kolmogorov-Smirnov or Cramer Von Mises. An alternative general approach, the one we follow here, is to build a larger model, or to perturb the original. This is a common approach and we highlight in particular, Box (1980), Cook (1986) and Critchley and Marriott (2004). Of particular interest is the observation in Cox (1986) who points out the importance of assessing that a parameter in a larger model has a meaning which is consistent with that in the smaller model. We want to make sure we are always comparing 'apples' with 'apples'. For that reason we will focus on attention on inference about quantities, such as population means, which have a 'model free' meaning.

In this initial, and exploratory, paper we will only look at the following class of models. We fully understand that, outside the classroom, it would be unusual for all of the regularity conditions to hold but it is common in practice that a substantial number will and hence we can learn a lot by exploring this basic class of models. Examples include maximum entropy and random graph models

Definition 1 All models in this paper satisfy the following regularity conditions: (a) all models are for discrete and finite random variables, (b) the observed data is independently and identically distributed, (c) the putative working model is a regular exponential family, (d) the parameter of inferential interest is the mean of a statistic, s, and (e) this statistic is part of the sufficient statistic.

Simple examples of such families include the distribution of a random vector X, where the probability vector $(P(X = x_i))_{i=0}^k$ is given by

$$\left(\pi_i^0 \exp(\phi s(x_i) - M(\phi))\right)_{i=0}^k \tag{1}$$

in which the inferential question of interest concerns $\mu = \mu(\phi) := E_\phi\left[s(X)\right]$.

We look at the sensitivity of the inferential answer to perturbations of Model (1). In particular we might, for example, perturb π_i^0 via

$$\pi_i^0 \to \pi_i^0 + \delta\omega_i =: \pi_i^0(\delta) \tag{2}$$

where $\sum_{i=0}^k \omega_i = 0$ and ω has unit length with respect to the Fisher information at π^0. We also look at extending the sufficient statistic via a larger model of the kind

$$\left(\pi_i^0 \exp(\phi_1 s(x_i) + \phi_2 s_2(x_i) - M(\phi_1, \phi_2))\right)_{i=0}^k \tag{3}$$

while keeping the inferential question about $E(s(X))$ fixed in both cases.

One property of perturbations (2) and (3) is that the maximum likelihood estimate of μ is always the sample mean $\frac{\sum_{j=1}^{N} s(x_j)}{N}$, where N is the sample size. Thus, for purely pointwise estimation, these perturbations play no role. We study them to investigate other aspects of inference, such as quantifying the uncertainty in the estimate and understanding the role of 'outliers'.

2 The Geometry of Model Sensitivity

In "Appendix 1: The Model Space, Cuts and Closures" we review some key concepts that are used in the analysis below, and give references for the interested reader. Nothing in there is new and those familiar with extended exponential families can move on without loss.

The idea of an inferential cut (Barndorff-Nielsen 1976; Barndorff-Nielsen and Koudou 1995) is key motivation for what follows. These are studied when we want to undertake inference on an interest parameter in the presence of nuisance parameters. Outside of the Bayesian inference approach this is a difficult question but important for understanding our perturbation approach to sensitivity analysis. Starting with a baseline model we will often be extending it, making it more flexible, at the cost of adding 'nuisance' parameters.

2.1 *Approximate Cuts*

Let

$$\mathcal{F} = \left\{ f_s(s;\phi) = \exp\left(s^T\phi - M(\phi)\right) f_s(s;0) \; : \; \phi \in \mathcal{P} \right\}$$

be a regular natural exponential family with respect to some fixed σ-finite measure ν on $\mathbb{R}^k$. The mean parameter function will be denoted by $\mu(\phi) := D_\phi M(\phi) = E[s;\phi]$. We will use the following notation

$$s = \left(s_1, s_{(1)}\right)^T, \qquad \mu = \left(\mu_1, \mu_{(1)}\right)^T, \qquad \phi = \left(\phi_1, \phi_{(1)}\right)^T,$$

where s_1, μ_1, ϕ_1 are of dimension r, the $\phi_{(1)}$ notation means exclude the elements in ϕ_1, so that $s_{(1)}, \mu_{(1)}, \phi_{(1)}$ are of dimension $k - r$. The following definition, and much more detail, can be found in Barndorff-Nielsen and Blaesild (1983).

Definition 2 (*Mixed Parameterisation*) For a regular exponential family $\mathcal{F}$, the map

$$\phi \mapsto \begin{pmatrix} \mu_1(\phi) \\ \phi_{(1)}(\phi) \end{pmatrix}$$

is a diffeomorphism on $\mathcal{P}$ with range $\mu_1(\mathcal{P}) \times \phi_{(1)}(\mathcal{P})$. The parameterisation $(\mu_1, \phi_{(1)})$ is called the mixed parameterisation of $\mathcal{F}$.

Definition 3 (*Cut*) The statistic s_1 is said to be a cut for the regular exponential family $\mathcal{F}$ if an only if

$$f_s(s; \mu_1, \phi_{(1)}) = f_{s_1}(s_1; \mu_1) f_{s_{(1)}|s_1}(s_{(1)} \mid s_1; \phi_{(1)})$$

for all s, μ_1, $\phi_{(1)}$.

In the presence of a cut, if μ_1 is of interest, we can make inferences about it (without knowledge of $\phi_{(1)}$) using only the marginal distribution of s_1. Analogously, if $\phi_{(1)}$ is of interest, we can make inferences about it (without knowledge of μ_1) using only the conditional distribution of $s_{(1)}$ given s_1.

The following structural result about the existence of a cut can be found in Barndorff-Nielsen and Koudou (1995).

Theorem 1 *Let $\mathcal{F}$ be a regular exponential family. The following are equivalent:*

1. *s_1 is a cut for $\mathcal{F}$*
2. *The variance of s_1 depends only on μ_1*
3. *s_1 follows a natural exponential family model on $\mathbb{R}^r$ with natural parameter given by $\phi_1^*(\mu_1)$*
4. *For some functions $\phi_1^* : \mathbb{R}^r \to \mathbb{R}^r$ and $H : \mathbb{R}^{k-r} \to \mathbb{R}^r$*

$$\phi_1(\mu_1, \phi_{(1)}) = \phi_1^*(\mu_1) + H(\phi_{(1)}) \tag{4}$$

5. *For some functions $k : \mathbb{R}^{k-r} \to \mathbb{R}^{k-r}$ and $h : \mathbb{R}^{k-r} \to \mathbb{R}^{(k-r)\times r}$*

$$\mu_{(1)}(\mu_1, \phi_{(1)}) = k(\phi_{(1)}) - h(\phi_{(1)})\, \mu_1 \tag{5}$$

6. *For some functions $M_1 : \mathbb{R}^r \to \mathbb{R}$ and $K : \mathbb{R}^{(k-r)} \to \mathbb{R}$*

$$M(\phi_1(\mu_1, \phi_{(1)}), \phi_{(1)}) = M_1(\phi_1^*(\mu_1)) + K(\phi_{(1)}). \tag{6}$$

An inferential cut, as defined in Definition 3, is a very useful tool allowing exact likelihood inference about a mean, say, independent of nuisance parameters. However, the existence of an exact cut is rather rare. Instead we might look to loosen its definition somewhat. Theorem 1 gives us a number of equivalent choices, any of which could be relaxed. In this paper we choose to focus on Condition (2) and define an approximate cut in the following, if rather informal, way. We also note related ideas in Christensen and Kiefer (1994, 2000).

Definition 4 (*Approximate cut*) In the notation of Theorem 1, the dependence of the variance of s_1 on nuisance parameters is a measure of the sensitivity of the model for inference about μ_1. When this dependence is small we say that we have an approximate cut and the corresponding nuisance parameter is called insensitive.

The motivation here is that we expect a small inferential change concerning μ_1 when we perturb the model in directions in which the conditions of Theorem 1 hold to a reasonable approximation. We have selected the variance characterisation to focus on due to computational advantages described below. This is clearly not the only choice from this theorem and other characterisations can be explored.

2.2 Directional Approach

We wish to investigate the sensitivity of inference on a mean with respect to the specification of a working, putative model. We first take a directional approach to perturbing the working model. That is, we define perturbations of the model using directions in the model space. Since the model space is affine the direction can be considered either as a tangent vector or as a vector in the tangent space. We shall see that many perturbation directions have effectively zero effect on the model performance – which and we call the *insensitive directions* – while other can have a large effect on inference. This second class is called the *sensitive directions.*

Definition 5 (*Directional perturbation*) Consider the perturbation given by Expression (2), where we treat $\delta\omega$ as the perturbation parameter. Defining the i^{rmth} component of s_2 by

$$s_{2i}(\delta) := \log\left(1 + \frac{\pi_i^0(\delta) - \pi_i^0}{\pi_i^0}\right) = \delta\frac{w_i}{\pi_i^0} + O(\delta^2)$$

we can write

$$\pi_i^0(\delta)\exp(\phi s_{1i} - M(\phi)) = \pi_i^0\exp(\phi s_{1i} + s_{2i}(\delta) - M(\phi,\delta)) = \pi_i^0\exp(\phi s_{1i} + \delta s_{2i} - M(\phi,\delta)) + O(\delta^2)$$

where $s_{2i} := \frac{w_i}{\pi_i}$. Thus, to first order in δ, the two perturbations schemes (2) and (3) are equivalent.

By combining Definitions 4 and 5 we look for the most sensitive directional perturbation by solving an optimisation problem. We are looking for the directional perturbation which gives the largest local effect on the variance. First, we note that if, as we do, we want to preserved the meaning of the parameter of interest under such a perturbation it will be convenient, although not essential, to perturb the distribution π^0 in ways which preserved the mean, i.e. satisfying

$$\sum_{i=0}^{k} s_{1i}\pi_i^0 = \sum_{i=0}^{k} s_{1i}(\pi_i^0 + \delta\omega_i) = \mu \Rightarrow \sum_{i=0}^{k} s_{1i}\omega_i = 0.$$

These ideas give rise to the following.

Theorem 2 *Consider the following optimisation problem:*

$$\max_{\omega\in\Omega}\sum_{i=0}^{k} s_{1i}^2\omega_i,$$

where

$$\Omega := \left\{\omega \mid \sum_{i=0}^{k}\omega_i = 0, \sum_{i=0}^{k} s_{1i}\omega_i = 0, \omega^T\Sigma_{\pi^0}\omega = \epsilon^2\right\},$$

Σ_{π^0} *is defined by the Fisher information at* π^0*, and* ϵ *is a small, user-selected tuning parameter.*

This has the solution that $\Sigma^{-1}\omega$ *is proportional to the* Σ^{-1}*-orthogonal projection of* $s_2^{(2)} := (s_{2i}^2)_{i=0}^k$ *on the space orthogonal to the space spanned by* 1*, and* s*. This problem could also be solved by standard linear programming methods.*

Proof This follows from a direct Lagrangian analysis of the problem. For specific details about the analytic calculation of this infinitesimal direction see "Appendix 3: Sensitive Infinitesimal Perturbations".

We note here that we have set up the optimization problem to find the direction which maximise the effect on the variance. This is motivated by the definition of the approximate cut. This results in a computationally tractable problem, even in high dimensional model spaces. This is, of course, not the only possible optimization problem that could be studied. There are many other measures – information theoretic or geometric – which could be used, and this is to be studied in future work.

Example 2 (*Revisited*) Returning to our running example we can solve the optimisation problem in Definition 2. Figure 3 shows the results of this based on the binomial working model. In Panel (a) we plot the statistic that we have added to the sufficient statistic, and can see that it is non-linear – and further analysis shows it is well approximated by a quadratic function. Panel (b) shows the corresponding ω-vector, while in (c) we show the effect on inference on the interest parameter μ. The solid line shows the deviance for μ from the unperturbed model, while the dash line is that from the profile likelihood for μ associated with Model in displayed Eq. (3). This gives us a way of making inference on μ in the presence of the new nuisance parameter ϕ. It can be shown that the profile likelihood is very close to the, 'model free', empirical likelihood in Fig. 2c. The main difference is that there is a small amount of skewness. It can be shown that adding one further element to the sufficient statistic, corresponding to a cubic – skewness – term gives almost exact agreement between the model-based and the model-free estimation.

For clarity, though, we note that we are not claiming that the model-free approach discussed above is the 'correct' inferential procedure. For instance, in Example 1, where there is a well-established theoretical model, it seems natural that the analyst use this model.

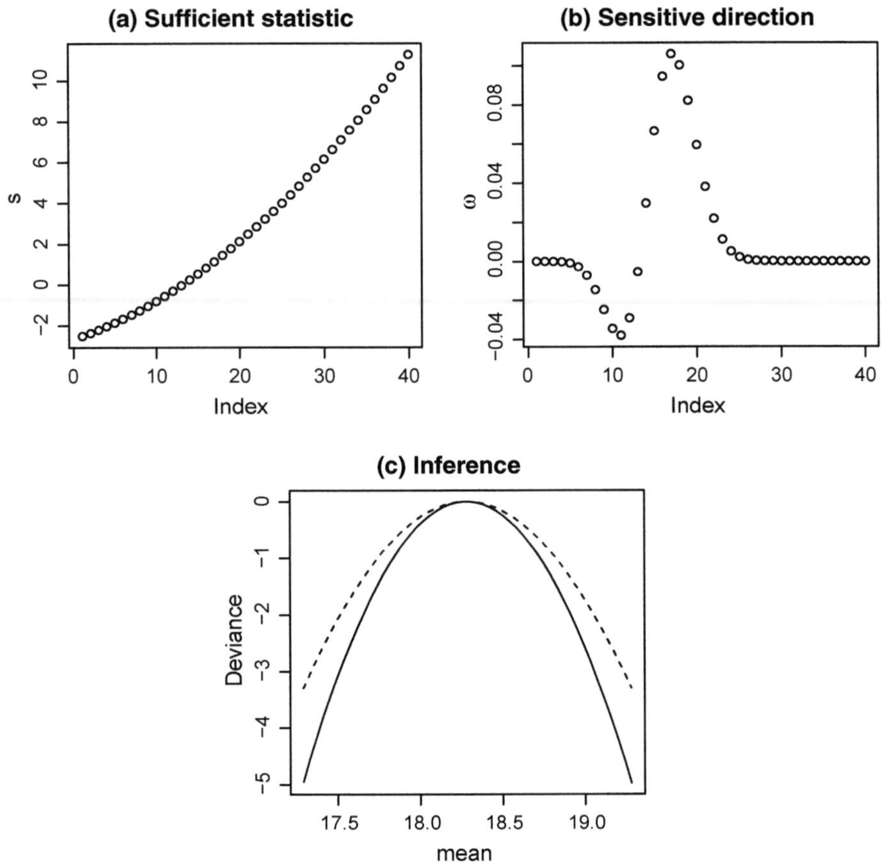

Fig. 3 Computing the direction using the Fisher matrix

The existence of a sensitive direction means a particular modelling choice has had a substantial impact on inference although, as discussed in Definition 1, not on the value of the point estimate. For example, we might measure that impact on inference by the difference in the plotted solid and dashed lines in Fig. 3c. If we have good reason to believe the model (e.g. Example 1) then that is not a problem, but if we are not sure about the model (Example 2) then we might want to use proportionally more information from the data.

It is illuminating to compare the results of this infinitesimal analysis with standard ways described in the literature to generalize the Binomial to a two-dimensional exponential family of the form

Definition 6 The following are extensions of the binomial model. They are all exponential families of the form

$$f(x;\phi_1,\phi_2) = \binom{K}{x} \exp(\phi_1 x + \phi_2 g(x) - M(\phi_1,\phi_2)), \quad x = 0, 1, \ldots, K.$$

where g is a function which the modeller selects. Important examples include (a) letting $g(x) = I(x = 0)$ giving the zero-inflated binomial Lambert (1992), (b) letting $g(x) = x(x - K)$ giving Altham's multiplicative binomial model Altham (1978), and (c) letting $g(x) = x \log\left(\frac{x}{K}\right) + (K - x)\log\left(1 - \frac{x}{K}\right)$ giving one of Efron's double binomial model Efron (1986).

For Example 2 we are adding a quadratic term to the sufficient statistic, this gives a model which is equivalent to Model (b) in Definition 6. If the sample size is large enough for first order asymptotic inference to be plausible then any model which is flexible enough to fit the sample mean and variance – that is the mode and the hessian of the log-likelihood – would essentially give the same inference.

The infinitesimal perturbation indicates that particular low dimensional exponential families in Δ^k might be of interest. To study these families, as is clear from the definition of an exact cut, Definition 3, the shape of the log-likelihood function in the mixed parameterisation, Definition 2, is critical in understanding the effect on μ-inference. Further, we need to understand the embedding of the families in the simplex and, in particular, the way that they meet the boundary.

Example 2 (*Revisited*) Figure 4 shows the geometry of a two dimensional family inside the simplex Δ^k. Panel (a) shows the function $g(x)$ which has been added, and this corresponds precisely to the term (s_i) in Definition 5. We note that all that matters when studying the sufficient statistics of exponential families is the span of the corresponding linear space. The form of the function Fig. 4a only differs from Fig. 3a by a linear function: they both have the same inferential effect.

Because the two-dimensional exponential families lies in a closed simplex we need to analyse its boundary. In Panel (c) this is shown in the mean parameters, while the corresponding convex polar – which shows the directions of recession for the natural parameters, see Appendix or Critchley and Marriott (2014b); Geyer (2009); Rinaldo et al. (2009) – is shown in Panel (b). We can extend the analysis by using the theory of approximate cuts as shown in Fig. 5. The fundamental result on exact cuts, Theorem 1, says that in the mixed parameterisation the Fisher information being independent of the nuisance parameter is a sufficient condition for an exact cut.

Figure 5a shows the contours of the log-likelihood in the natural parameters.We have also added the directions of recession, see Fig. 4b, for later analysis. We see that the log-likelihood function is close to, but is not exactly, a quadratic function in these parameters. The solid horizontal line through zero is the working model in these parameters. Panel (c) shows the same information, but now in the mean parameters. The dash line corresponds to the boundary of the exponential family shown in detail in Fig. 4c. The solid curve is the null model – not straight here since it is an exponential family and these are the mean parameters. Again the log-likelihood is not a quadratic function of these parameters. Panel (b) shows the same information in the mixed parameters, noting the vertical axes of (a) and (b) and the horizontal axes of (b) and (c) agree.

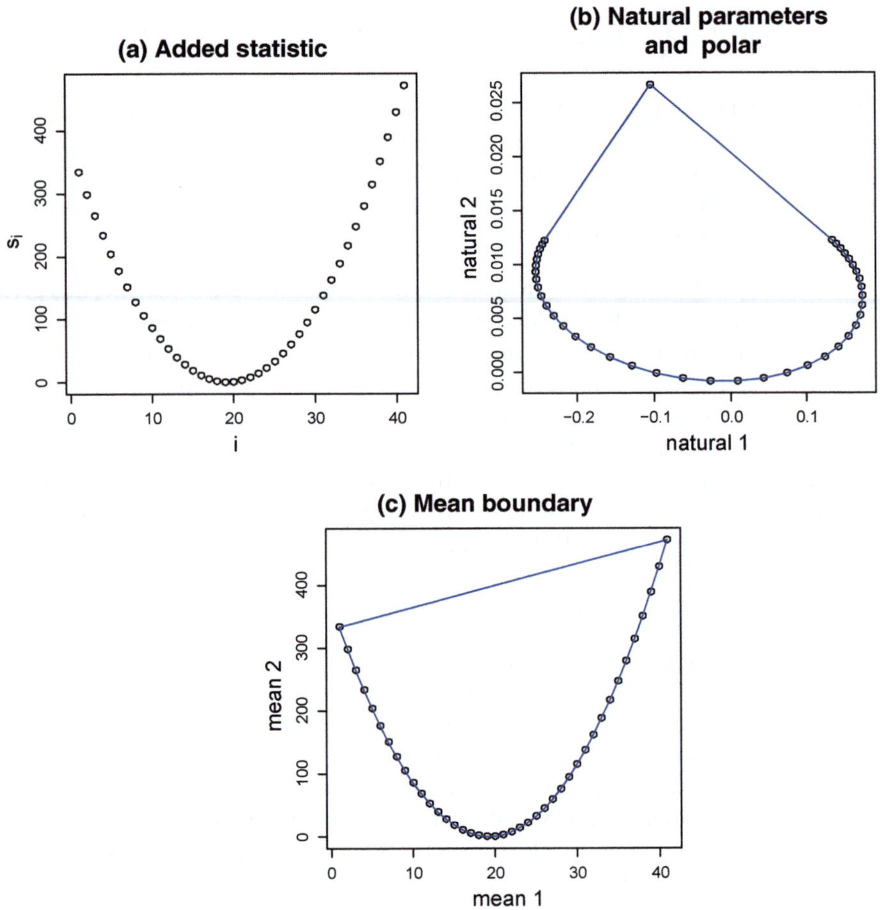

Fig. 4 The geometry of the two dimensional full exponential family

If we perturb the model in the nuisance direction – which corresponds in Panel (b) to a translation of the base model vertically – we want to see how this affects inference about the interest parameter – the horizontal axis in (b). We see from the shape the of log-likelihood in (b) that a vertical shift will change the hessian of the log-likelihood– i.e. the Fisher information – thus strongly affects inference.

In more generality we can use the shape of the log-likelihood function in the mixed parameters to assess the effect of a perturbation of the model in a given direction. Theorem 1 essentially states that if the log-likelihood was quadratic – i.e. had a fixed hessian for different horizontal slices – then we would be close to an exact cut. Hence perturbations in that direction would not have much effect on μ-inference. We illustrate this in the example below.

The infinitesimal analysis of of this section indicates that perturbations which add a quadratic, and to a small extent a cubic, sufficient statistic to the model are going

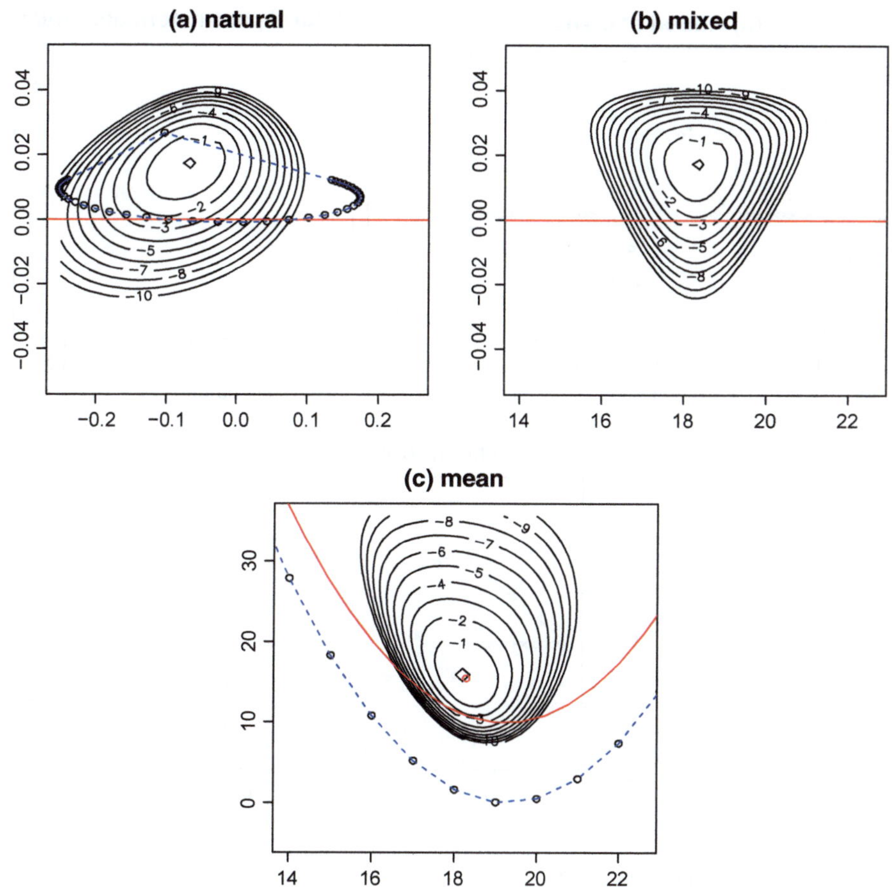

Fig. 5 Using approximate cut theory

to have an effect on inference, but that there will be many perturbations that have no effect. We see this below.

Example 2 (*Revisited*) We choose an arbitrarily selected perturbation direction, shown in Fig. 6a. The only constraint we put on the selection was that it only has weight in cells which have a positive observed count, see Fig. 2a for the data. We will make clearer in the following section why we add this constraint, but intuititively it seems sensible to first focus on directions where the data is informative. Panels (b) and (c) show, as before, the geometry of the natural and mean parameters taking into account their boundaries. We see here that the two dimensional model meets five distinct vertices of the simplex.

As in the previous example we show the contours of the log-likelihood in three parameterisations (a) natural, (b) mixed and (c) mean. The key plot in Fig. 7 is the middle one where the log-likelihood looks very close to being a quadratic function.

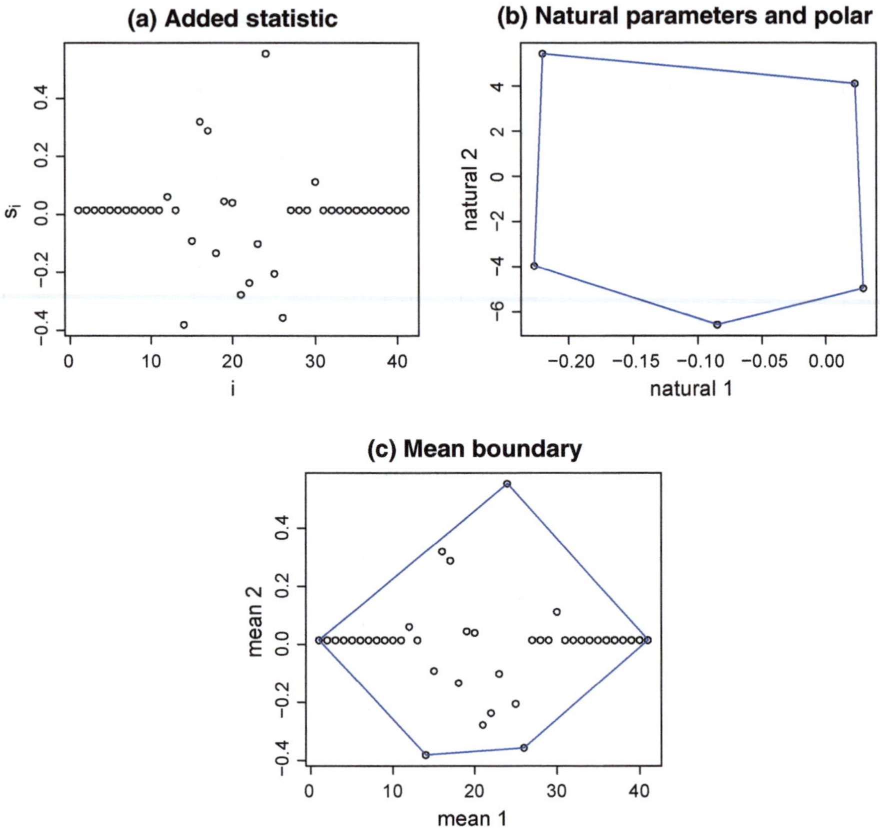

Fig. 6 Using approximate cut theory

This means that perturbations of the based model – illustrated with the solid horizontal line in (b) – have almost no effect on inference on μ. This direction is – as expected – very insensitive as far as μ inference is concerned. It can be shown that this is true of most 'randomly selected' directions for this example.

At this point in the analysis, of our simple problem, it seems that just a couple of low order polynomials will completely determine the sensitive directions. In fact, as we now show, we can find other interesting directions by considering perturbations of the data rather than the model. This is a form of robustness analysis where some data points might be considered 'outliers' by the analyst so are not representative of the model and can be down-weighted or removed. If we look at the models described in Definition 6 we see that Model (a) allows the changing of the weight in a single cell. In that case the zero cell. Of course we can perturb the weight of other cells in particular ones which we may have identified as containing 'outliers'.

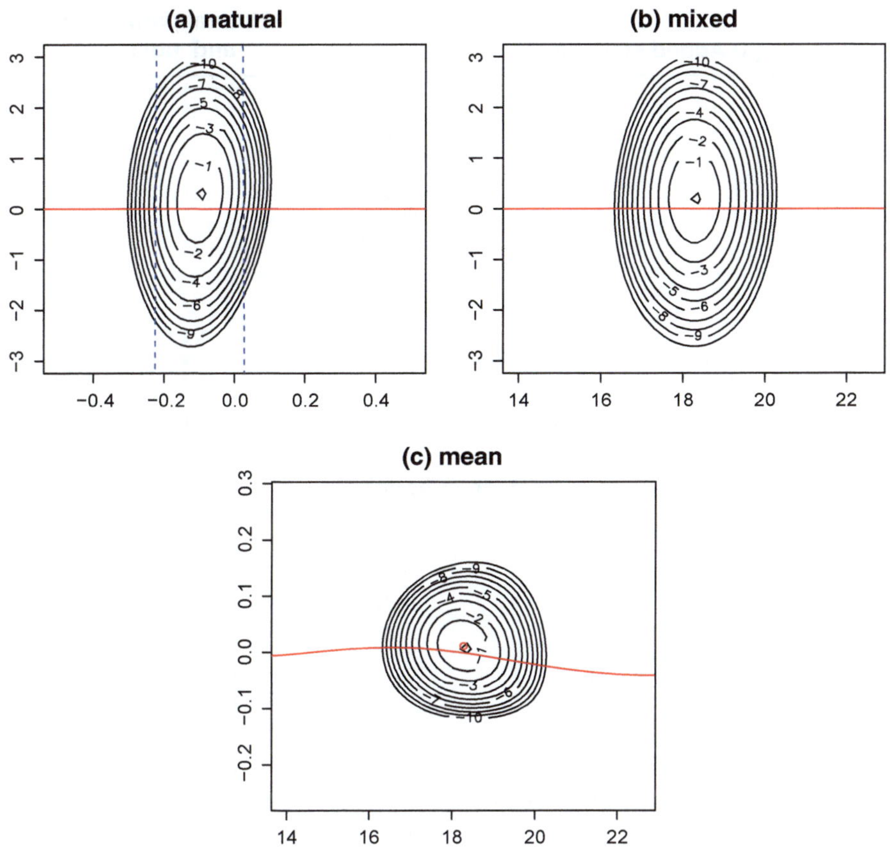

Fig. 7 The geometry of the two dimensional full exponential family

Example 2 (*Revisited*) Let us start with a perturbation of a cell which is clearly not an outlier and lies right at the centre of the observed data. The geometry of the family is shown in Fig. 8 as before. Panel (a) shows the perturbation vector, which is is reweighing of cell 18, the sample mean of the data. The boundary and corresponding polar dual are shown in (b) and (c).

We can see the inferential effect of this perturbation in Fig. 9. From Panel (c) we see that the boundary of the family is having an effect on the shape of the log-likelihood in the mean parameterisation, with the relavant part of the boundary being the horizontal dashed line and the working model the curved solid line. In Panel (a) we see some distortion in the corresponding direction of recession. In the mixed paremeters, Panel (b), we see that the log-likelihood is close to quadratic indicating a small effect on μ-inference.

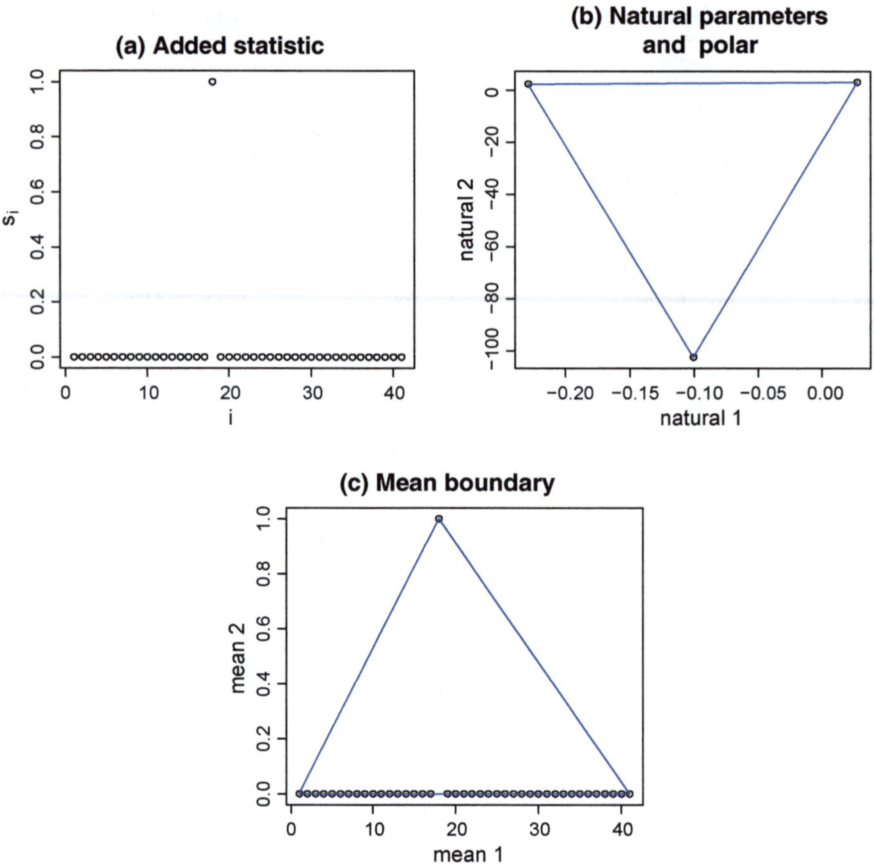

Fig. 8 The geometry of the two dimensional full exponential family

We can contrast this with the following case illustrated in Figs. 10 and 11. Here the perturbation is on the cell which contains the largest observed value. This *might*, in our example, be a candidate for being considered an 'outlier'. The perturbation vector is shown Fig. 10a and the corresponding boundary geometry in Panels (b) and (c).

Figure 11 shows the effect on μ-inference. Panel (c) shows that the boundary and the model are very close – on a scale defined by the size of log-likelihood based inference – indeed the solid and dash lines are almost on top of one another in the panel. This effect is mirrored in Panel (a) where the contours of the log-likelihood function are being pulled in the direction of the corresponding direction of recession. In Panel (b) the mixed parameterisation shows the effect on μ-inference is very strong. The log-likelihood is very far from being a quadratic function and so keeping or removing this single point can strongly change our inferential conclusions.

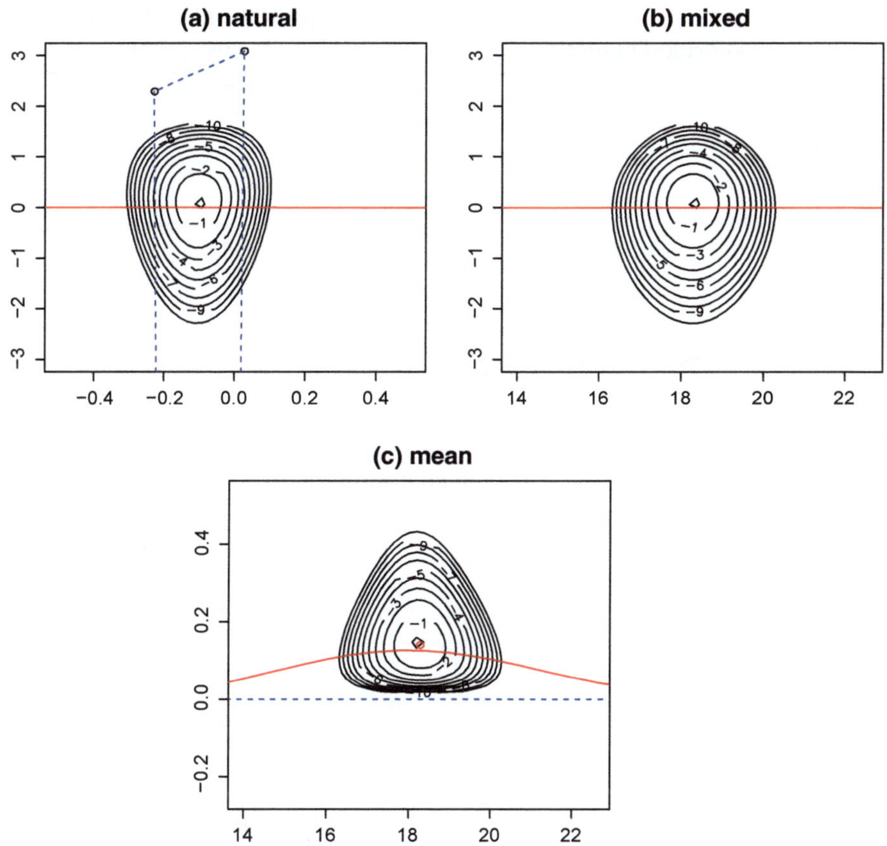

Fig. 9 The geometry of the two dimensional full exponential family

If the distance to the boundary was small we would expect that the shape of the likelihood would be distorted and so it would be unlikely that there would be even approximately a cut in that direction. We therefore, as part of our search for sensitivity directions should look for directions where the distance is small.

Definition 7 We can define the distance – as measured by the Fisher information – between π^0 and π where π lies on the face defined by the set of indexes $\mathcal{I}$ i.e.

$$\{\pi | \pi_i = 0 \iff i \in \mathcal{I}\}.$$

We look at the squared -1-distance, in the notation of Amari (1985), with a fixed metric at π^0 which is

$$Q(\pi) := \sum_{i=0}^{k} \frac{(\pi_i - \pi_i^0)^2}{\pi_i^0}.$$

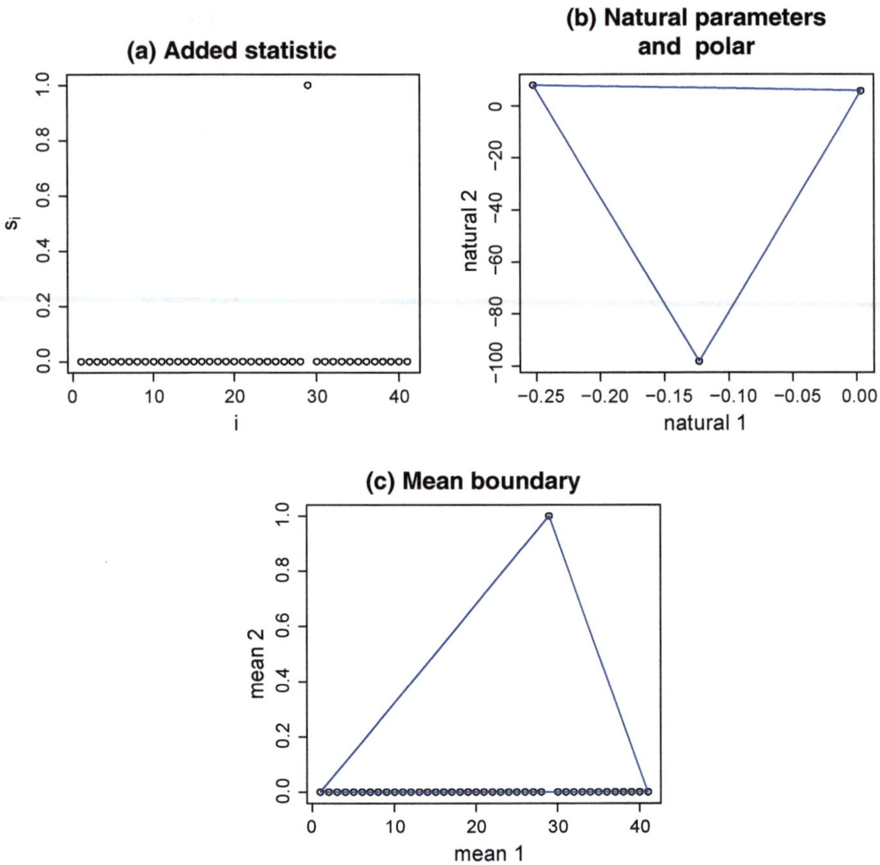

Fig. 10 The geometry of the two dimensional full exponential family

Theorem 3 *The minimum squared distance* $Q(\pi)$ *is* $\frac{\pi_{\mathcal{I}}^0}{1-\pi_{\mathcal{I}}^0}$. *where* $\pi_{\mathcal{I}}^0 := \sum_{i\in\mathcal{I}} \pi_i$. *Further, the set of directions which are close to the boundary form a union of cones.*

Proof This follows from direct calculation.

2.3 *Global Perturbations: Region of Interest*

One feature of the analyses shown in Figs. 9 and 11 is that the 'distance' to the boundary seems to play an important role when looking for sensitive directions. It is for this reason that the previous analyses – which are fundamentally based on infinitesimal arguments – can be complemented with more global ones, briefly explored in this section.

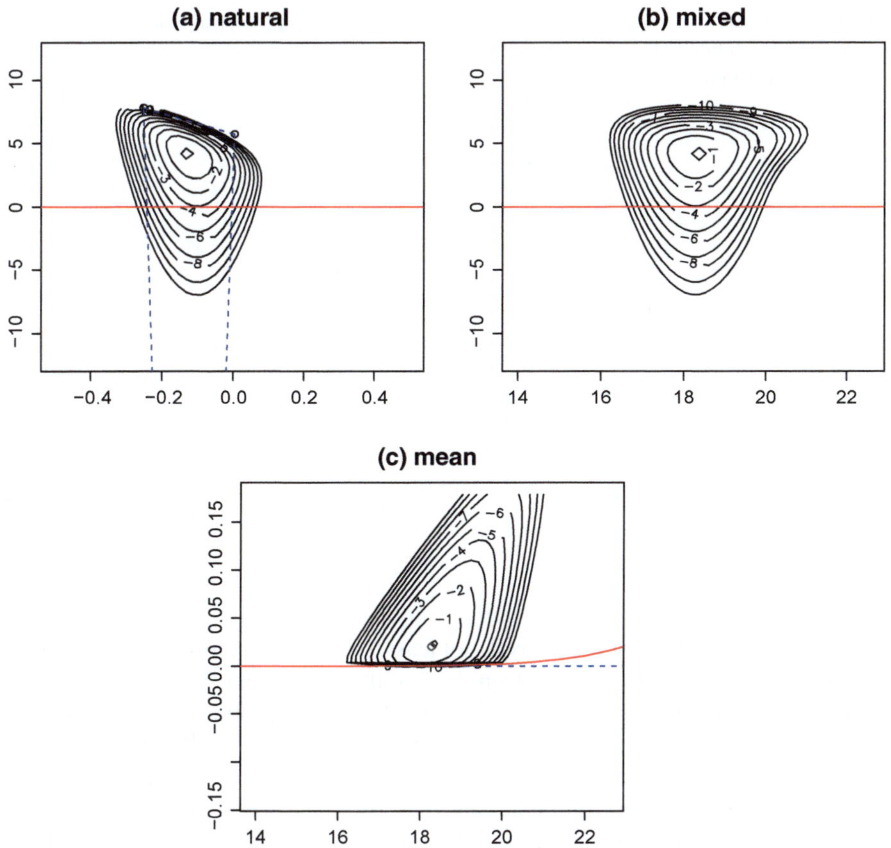

Fig. 11 The geometry of the two dimensional full exponential family

One major advantage of the infinitesimal approach is that, because of Definition 5 perturbations of the form (2) and (3) are the same. The corresponding direction of perturbation is just a tangent vector and can be represented in the $+1$ or -1 form in the notation of Amari (1985). In this section we take a more global approach and focus on perturbations of the form (2) i.e. $\pi_i^0 \rightarrow \pi_i^0 + \omega_i$. Looking for interesting perturbation vectors would then involve the following optimisation problem.

$$\max_{\omega} \sum_{i=0}^{k} t_i^2 w_i \text{ such that} \sum_{i=0}^{k} \omega_i = 0, \sum_{i=0}^{k} t_i \omega_i = 0, \pi_i + \omega_i \geq 0 \tag{7}$$

Assuming that all values of t_i are distinct, all but 2 values of $\pi_i + w_i$ will be zero. Using this, or simply using linear programming to solve the problem numerically, gives the solution shown in Fig. 12. This is simply the distribution which has the empirical mean of the data and the maximum variance. Of course Fig. 12 is not going

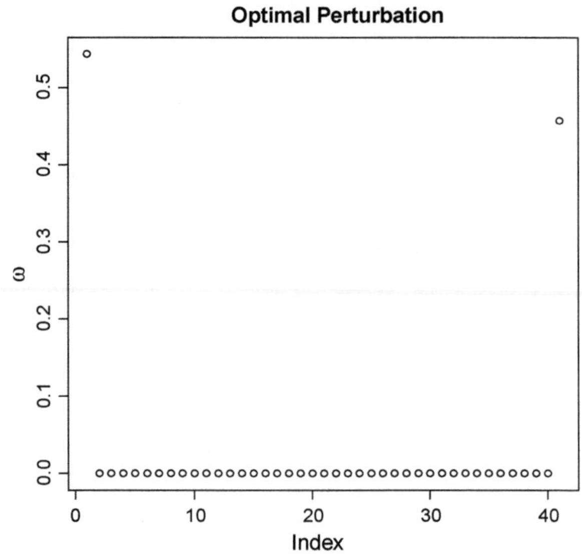

Fig. 12 The linear programming solution to the unconstrained optimisation problem

to be a plausible model for the analyst, not least because it is completely inconsistent with empirical distribution, aside from having the right mean. We propose that we need to restrict the search space of ω in the optimisation problem to a subset of Δ^k of distributions which are 'consistent' with the data. We call such a subset a *region of interest* in Δ^k. There are a number of ways to do this and we explore just one here.

We first note that, for computational reasons, since we are trying to solve an optimisation problem, it would be advantageous to look for convex regions of interest and the simplest choice is to have the convexity in the (-1)-affine structure of Δ^k. We can then still apply the method of linear programming which can work in very high dimensional problems.

The region of interest is designed to represent models which are consistent with the data. For any set of indices, $\mathcal{I}$, we have a corresponding count set $\sum_{i\in\mathcal{I}} n_i$ and we can use these aggregate counts to define linear inequalities on probabilities;

$$l_{\mathcal{I}}\left(\sum_{i\in\mathcal{I}} n_i\right) \leq \sum_{i\in\mathcal{I}} \pi_i \leq u_{\mathcal{I}}\left(\sum_{i\in\mathcal{I}} n_i\right), \tag{8}$$

which would show that the empirical and model based probability masses are consistent after selecting the lower bound function l and the upper bound function u. How to choose the index subsets is a matter of choice, since there are exponentially many in k to select. For simplicity here, and to explore the problem, we only look at contiguous subsets. A simple way to select upper and lower bounds is to treat each count $\sum_{i\in\mathcal{I}} n_i$ as an observation from a binomial distribution and use a corre-

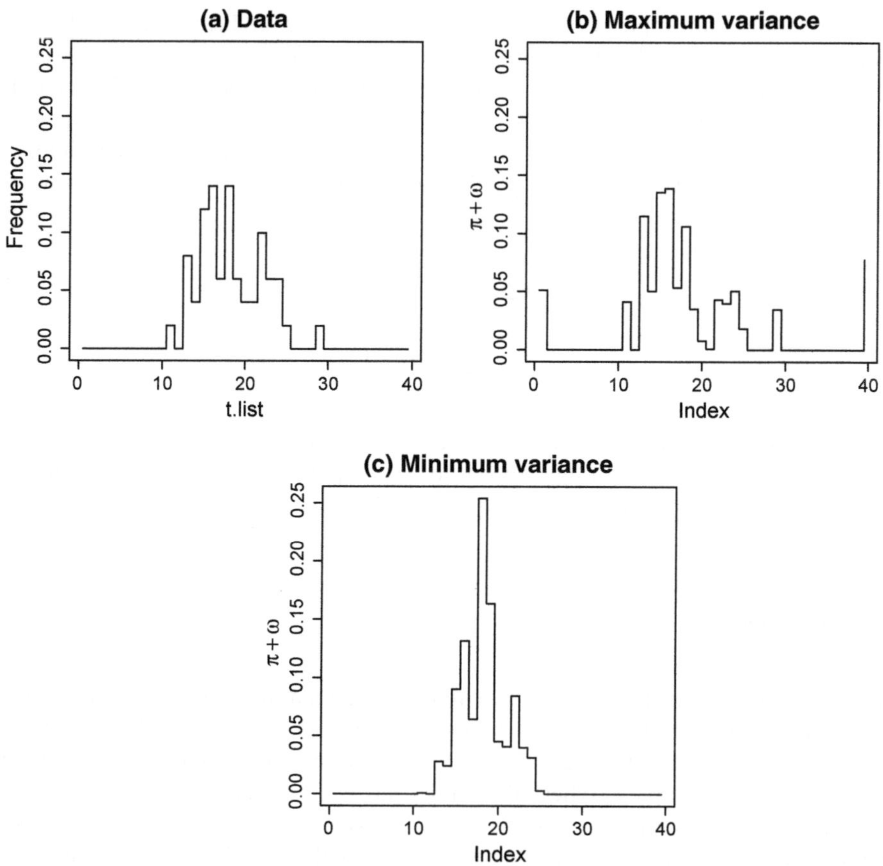

Fig. 13 The data and distribution with maximum and minimum variance in the region of interest

sponding binomially based confidence interval. This will result is a very conservative region of interest.

Figure 13 shows the results of the corresponding linear programming problem which results when constraint (8) is added to (7). The solutions show in Fig. 13b, c are much closer to the data than that of Fig. 12. We see that in Panel (b) the optimisation is still putting some probability in the tails of the solutions in particular in the extreme bins.

One of the advantages of using the extended exponential family Δ^k, rather than the more common multinomial, is that there is a very close relationship between the space of models and the sample space. In particular we note that in data shown in Fig. 2 for Example 2 there many bins where there are zero counts. We can define the two sets

$$\mathcal{P} = \{i|n_i > 0\}, \ \mathcal{Z} = \{i|n_i = 0\},$$

and we call the subsimplex of Δ^k indexed by $\mathcal{P}$ the observed face. The way that this decomposition effects the shape of the likelihood function across Δ^k is discussed in Critchley and Marriott (2014a), which points out that there are many directions in which the log-likelihood is flat – i.e. we can learn nothing from this particular set of data. Using this decomposition gives a different way to construct a region of interest of Δ^k which looks at points with high likelihood values. That is we add the following constraint

$$\sum_{i \in \mathcal{P}} n_i \log \pi_i \geq C_1 \tag{9}$$

$$\sum_{i \in \mathcal{Z}} \pi_i \leq C_2 \tag{10}$$

for suitably chosen values of C_1, C_2. We do not have space here to describe the solutions except to note that they appear to be computational tractable and can give attractive solutions to simple problems such as Example 2.

3 Discussion

This paper is an example of what we call computational information geometry and gives an illustration of how it can be used in the foundational problem of understanding the way that selecting a statistical model affects a given inference problem. We contrasted Example 1, where the model has been selected by theoretical considerations, with Example 2, where a more empirical approach has been taken. We believe that ideas of this paper could be applied to both examples, but have particular importance in the second.

We note that using the methods of CIG we can find sensitive perturbations directions which generate a range of inferences about μ. These include ones where the model is treated as completely correct to ones where the model has been extended so that the resultant inferences agree with ‘model free’ inferences. What was perhaps surprising was that the space of sensitive perturbations was very small.

We also showed that there are different types of perturbations – which are based on more global considerations – which explore the robustness aspects of the inference problem. While the ‘model free’ inference might seem to have less assumptions they do put much more weight on the observed data being exactly as expected, without ‘outliers’. The sensitive directions discovered allows the analyst to understand the different choices available to them, balancing belief between the model and the data.

There are a number of computational issues which have naturally arisen in our analysis. These include the potential high dimensionality of Δ^k, so that optimisations based on methods which work in high dimensions have been focused on, such as convex and linear optimisation. Further, the role of boundaries in the mean parameters and the polar dual of such boundaries, turned out to be critical in the analysis. In the examples shown in this paper the computation of the boundary polytopes are completely straightforward and there are many cases where the key step, computing the convex hull of a finite number of points in $\mathbb{R}^k$, can be done with standard software. In general, however, as the number of parameters and the sample size grows, complete enumeration of the boundary becomes computationally infeasible, see Fukuda (2004) and the corresponding computational issues will be the subject of further work. Finally the role of a mixed parameterisation, which has aspects of both the ± 1 geometries of exponential families was highlighted. In general these can only be computed numerically and further research will be done on efficient ways to do this, particularly in the case of non-trivial boundaries.

Appendix 1: The Model Space, Cuts and Closures

Model Space

A key concept in building the perturbation space is to first represent statistical models – sample spaces, together with probability distributions on them – and associated inference problems, inside adequately large but finite dimensional spaces, see Critchley and Marriott (2014a) for details. Consider the general k–dimensional extended multinomial model

$$\Delta^k := \left\{ \pi = (\pi_0, \ldots, \pi_k)^T, \pi_i \geq 0, \sum_{i=0}^{k} \pi_i = 1 \right\}. \tag{11}$$

The multinomial family on $k+1$ categories can be identified with the (relative) interior of this space, $int(\Delta^k)$, while the extended family, (11), allows the possibility of distributions with different support sets. This paper looks at (extended) exponential families embedded in Δ^k and uses the following notation.

Definition 8 Let $\pi^0 = (\pi_i^0) \in int(\Delta^k)$, and V be a $(k+1) \times p$ matrix of the form $(v^{(1)}|\ldots|v^{(p)}) = (v_0|\ldots|v_k)^T$ with linearly independent columns and chosen such that $1_{k+1} := (1, \ldots, 1)^T \notin \text{Range}(V)$. With these definitions there exists a p-dimensional full exponential family in Δ^k, denoted by $\pi(\phi) = \pi_{(\pi^0, V)}(\phi)$ with general element:

$$\pi_i(\phi) = \pi_i^0 \exp\{v_i^T \phi - M(\phi)\}, \tag{12}$$

$i = 0, \ldots, k$ with normalising constant

$$\exp\{M(\phi)\} := \sum_{i=0}^{k} \pi_i^0 \exp\{(V\phi)_i\} = \sum_{i=0}^{k} \pi_i^0 \exp\{v_i^T \phi\},$$

for all $\phi \in \mathbb{R}^p$.

Using this formalism selecting a one dimensional model to undertake inference about $\mu = E(V)$, as in Examples (1) and (2), requires selecting a sufficient statistic V and a basepoint π^0. Initially we concentrate on the case where the choice of model contributes the minimal amount of information to the inference problem. We call these least informative models.

Definition 9 (*Least informative model*) Let X be the random variable over the $k+1$ categories of Δ^k which takes values x_i in category i. The model $\pi(\phi) = \pi_{(\pi^0,V)}(\phi)$ is a one dimensional least informative model for the estimation of $E(X)$ when V is $(k+1) \times 1$ and $v^{(1)} \propto (x_i)$.

Both the models considered in Examples (1) and (2) are least informative for the parameter of interest. Choices between different least informative models then correspond to selecting different base measures $\pi^0 \in \Delta^k$. We can think of these geometrically as translations of exponential families in the affine geometry defined by the natural parameters.

Closures of Exponential Families

In this section we consider the closure of discrete p-dimensional exponential families which are subsets of Δ^k For more general results on closures of exponential families see Barndorff-Nielsen (1978), Brown (1986), Lauritzen (1996) and Csiszar and Matus (2005). In the discrete case considered here, we can understand boundary behaviour in extended exponential families by considering the polar dual (Critchley and Marriott 2014b) or alternatively the directions of recession, Geyer (2009), Rinaldo et al. (2009) and described in detail in Anaya-Izquierdo et al. (2014).

We want to consider the limit points of the p-dimensional exponential family, so we consider the limiting behaviour of the path $\phi(\lambda) := \lambda q$ as $\lambda \to \infty$ where $q \in \mathbb{R}^p$, and $\|q\| = 1$. The support of the limiting distribution is determined by the maximal elements of the set

$$\{s_0^T q, \ldots, s_k^T q\}$$

where $s_i := \left(S_0(i), \ldots, S_p(i)\right)^T$. There exist a correspondence between the limiting behaviour of exponential families in a certain direction – the direction of recession – and the set of normals to faces of a convex polygon, the polar dual, Tuy (1998).

Appendix 2: Empirical Likelihood for the Mean Parameter in a Multinomial Setting

Let T be a discrete random variable with $k+1$ values $\{t_0, \ldots, t_k\}$ so that the probability mass function is $P[T = t_i] = \pi_i$ for $i = 0, 1, \ldots, k$, where $\sum_{i=0}^{k} \pi_i = 1$ and $\pi_i \geq 0$. The distribution of T depends on k free parameters and we are interested in making inferences about the expectation parameter

$$\phi = \sum_{i=0}^{k} t_i \pi_i = t_0 + \sum_{i=1}^{k} \pi_i (t_i - t_0)$$

in the presence of the other $k - 1$ nuisance parameters.

Theorem 4 *For a given random sample of size N from T, let t_- be the minimum observed value of T and t_+ be the maximum observed value of T, and we work in the generic case where all t_i's are distinct. Then for any $\phi \in (t_-, t_+)$ the profile likelihood for the mean parameter ϕ is given by*

$$\hat{\pi}_i(\phi) = \frac{n_i}{N + \hat{\delta}_\phi (\phi - t_i)}, \qquad i \in \mathcal{P}$$

Here, n_j is the number of times that t_j appears in the sample so that $N = \sum_{i=0}^{k} n_i$ and $\hat{\delta}_\phi$ is the unique solution to the equation

$$\sum_{i \in \mathcal{P}} \frac{n_i (t_i - \phi)}{N + \hat{\delta}_\phi (\phi - t_i)} = 0$$

in the interval $\left(\frac{N}{t_- - \phi}, \frac{N}{t_+ - \phi}\right)$.

Proof The empirical (profile) likelihood for ϕ can be found by solving the following optimization problem

$$\max_{\pi} \sum_{i \in \mathcal{P}} n_i \log \pi_i \text{ s.t. } \sum_{i \in \mathcal{P} \cup \mathcal{Z}} \pi_i = 1, \sum_{i \in \mathcal{P} \cup \mathcal{Z}} t_i \pi_i = \phi$$

where, we recall, $\mathcal{P} = \{i : n_i > 0\}$ and $\mathcal{Z} = \{i : n_i = 0\}$. Since the t_i's are distinct and we can also assume without loss that $\pi_i > 0$ for $i \in \mathcal{P}$ because otherwise $\ell = -\infty$. The Lagrangian is given by

$$\mathcal{L} = \sum_{i \in \mathcal{P}} n_i \log \pi_i + \lambda \left(\sum_{\in \mathcal{P} \cup \mathcal{Z}} \pi_i - 1 \right) + \delta \left(\sum_{\in \mathcal{P} \cup \mathcal{Z}} \pi_i t_i - \phi \right)$$

and the key turning point equations are given by

$$i \in \mathcal{P}, \frac{\partial}{\partial \pi_i} = 0 \Rightarrow n_i + \hat{\lambda} \hat{\pi}_i + \hat{\delta} t_i \hat{\pi}_i = 0$$

$$i \in \mathcal{Z}, \frac{\partial}{\partial \pi_i} = 0 \Rightarrow \hat{\lambda} + \hat{\delta} t_i = 0$$

which give the solutions

$$\hat{\pi}_i = \frac{n_i}{N + \hat{\delta}(\phi - t_i)}, \; N + \hat{\delta}(\phi - t_i) > 0$$

with $\hat{\delta}_\phi$ defined as the solution $H_\phi(\delta) = 0$ where

$$H_\phi(\delta) := \sum_{i \in \mathcal{P}} \frac{n_i(t_i - \phi)}{N + \delta(\phi - t_i)}.$$

Calculations show that

$$\delta_{min} = \frac{N}{t_- - \phi} < \hat{\delta} < \frac{N}{t_+ - \phi} = \delta_{max}.$$

giving

$$H'_\phi(\delta) = \sum_{i \in \mathcal{P}} \frac{n_i(t_i - \phi)^2}{(N + \delta(\phi - t_i))^2} > 0$$

so that $H_\phi(\delta)$ is a strictly increasing function. Also

$$\lim_{\delta \to \delta_{\min}} H_\phi(\delta) = -\infty, \qquad \lim_{\delta \to \delta_{\max}} H_\phi(\delta) = \infty$$

so that $H_\phi(\delta) = 0$ has a unique solution in the interval $(\delta_{min}, \delta_{max})$.

Appendix 3: Sensitive Infinitesimal Perturbations

We proceed from the minimal exponential family representation of the multinomial for the observed counts $n = (n_1, \ldots, n_k)^T$

$$f_n(n; \eta) = \exp\left(n^T \eta - \varphi(\eta)\right) h(n)$$

where the relation with the probability parameter π is given by $\eta_i(\pi) = \log\left(\frac{\pi_i}{1 - \sum_{r=1}^k \pi_r}\right)$, $\pi_i(\eta) = \frac{e^{\eta_i}}{1 + \sum_{i=1}^k e^{\eta_i}}$ for $i = 1, \ldots, k$, $\varphi(\eta) = N \log (1 + \sum_{i=1}^k e^{\eta_i})$, and $h(n)$ is the multinomial coefficient.

We define the following coordinate system in $\mathcal{N}$, the natural parameter space. Consider a fixed point $\eta_0 \in \mathbb{R}^k$ and $d^T := (t_1 - t_0, \ldots, t_k - t_0)/N$. Let $\{v_1, \ldots, v_{k-1}\}$ be an orthogonal basis for the orthogonal complement of d. If we take $A = (d, v_1 \ldots, v_{k-1})$, then for any $\eta \in \mathbb{R}^k$ we can write $\eta = \eta_0 + A\phi$ for some $\phi \in \mathbb{R}^k$. So ϕ defines a new parameterisation for the multinomial. By defining $s := A^T n + c$ with $c^T = (t_0, 0, \ldots, 0)$ we have

$$f_s(s;\phi) = \exp\left(s^T\phi - M(\phi)\right) f_n((A^T)^{-1}(s-c);\eta_0)$$
$$= \exp\left(s^T\phi - M(\phi)\right) f_s(s;0)$$

where

$$M(\phi) = \varphi(\eta_0 + A\phi) - \varphi(\eta_0) - c^T\phi.$$

This is of course, the same regular natural exponential family but now with natural parameter ϕ and expectation parameter

$$\mu(\phi) = D_\phi M(\phi) = E[s;\phi] = A^T E[n] + c.$$

We are interested in making inferences about $\mu_1 = E[s_1] = \sum_{i=0}^k t_i\,\pi_i = \phi$.

According to the variance Condition 2 in Theorem 1: $s_1 = n^T d + t_0$ is an exact cut for the regular exponential family

$$\mathcal{F} = \left\{f_s(s;\phi) = \exp\left(s^T\phi - M(\phi)\right) f_s(s;0) \,:\, \phi \in \mathcal{P}\right\}$$

if and only if its variance depends only on μ_1. If such exact cut exists, we can then make exact marginal inferences for μ_1 using the marginal distribution of s_1 given by

$$f_{s_1}(s_1;\mu_1) = \exp(s_1\,\phi^*(\mu_1) - \psi(\phi_1^*(\mu_1)))\,h^*(s_1)$$

for some real valued functions h^* and ψ. We define

$$\pi(\mu) = N^{-1}(A^T)^{-1}(\mu - c)$$

and then we have

$$Var(s_1;\mu) = d^T Var(n;\pi(\mu))d$$
$$= N\,d^T[diag(\pi(\mu)) - \pi(\mu)\pi(\mu)^T]d$$

so we can check how much this vary as a function of $\mu_{(1)}$. For any fixed μ_1^0 we would like to explore the variation of $V(s_1;\mu)$ in the subspace of densities given by $\mu_1 = \mu_1^0$. We would like to find a direction in such space such that $Var(s_1;\mu)$ changes the most.

We define the following inner products for $u, v \in \mathbb{R}^k$

$$\langle u, v\rangle_\mu := u^T I(\mu)v,\ \langle u, v\rangle_\phi := u^T I(\phi)v,\ \langle u, v\rangle_\pi := u^T I(\pi)v,\ \langle u, v\rangle_\eta := u^T I(\eta)v$$

and orthogonal projections matrices are

$$P_\eta^\perp(v;u) := v - \left[\frac{\langle u, v\rangle_{\eta_0}}{\langle u, u\rangle_{\eta_0}}\right] u$$

If ω is such that $\omega_1 = 0$ and $\mu_0 = \mu(\phi)$ with $\phi = 0$ then

$$Var(s_1; \mu_0 + \lambda\omega) = Var(s_1; \mu) + \lambda\langle\omega, I(\phi_0)A^{-1}d^{(2)}\rangle_{\mu_0}$$

so the directional derivative at μ_0 along the vector ω is given by $\langle\omega, I(\phi_0)A^{-1}d^{(2)}\rangle_{\mu_0}$. To explore the variation of $Var(s_1; \mu_0 + \lambda\omega)$ we define the following optimisation problem

$$\max_{w}\langle\omega, I(\phi_0)A^{-1}d^{(2)}\rangle_{\mu_0} \text{ s.t. } \langle w, w\rangle_{\mu_0} = 1, \langle w, I(\phi_0)e_1\rangle_{\mu_0} = 0$$

where $e_1^T = (1, 0, \ldots, 0)$.

The solution is given by $\hat{\omega} = \hat{u}/\|\hat{u}\|_{\mu_0}$ where

$$\hat{u} = P_{\mu_0}^{\perp}(I(\phi_0)A^{-1}d^{(2)}; I(\phi_0)e_1)$$

that is, the normalised projection of $I(\phi_0)A^{-1}d^{(2)}$ orthogonal to $I(\phi_0)A^{-1}d$ in the metric $I(\mu_0)$. Note that $A^{-1}d = e_1$ and also $\|\hat{u}\|_{\mu_0} = \|P_{\eta_0}^{\perp}(d^{(2)}; d)\|_{\eta_0}$. We can write $\hat{\omega}$ as

$$\hat{\omega} = I(\phi_0)A^{-1}\frac{P_{\eta_0}^{\perp}(d^{(2)}; d)}{\|P_{\eta_0}^{\perp}(d^{(2)}; d)\|_{\eta_0}}$$

The objective function evaluated at the maximum is

$$\langle\hat{\omega}, I(\phi_0)A^{-1}d^{(2)}\rangle_{\mu_0} = \left\|P_{\eta_0}^{\perp}(d^{(2)}; d)\right\|_{\eta_0}$$

This has a nice interpretation. If we take $\eta_0 = \eta(\hat{\pi}_{Global}) = \eta(n/N)$ we have

$$\langle\hat{\omega}, I(\phi_0)A^{-1}d^{(2)}\rangle_{\mu_0} = \|d\|_{\eta_0}^2\, C_{+1}(\hat{\phi}_{Global})$$

then it can be interpreted as $\|d\|_{\eta_0}^2$ times the $+1$ curvature of the profile likelihood curve for ϕ at $\phi = \hat{\phi}_{Global}$. The profile likelihood curve defines a curved exponential family embedded in the multinomial. We have

$$N\frac{\partial\eta}{\partial\phi}(\hat{\phi}) = \frac{1}{\|d\|_{\eta_0}^2}d$$

$$N^2\frac{\partial^2\eta}{\partial\phi^2}(\hat{\phi}) = \frac{1}{\|d\|_{\eta_0}^4}\left[P_{\eta_0}^{\perp}(d^{(2)}; d)\right] - \frac{S}{\|d\|_{\eta_0}^6}d$$

so the $+1$ embedding curvature of the profile likelihood curve at $\phi = \hat{\phi}$ is given by

$$C_{+1}(\hat{\phi}_{Global}) = \frac{\left\|P_{\eta_0}^{\perp}(d^{(2)}; d)\right\|_{\eta_0}}{\|d\|_{\eta_0}^2}$$

The solution $\hat{\omega}$ determines a direction in the -1 space of the exponential family $\mathcal{F}$. If variation in this direction is small we can consider s_1 as an approximate cut.

References

Altham, P. M. (1978). Two generalizations of the binomial distribution. *Applied Statistics*, *27*(2), 162–167.

Amari, S.-I. (1985). *Differential-geometrical methods in statistics*. New York: Springer.

Anaya-Izquierdo, K., Critchley, F., & Marriott, P. (2014). When are first-order asymptotics adequate? a diagnostic. *Stat*, *3*(1), 17–22.

Anaya-Izquierdo, K., Critchley, F., Marriott, P., & Vos, P. (2013). Computational information geometry in statistics: Foundations. *Geometric science of information* (pp. 311–318). New York: Springer.

Barndorff-Nielsen, O. (1976). Factorization of likelihood functions for full exponential families. *Journal of the Royal Statistical Society. Series B (Methodological)*, *38*(1), 37–44.

Barndorff-Nielsen, O. (1978). *Information and exponential families in statistical theory*. New Jersey: Wiley.

Barndorff-Nielsen, O., & Blaesild, P. (1983). Exponential models with affine dual foliations. *Annals of Statistics*, *11*(3), 753–769.

Barndorff-Nielsen, O., & Koudou, A. (1995). Cuts in natural exponential families. *Theory of Probability and Its Applications*, *40*, 220–229.

Box, G. (1976). Science and statistics. *Journal of the Acoustical Society of America*, *71*, 791–799.

Box, G. (1980). Sampling and Bayes' inference in scientific modelling and robustness (with discussion). *Journal of Reliability and Statistical Studies*, *B 143*, 383–430.

Brown, L. (1986). *Fundamentals of statistical exponential families: With applications in statistical decision theory*. Hayward: Institute of Mathematical Statistics.

Christensen, B. J., & Kiefer, N. M. (1994). Local cuts and separate inference. *Scandinavian Journal of Statistics*, *21*(4), 389–401.

Christensen, B. J., & Kiefer, N. M. (2000). Panel data, local cuts and orthogeodesic models. *Bernoulli*, *6*(4), 667–678.

Cook, R. D. (1986). Assessment of local influence. *Journal of the Royal Statistical Society, Series B, Methodological*, *48*, 133–155.

Cox, D. (1986). Comment on 'Assessment of local influence' by R. D. Cook. *Journal of the Royal Statistical Society. Series B (Methodological)*, 133–169.

Cox, D., & Reid, N. (1987). Parameter orthogonality and approximate conditional inference. *Journal of the Royal Statistical Society, Series B: Methodological*, *49*, 1–18.

Critchley, F., & Marriott, P. (2004). Data-informed influence analysis. *Biometrika*, *91*, 125–140.

Critchley, F., & Marriott, P. (2014a). Computational information geometry in statistics: Theory and practice. *Entropy*, *16*(5), 2454–2471.

Critchley, F., & Marriott, P. (2014b). Computing with fisher geodesics and extended exponential families. *Statistics and Computing*, 1–8.

Csiszar, I., & Matus, F. (2005). Closures of exponential families. *The Annals of Probability*, *33*(2), 582–600.

Efron, B. (1986). Double exponential families and their use in generalized linear regression. *Journal of the American Statistical Association*, *81*(395), 709–721.

Fukuda, K. (2004). From the zonotope construction to the Minkowski addition of convex polytopes. *Journal of Symbolic Computation*, *38*, 1261–1272.

Geyer, C. J. (2009). Likelihood inference in exponential families and directions of recession. *Electronic Journal of Statistics*, *3*, 259–289.

Lambert, D. (1992). Zero-inflated poisson regression, with an application to defects in manufacturing. *Technometrics*, *34*(1), 1–14.
Lauritzen, S. (1996). *Graphical models*. Oxford: Oxford University Press.
Owen, A. B. (1988). Empirical likelihood ratio confidence intervals for a single functional. *Biometrika*, *75*(2), 237–249.
Rinaldo, A., Fienberg, S. E., & Zhou, Y. (2009). On the geometry of discrete exponential families with applications to exponential random graph models. *Electronic Journal of Statistics*, *3*, 446–484.
Tuy, H. (1998). *Convex analysis and global optimization*. London: Klumer academic publishers.

On the Geometric Interplay Between Goodness-of-Fit and Estimation: Illustrative Examples

Karim Anaya-Izquierdo, Frank Critchley, Paul Marriott and Paul Vos

1 Introduction

In statistical analysis, it is common practice to end the model building phase when one, or more, goodness-of-fit tests no longer reject the hypothesis that the data generation process lies in a given parametric model. This model is, often, then treated as known, and parametric inference theory, within it, is assumed to be sufficient to describe the uncertainty in the problem. As a corollary of this, only information captured by the sufficient statistics for the final model is used in the inference. The excellent papers Eguchi and Copas (2005), Copas and Eguchi (2010), summarised below, take a first order geometric approach, which defines the envelope likelihood and a 'double the variance' rule, which are designed to capture the actual model uncertainty.

F. Critchley—This work has been partly funded by EPSRC grant EP/L010429/1 and code which generates the figures in this paper is available at http://users.mct.open.ac.uk/rs23854/EGSS/software.php.
P. Marriott—This work has been partly funded by NSERC discovery grant 'Computational Information Geometry and Model Uncertainty'.

K. Anaya-Izquierdo
Department of Mathematical Sciences, University of Bath, Bath BA2 7AY, UK
e-mail: kai21@bath.ac.uk; K.Anaya-Izquierdo@bath.ac.uk

F. Critchley
The Open University, Walton Hall, Milton Keynes, Buckinghamshire MK7 6AA, UK
e-mail: f.critchley@open.ac.uk

P. Marriott (✉)
University of Waterloo, 200 University Avenue West, Waterloo, ON, Canada
e-mail: pmarriot@uwaterloo.ca

P. Vos
East Carolina University, Greenville, NC 27858-4353, USA
e-mail: VOSP@ecu.edu

F. Nielsen et al. (eds.), *Computational Information Geometry*,
Signals and Communication Technology, DOI 10.1007/978-3-319-47058-0_3

This paper examines the same problem, but uses a global, rather than local, geometric approach. We show how 'rotations' and 'translations' of working parametric models – which we define using Information Geometric ideas – affect estimation results in ways analogous to those shown by Copas and Eguchi. Further, through a form of bias-variance trade-off, see Hastie et al. (2001), we define, what we call, least-informative families, these being families which, in some sense, add the least amount of information to the estimation problem. These, we show, are connected to ideas from Maximum Entropy theory, Jaynes (1978), Skilling (2013), Schennach (2005), and non-parametric inference methods, with Efron (1981) and Owen (2001) being important references.

Copas and Eguchi (2010) note that in practice the choice of statistical model, made by an analyst, can be rather arbitrary. There may well be other models which fit the data equally well, but give substantially different inferences. We concur with this conclusion which can be summarised as the main theme of the paper: namely, that goodness-of-fit is necessary but not sufficient for model selection. Of course, this is not a new conclusion. In the extreme case, over-fitting of sample data, giving a poor representation of the population, is an extremely well documented phenomenon, Hastie et al. (2001). Rather, this paper points to new, geometrically-based, methodologies to deal with the consequences of this conclusion.

Copas and Eguchi (2010) define *statistically equivalent* models, f and g, to mean that hypothesis tests that the data were sampled from g rather than f would result in no significant evidence one way or another. So, if one model passed a goodness-of-fit test, the other would too. They define the class of statistically equivalent models using first-order asymptotic statistical theory, and hence local linear geometry. They then, building on earlier results in Eguchi and Copas (2005) and similar ideas in Kent (1986), build an envelope of likelihood functions, which gives a conservative inferential framework across the set of statistically equivalent models. A similar idea, in Eguchi and Copas (2005), again using an elegant first order asymptotic and geometric argument, results in the idea of doubling the Fisher information from a single model before calculating confidence intervals to correct for the existence of statistically equivalent models.

One reason that two different analysts may select two different models, both close to the data but giving different inferences about the same parameter, may come from the fact that the models are built with different a priori information. One of the ideas that this paper starts to explore is to what extent can different a priori assumptions be encoded geometrically in some 'space of all models'. We can then think of the least-informative model as one which, in some sense, uses a minimal amount of extra-data information.

The paper takes an illustrative approach throughout by using low-dimensional models which are simple enough that figures can convey, without misleading, general truths. In particular, we discuss simple *thought experiments*, see Sect. 2, which, while undoubtably 'toy', illustrate clearly the essential points we wish to make. The paper is organised as follows: in Sect. 2 we use some very basic geometric transformation in the 'space of models' to explore the relationship between goodness-of-fit tests and parametric inference. In Sect. 3 we formalise these ideas and define the least informative family. We conclude with a discussion which links the least informative idea with similar ideas in non-parametric inference.

2 Thought Experiments

2.1 Introduction

This section looks at simple geometric concepts, such as translation and rotation, of models in a, in practice, high-dimensional, space of models. We show how these geometric ideas are related to inferential concepts, such as efficiency, information about a parameter, and bias-variance trade-off. The ideas also illustrates relationships between parametric and non-parametric approaches to inference.

We are interested in the global relationship between mean and natural parameters, which we denote by (-1) and $(+1)$-affine parameters following Amari (1985) and described in Critchley and Marriott (2014). This global approach, explicitly using affine geometries and convex sets, complements that of Copas and Eguchi which is local and first order asymptotic.

To start the discussion, since we want to understand the geometry of sets of models, we first give a formal definition of what we mean by the space of all models, at least in the finite, discrete case. We consider, purely for illustrative reasons, a very simple example where we have a discrete sample space of 3 values: $\{t_0, t_1, t_2\}$. It might be considered natural to consider the space of models as being represented by the multinomial distribution parameterized in the mean parameters by the simplex

$$\Delta_{int} = \left\{ (\pi_0, \pi_1, \pi_2) | \pi_i > 0, \sum_{i=0}^{2} \pi_i = 1 \right\},$$

and in the natural parameters by

$$\Delta_{int}^* = \left\{ (\eta_1, \eta_2) | \eta_i = \log(\pi_i / \pi_0) \right\}.$$

In fact, both geometrically and statistically, it is far neater to work on the closure of the simplex. The global relationship between the (± 1)-parameters is much easier to understand in the closure. Furthermore, when considering non-parametric approaches such as the empirical likelihood, Owen (2001), it is natural to consider the boundary of the closure.

The closure of the multinomial is called the extended multinomial distribution, Critchley and Marriott (2014), with mean parameter space

$$\Delta = \left\{ (\pi_0, \pi_1, \pi_2) | \pi_i \geq 0, \sum_{i=0}^{2} \pi_i = 1 \right\}.$$

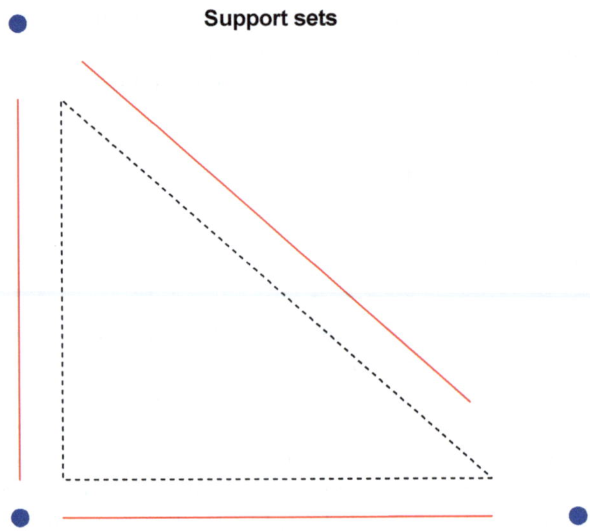

Fig. 1 The extended exponential family as a union of exponential families with different support sets

To define the structure of the 'natural parameters' of the closure, Δ^*, we need to use the concept of the polar dual of the boundary of the simplex to define the limiting behaviour, again see Critchley and Marriott (2014). The boundary is a union of exponential families each with corresponding natural $(+1)$-parameters. The different support sets are illustrated in Fig. 1. To sum up, the space of all models we call a structured extended multinomial (SEM), denoted by $\{\Delta, \Delta^*, (t_0, t_1, t_2)\}$, where the t_i are numerical labels associated with the categories of the extended multinomial. Without loss of generality we assume $t_0 \leq t_1 \leq t_2$. For illustration, in this paper, we take $(t_0, t_1, t_2) = (1, 2, 3)$.

2.2 *First Thought Experiment*

In our first thought experiment, suppose we are trying to estimate the mean, μ_T, of the random variable T which takes values t_i with probability π_i, $i = 0, 1, 2$. In the simplex, shown in Fig. 2a, sets of distributions with the same mean are (-1)-geodesics and are straight lines in this parameterization. The same sets, in the $(+1)$-affine parameters of the relative interior of the simplex, are shown in Panel (b), and are clearly non-linear. The global structure of the (-1)-geodesics in the $(+1)$-affine parameterization is determined by the limit sets defined by the closure and given by the polar dual of the simplex. These can be expressed in terms of the, so-called, directions of recession, Geyer (2009), Feinberg and Rinaldo (2011). The directions associated with the polar dual are illustrated with dashed lines in the panel, again see Critchley and Marriott (2014) for details.

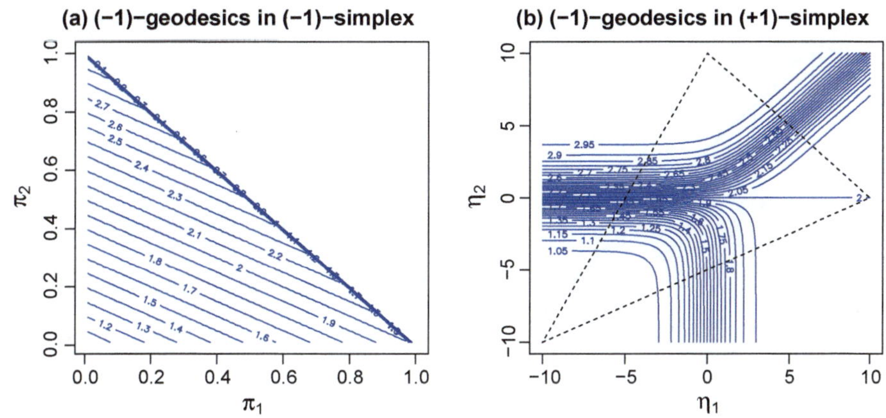

Fig. 2 **a** The extended multinomial in the (-1)-affine parameters, **b** the relative interior of the extended multinomial in the $(+1)$-affine parameters. The *dash lines* represents the boundary of the closure 'at infinity'

Consider, then, Fig. 3. In this thought experiment there is a set of models, all one dimensional exponential families, which intersect at the true data generation process, shown as the red circle in panels (a) and (b). These models are shown in panel (b) as straight lines in the (+1)-affine parameters, becoming curves in the (-1)-affine parameters of panel (a). We can, then, think of these models as a set of lines rotating around the data generation process. The red line is the $(+1)$-geodesic which is Fisher orthogonal to the (-1)-geodesics of interest. In Sect. 3, we will define this as the least-informative model. In our figure we show a set of models, with the most extreme plotted in green for clarity.

For each model the corresponding deviance (twice the normalised log-likelihood) for μ_T, corresponding to each model, and the counts (50, 10, 40), is shown in Fig. 3c. The colour coding here is the same as in panel (b). It is clear that there are considerable differences in inference across this range of models. The vertical scale in panel (c) is selected to show the part of the parameter space of reasonable inferential interest. Since the data generation process lies in each of the models, each should pass any reasonable goodness-of-fit test. Hence, the thought experiment establishes the main theme of the paper: namely, that goodness-of-fit is necessary but not sufficient for model selection.

As discussed above, Fig. 3c shows the set of deviances for our set of models. This is similar to the set of deviances defined in the papers Eguchi and Copas (2005), Copas and Eguchi (2010), where they recommend to use a conservative 'envelope' approach. The likelihood which gives the most conservative inference is shown in red and corresponds to the *least informative model* defined and discussed in Sect. 3, and shown by the red curves in (a) and (b). We might extend our thought experiment and imagine the case where one scientist has clear extra-data information which informs the model choice, while a second does not have such information so makes the most conservative choice possible.

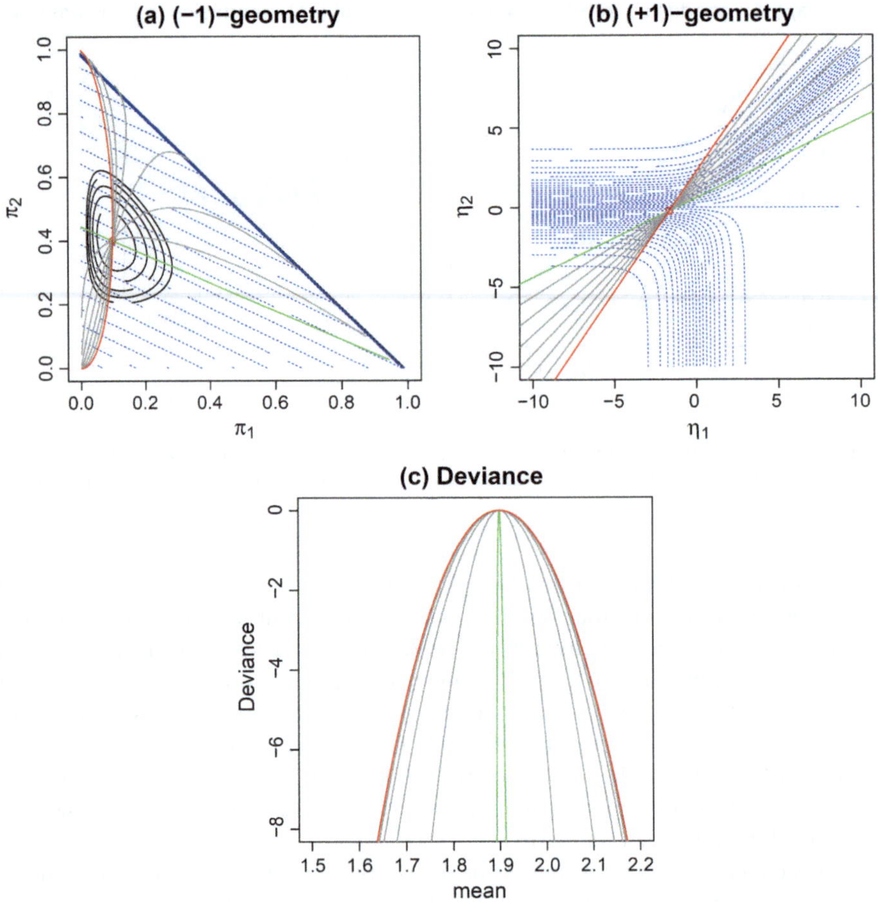

Fig. 3 The *black lines* in panel **a** are the level sets of the log-likelihood for a sample, (50, 10, 40), drawn from the true data generation process, shown as the *red circle* in panels **a** and **b**. **a** The set of models, in the (−1)-affine representation; the *blue lines* are sets of constant parameter values. **b** Same structure in the (+1)-affine representation. **c** The deviance function for the set of models (color figure online)

We also note that adding complexity penalties, as the Akaike information criterion (AIC) or other information criteria do, does not help in the thought experiment since all considered models have the same complexity.

2.3 Second Thought Experiment

The first thought experiment concerns rotations in the (+1)-affine parameters, the second concerns translations, illustrated by Fig. 4. In panel (b) we show a set of (+1)-geodesics. These all have the same sufficient statistic, T, and so their expectation

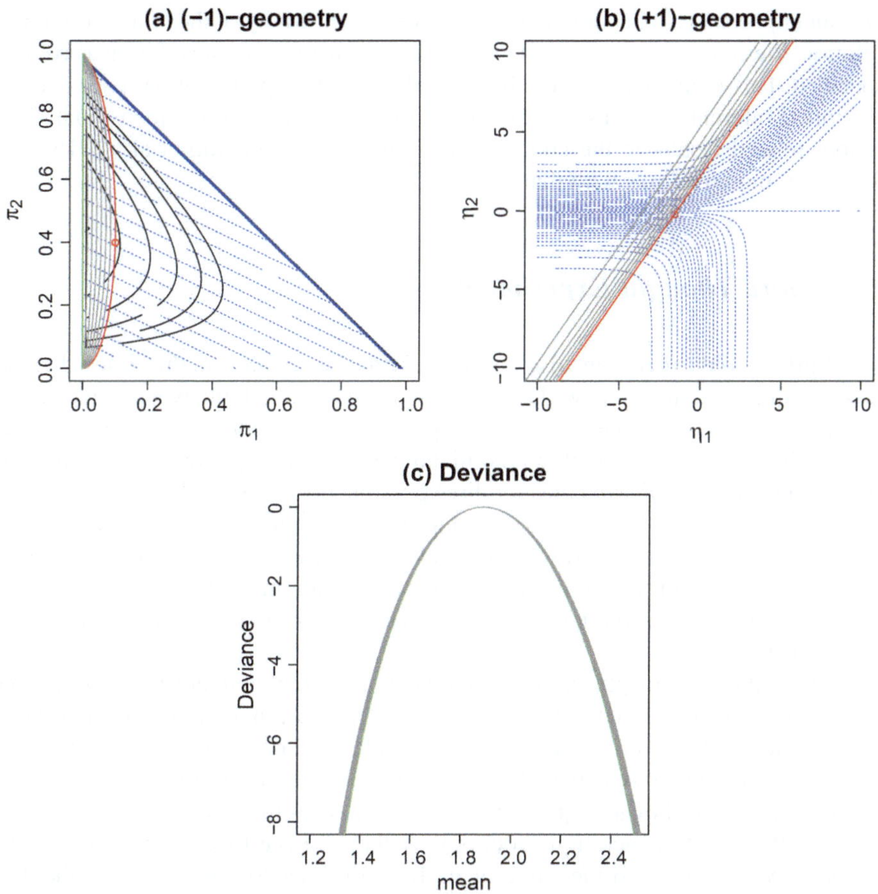

Fig. 4 The data considered here are the counts $(10, 0, 8)$. **a** The set of translated models in the (-1)-affine representation. The *black lines* are the level sets of the log-likelihood, the *blue lines* sets of constant parameter values. **b** The same structure in the $(+1)$-affine representation. **c** The deviance function for the set of models. Note that the ten different deviance plots are so similar that they are superimposed (color figure online)

parameter is μ_T, the parameter of interest. Since they share a common sufficient statistic, they are all $(+1)$-parallel in the $(+1)$-affine parameters. Their corresponding representation in (-1)-affine parameters is shown in panel (a). Here we note that the green line is the limit of this set of translations and lies in the boundary, and so corresponds to a change of support. The data for this example are counts $(10, 0, 8)$ and so the non-parametric maximum likelihood estimate also lies in the boundary. The black curves in Fig. 4a are, as in Fig. 3a, the likelihood contours in the simplex. Panel (c), in Fig. 4, shows the deviance plots for the parameter of interest for this set of models, and it is clear that they all giving essentially the same inference. It can be easily shown that the empirical likelihood for μ_T would also give very similar inference, Owen (2001). In this example we see that the goodness-of-fit does *not* play an important role in our understanding of the sensitivity of the related inference solution. We call such directions in $(+1)$-space insensitive, for the inference problem specified. This example illustrates the general fact that the perturbation space of data-supported inferentially sensitive directions may, indeed, be low dimensional. This echoes related results in the companion paper Anaya-Izquierdo et al. (2016): see Sect. 4 for more on links with that paper.

2.4 Third Thought Experiment

The third thought experiment also considered translations shown, this time schematically, in Fig. 5. Assume that we have a *fixed* one dimensional full exponential family, which we plot in Fig. 5 in the (-1)-affine parameters, as a curved line. For example, it might be that an analyst feels a binomial model was appropriate and we denote its sufficient statistic by T. We consider three possible positions for the data generation process, A, B and C. These are selected to lie on the same (-1)-geodesic which is Fisher orthogonal to the working binomial model. This would mean that for each position the pseudo-true value (i.e. the one which minimises the Kullback–Leibler divergence between the model and the data generation process) will be the same, i.e. the point A.

In this example, though, suppose the object of inferential interest was not μ_T but $\mu_S := E(S)$, where S is a different random variable to T. In our thought experiment, the parameter of interest might be, for example, one of the bin probabilities π_i. The sets of distributions which share a common value of $E(S)$ are parallel (-1)-geodesics, and are shown in green. We have three cases to consider.

First, if the data generation process is A then we see that the working model is correctly specified and the value of μ_S from the pseudo-true value and the data generation process are obviously the same. It is the job of the analyst to quantify the variability of the estimate of μ_S and they, naturally, want to use the model that

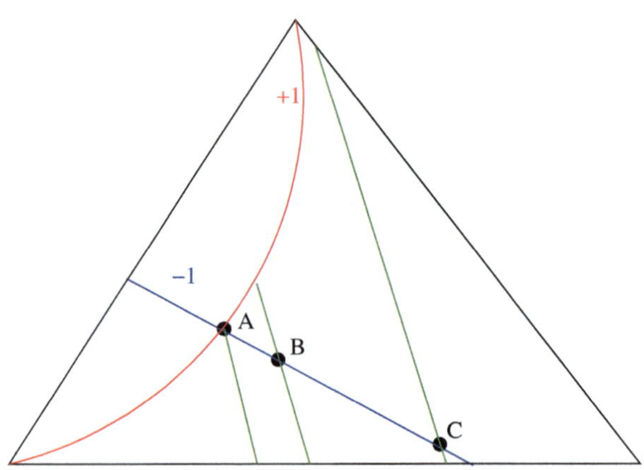

Fig. 5 A schematic view of bias-variance trade-off

they selected to do this. Given the model we can write μ_S as a, typically non-linear, function of the mean parameter μ_T. By the arguments in the following section, the Fisher information about μ_S is increased by using this approach and the model specification is informative about the inferential question of interest.

Second, consider the case where the data generation process is located at the point B. Above, the model has been actively informative about inference for the parameter of interest. Here, there is a cost. We see that there is considerable bias in the estimate of $E(S)$. Lines of constant values of $E(S)$ are shown in green and the 'true' line passes through B. However the pseudo true value is still at A. So, we have a situation where there is a reduction in estimation variance but there is now bias. It is here that goodness-of-fit plays a role. The 'distance' between A and B is one of the things that a goodness-of-fit statistic measures and the smaller this is, the smaller the bias.

Third, consider the case where the data generation process is at the point C, considerably further away. This results in an odd situation. We see that the 'true' value of μ_S, shown by the green line through C, does not intersect the model at all. Hence, the model is so poorly specified that it cannot estimate the true value of $E(S)$ at all. This is a case where we would expect that goodness-of-fit to play a really important role. Hopefully, it could rule out this working model completely.

3 Least-Informative Models

The three thought experiments have been designed to illustrate, in a visual way the, rather weak, way that goodness-of-fit testing controls the effect of model choice on inference. These models are, of course, toy but the plots represent much more general truths. In this section we move to a much more general statistical analysis of the same problem.

We have explored the effect of different choices of low dimensional, exponential family models in some large model space. We have shown that pure goodness-of-fit tests, and indeed penalty methods, will not give enough information to give unambiguous inference and, in general, methods for taking account of the model uncertainty need to be used. This is in complete agreement with, and gives a global extension to, the results of Eguchi and Copas (2005), Copas and Eguchi (2010). Two question which naturally arises are, then: why is it that in practical statistical modelling it is very common that low dimensional exponential families are used? and, what sort of information, if not data based, can be used to justify the choice of such low dimensional models?

One justification is through limit theory. The analyst assumes that enough regularity holds such that central or Poisson limit theorems, or similar, hold and uses these to justify the low dimensional assumptions. Such arguments mean that, in the geometry of model spaces, there will be particular directions which are special and some directions in the space of rotations are preferred. Similar arguments can come through assumption that particular equilibrium distributions are appropriate. An example would be assuming a binomial model in Hardy–Weinberg equilibrium theory, Hardy (1908). Another argument, which also produces low dimensional models is maximum entropy theory, Jaynes (1978), Skilling (2013), Schennach (2005), where distributions are selected which maximise the entropy subject to a set of moment constraints. Again, this would imply that in the $(+1)$-affine structure, certain directions are special.

In this section, we present a new approach which may give some guidance for the selection of models. We call this the *least-informative model approach.* We work in the general space of k-dimensional discrete distributions, the k-dimensional extended multinomial models, see Critchley and Marriott (2014).

We consider here the one-dimensional subfamily generated by exponential tilting of a fixed distribution $\{\pi_0^0, \pi_1^0, \ldots, \pi_k^0\}$ via a real-valued function g. It is easy to see that, as is necessary for ϕ_1^* to parameterise this subfamily, the map defined on $\mathbb{R}$ by

$$\phi_1^* \mapsto \{\pi_i^0 \, e^{\phi_1^* g(t_i) - \psi(\phi_1^*)}\}_{i=0}^k \qquad \text{where} \qquad \psi(\phi_1^*) := \log\left(\sum_{i=0}^k \pi_i^0 e^{\phi_1^* g(t_i)}\right)$$

is one-to-one if and only if the $\{g(t_i)\}_{i=0}^k$ are not all equal, which we now assume. The general member of this subfamily assigns cell i $(i = 0, \ldots, k)$ the probability denoted by:

$$f_g(t_i; \phi_1^*) = P[T = t_i; \phi_1^*] = \pi_i(\phi_1^*) = e^{\phi_1^* g(t_i) - \psi(\phi_1^*)} \, f_g(t_i; 0)\,, \tag{1}$$

so that the original distribution corresponds to $\phi_1^* = 0$, in which case $\mu_T = \sum_{i=0}^k t_i \, \pi_i^0$. This is an exponential family with natural parameter ϕ_1^*, mean parameter $\mu := E[g(T)]$ and sufficient statistic

$$s = \sum_{i=0}^k n_i g(t_i)$$

where n_i is the number of times t_i appears in a sample of size $N = \sum_{i=0}^k n_i$ from T.

Our set of rotations, from the above discussion, corresponds to different choices of the function g, according to what the analyst thinks is important, or, equally importantly, has available as a sufficient statistic. Again, the set of translations is basically the choice of the base-line distribution $\{\pi_0^0, \pi_1^0, \ldots, \pi_k^0\}$.

In this context, we are considering the case where the inferential problem of interest concerns not the mean of $g(T)$, but the mean of T. For any member of the subfamily (1), this mean is

$$\mu(\phi_1^*) = \sum_{i=0}^k t_i \, \pi_i(\phi_1^*). \tag{2}$$

If the function $g(t_i) = a\, t_i + b$ $(a \neq 0)$ is affine and invertible, then the map $\phi_1^* \mapsto \mu(\phi_1^*)$ is one-to-one since $\mu_1(\phi_1^*) = a\, \mu(\phi_1^*) + b$ and $\phi_1^* \mapsto \mu_1(\phi_1^*)$ is one-to-one.

Suppose now that g is *any* function such that $\phi_1^* \mapsto \mu(\phi_1^*)$ is one-to-one. Now let $\mu \mapsto \phi_1^*(\mu)$ be the inverse map. Then the expected Fisher information about μ in a sample of size one is given by

$$I_g(\mu) = \psi''(\phi_1^*(\mu))[(\phi_1^*)'(\mu)]^2 = \mathrm{Var}_\mu(g(T))[(\phi_1^*)'(\mu)]^2$$

where the subscript μ in Var_μ means the variance is calculated with respect to (1), and $'$ denotes the derivative. But, differentiating (2) with respect to μ we obtain

$$(\phi_1^*)'(\mu) = \frac{1}{\mathrm{Cov}_\mu(T, g(T))},$$

where we make a further regularity assumption that this is also finite. This gives

$$I_g(\mu) = \frac{\mathrm{Var}_\mu(g(T))}{\mathrm{Cov}_\mu^2(g(T), T)}.$$

Denoting by h any invertible affine function of T, as considered above, then

$$I_h(\mu) = \frac{1}{\mathrm{Var}_\mu(T)}.$$

Thus, Cauchy-Schwarz gives at once

$$I_h(\mu) \leq I_g(\mu)$$

and equality holds if and only if g is of the form h. For this reason we call the family (1) the least-informative family for estimation of μ.

We can reconsider the thought experiments in the light of this concept. In Figs. 3 and 4 panels (b) the least informative model corresponds to the red (+1)-geodesic. Under rotations these will have the smallest Fisher information about the parameter of interest, and this can be seen in Fig. 3c. Under translation, as shown in Fig. 4c, there is relative stability in the inferences. Further, there is very good agreement with the empirical likelihood, a model free inference method, Owen (2001).

Each possible choice of one dimensional model introduces information about the parameter of interest that has not come from the data. Therefore one argument would be, if you have no reason to prefer any of one of the set of data supported models select the model which introduces the least amount of extra-data information. In terms of the size of the confidence interval this would be a conservative approach. It is not as conservative as the envelope method, which gives all models in the rotation set equal weight, even if there was no scientific reason for justifying the dimensional reduction for a particular one. We note that the notion of using the most conservative model, rather than averaging inference over sets of models, was advocated by Tukey (1995) in his discussion of Draper (1995).

The third thought experiment, shown in Fig. 5, gives an interesting illustration of what not using least-informative models means. If we have good extra-data reasons for using a non least-informative model – for example a limit result or scientific theory such as Hardy–Weinberg equilibrium – then that information enters the inference problem through an increase in the Fisher information, and thus, smaller confidence intervals and more precise inference. However, there is a cost to this. When the models selected is misspecified then there is a bias introduced into the estimation problem. Hence there is, in this sense a bias-variance trade-off in the model selection choice.

4 Discussion

The concept of least-informative models is related to similar ideas in the literature. For example, moment constrained maximum entropy. A good introduction can be found in the book Buck and Macaulay (1991) and we also note the work in Jaynes (1978), Skilling (2013) and Schennach (2005). In the simplex, we can look for the distribution which maximises entropy, $-\sum_{i=0}^{k} \pi_i \log \pi_i$, with a given value of the mean of $\mu_T = E(T)$. As the value of μ_T changes the solution set is a least informative family for μ_T which passes though the uniform distribution at the centre of the simplex. In our definition of least informative model, we do not insist that the uniform distribution is part of the model. In practice, goodness-of-fit with the observed data might be an alternative way of selecting which of the $(+1)$-parallel least informative models to use. Indeed, it would be interesting to explore if this is the actual role that, in some sense, goodness-of-fit should play in model selection. Note that there is also a difference here in that the maximum entropy principle focuses on properties of the underlying distribution, while we are primarily interested in inference about a given interest parameter. It would also be interesting to explore the link between least informative models, which by design cut the level sets of the interest parameter Fisher orthogonally, and the minimum description length approach to parametric model selection, see for example, Balasubramanian (2005).

An obvious question, about selecting low dimensional parametric models, is the following. If you are unsure about the parametric model, why not use non-parametric approaches? In fact, there are close connections between non-parametric approaches and least-informative models. For example, consider Efron (1981), which investigates a set of common nonparametric methods including the bootstrap and jackknife among other methods. It defines the corresponding confidence intervals by basing them on an exponential tilting model. This model, which they called 'least favourable' is essentially a least informative model. The least favourable method is also discussed in DiCiccio et al. (1989). Another link with nonparametric methods is through the empirical likelihood function as described in the second thought experiment, see also Murphy and Van der Vaart (2000). The empirical likelihood in this case can be interpreted as a real likelihood on a model which lies in the boundary of the closure of the simplex.

In this paper we have used extensively the (-1)-affine geometry of the extended multinomial model. This, like all affine structures, is global. Our analysis is complementary to the local, and asymptotically based, work of Copas and Eguchi (2010). The choice of the (-1)-structure, aside from its global nature, has some other advantages due to the interpretation of its parameters as expectations, which are model free concepts. As Cox (1986) points out, when we are looking at perturbations of models – for example, as described here in the thought experiments – there are two ways of defining what the 'same' parameter means in different models. Firstly, that the parameters have the same real world meaning in different models. We have exploited the fact that (-1)-affine expectation parameters have this property, while $(+1)$-affine parameters, in general, do not. The second way of connecting parameters in different models, Cox (1961, 1986), regards them merely as model labels and not of intrinsic interest themselves. It operates via minimising some 'natural measure of distance', using Cox's words, between points in the different models. This measure may be naturally suggested by the fitting criterion, as Kullback–Leibler divergence is by maximum likelihood. This would be a natural approach to take if an affine structure other than the (-1) one was used. Further, it might be possible to generalising our results to non-exponential families by using the approximating exponential families of Barndorff-Nielsen and Jupp (1989).

As noted in the second thought experiment (Sect. 2.3), this paper can be viewed as complementary to another paper in this volume, Anaya-Izquierdo et al. (2016). That paper shows, within the space of extended multinomial models, how to iteratively construct a – surprisingly simple (low-dimensional) – space of all important perturbations of the working model, where important is relative to changes in inference for the given question of interest. The iterative search first looks for the directions of most sensitivity. It also carefully distinguishes between possible modelling choices that are empirically answerable and those which must remain purely putative. Unlike the approach taken in this paper, the iterative steps changes the dimension of the model by adding 'nuisance parameters' whose role is to inform the inference on the interest parameter.

All examples in this paper are finite discrete models and it is natural to consider extensions to the infinite discrete and continuous cases. The underlying IG for the infinite case is considered in detail in Critchley and Marriott (2014). Section 3 of that paper explores the question of whether the simplex structure, which describes the finite dimensional space of distributions, extends to the infinite dimensional case. Overall the paper examines some of the differences from the finite dimensional case, illustrating them with clear, commonly occurring examples.

The fundamental approach of computational information geometry is, though, inherently discrete and finite, if only for computationally operational reasons. Sometimes, this is with no loss at all, the model used involves only such random variables. In general, suitable finite partitions of the sample space can be used in constructing these computational spaces. While this is clearly not the most general case mathematically speaking (an obvious equivalence relation being thereby induced), it does provide an excellent foundation on which to construct a computational theory. Indeed it has been argued, Pitman (1979)

> ⋯ statistics being essentially a branch of applied mathematics, we should be guided in our choices of principles and methods by the practical applications. All actual sample spaces are discrete, and all observable random variables have discrete distributions. The continuous distribution is a mathematical construction, suitable for mathematical treatment, but not practically observable.

Since real world measurements can only be made to a fixed precision, all models can, or should, be thought of as fundamentally categorical. The relevant question for a computational theory is then: what is the effect on the inferential objects of interest of a particular selection of such categories?

In summary of this and related papers, it will be of great interest to see how far the potential conceptual, inferential and practical advantages of computational information geometry can be realised.

References

Amari, S.-I. (1985). *Differential-geometrical methods in statistics*. New York: Springer.

Anaya-Izquierdo, K., Critchley, F., Marriott, P., & Vos, P. (2016). The geometry of model sensitivity: an illustration. In *Computational information geometry: For image and signal processing*.

Balasubramanian, V. (2005). Mdl, bayesian inference, and the geometry of the space of probability distributions. In *Advances in minimum description length: Theory and applications* (pp. 81–98).

Barndorff-Nielsen, O., & Jupp, P. (1989). Approximating exponential models. *Annals of the Institute of Statistical Mathematics*, *41*, 247–267.

Buck, B., & Macaulay, V. A. (1991). *Maximum entropy in action: a collection of expository essays*. Oxford: Clarendon Press.

Copas, J., & Eguchi, S. (2010). Likelihood for statistically equivalent models. *Journal of the Royal Statistical Society: Series B (Statistical Methodology)*, *72*(2), 193–217.

Cox, D. (1986). Comment on 'Assessment of local influence' by R. D. Cook. *Journal of the Royal Statistical Society. Series B (Methodological)*, 133–169.

Cox, D. R. (1961). Tests of separate families of hypotheses. In *Proceedings of the fourth Berkeley symposium on mathematical statistics and probability*, *1*, 105–123.

Critchley, F., & Marriott, P. (2014). Computational information geometry in statistics: theory and practice. *Entropy*, *16*(5), 2454–2471.

DiCiccio, T. J., Hall, P., & Romano, J. P. (1989). Comparison of parametric and empirical likelihood functions. *Biometrika*, *76*(3), 465–476.

Draper, D. (1995). Assessment and propagation of model uncertainty. *Journal of the Royal Statistical Society. Series B (Methodological)*, 45–97.

Efron, B. (1981). Nonparametric standard errors and confidence intervals. *The Canadian Journal of Statistics*, *9*(2), 139–158.

Eguchi, S., & Copas, J. (2005). Local model uncertainty and incomplete-data bias. *Journal of the Royal Statistical Society, Series B, Methodological*, *67*, 1–37.

Feinberg, S., & Rinaldo, A. (2011). Maximum likelihood estimation in log-linear models: Theory and algorithms. arxiv:1104.3618v1.

Geyer, C. J. (2009). Likelihood inference in exponential families and directions of recession. *Electronic Journal of Statistics*, *3*, 259–289.

Hardy, G. H. (1908). Mendelian proportions in a mixed population. *Science*, 49–50.

Hastie, T., Tibshirani, R., & Friedman, J. (2001). *The elements of statistical learning*. Springer series in statistics (Vol. 1). Berlin: Springer.

Jaynes, E. T. (1978). Where do we stand on maximum entropy? *The maximum entropy formalism conference*, MIT, 15–118.

Kent, J. T. (1986). The underlying structure of nonnested hypothesis tests. *Biometrika*, *73*(2), 333–343.

Murphy, S. A., & Van der Vaart, A. W. (2000). On profile likelihood. *Journal of the American Statistical Association*, *95*(450), 449–465.

Owen, A. B. (2001). *Empirical likelihood*. Boca Raton: CRC Press.

Pitman, E. (1979). *Some basic theory for statistical inference*. London: Chapman and Hall.

Schennach, S. M. (2005). Bayesian exponentially tilted empirical likelihood. *Biometrika*, *92*(1), 31–46.

Skilling, J. (2013). *Maximum entropy and bayesian methods: Cambridge, England, 1988* (Vol. 36). Springer Science & Business Media.

Tukey, J. W. (1995). Comment on 'Assessment and propagation of model uncertainty' by D. Draper. *Journal of the Royal Statistical Society. Series B (Methodological)*, 45–97.

Spontaneous Learning for Data Distributions via Minimum Divergence

Shinto Eguchi, Akifumi Notsu and Osamu Komori

1 Introduction

The theory of statistical estimation has been deeply researched in a situation where the true distribution is assumed to be in a parametric model since Fisher's early work Fisher (1912, 1922). The information geometry offers an intrinsic perspective for the estimation theory, and more profound understandings for statistics and machine learning, cf. Amari (1985); Amari and Nagaoka (2000). The minimum divergence estimation is discussed from a viewpoint of information geometry, cf. Eguchi (1983, 1992, 2008). In principle all the minimum divergence estimators satisfy Fisher consistency. In the class the robustness and efficiency properties are investigated in a wide perspective with redescending influence function, gross error sensitivity and so forth, cf. Basu et al. (1998); Minami and Eguchi (2002); Murata et al. (2004); Fujisawa and Eguchi (2008). We observe that the degree of robustness depends on the choice of the divergence measure with a trade-off for the efficiency. There is a subclass of minimum divergence estimators that satisfy much robustness for a heavy contamination at the cost of the efficiency, cf Scott (2001).

We focus on the behavior of minimum divergence estimators in which the theory is collapsed when the true distribution rather deviates from the parametric model. In effect we consider extreme deviation from the model beyond the break-down point, in which any idea for robustness is already a useless issue since there is no guarantee for any closeness between the true distribution and model distribution. We envisage in the extreme situation that the true distribution would have multiple modes; the

S. Eguchi (✉)
Institute of Statistical Mathematics, Tachikawa 190-8562, Japan
e-mail: eguchi@ism.ac.jp

A. Notsu
Oita University of Nursing and Health Sciences, Oita 870-1201, Japan

O. Komori
University of Fukui, Fukui 910-8507, Japan

F. Nielsen et al. (eds.), *Computational Information Geometry*,
Signals and Communication Technology, DOI 10.1007/978-3-319-47058-0_4

working model is supposed to be unimodal. Under this context an optimal estimator is searched to suggest appropriately the multimodality of the true distribution in a variety of candidate estimators. In accordance with this, we aim to extract information of the true distribution, called *spontaneous data learning* (SDL) in the nonparametric perspective.

We begin with estimating a mean of a normal distribution, which is one of the most elementary tasks in statistics. The typical solution is the maximum likelihood, or the sample mean, which is supported as the uniformly minimum variance unbiased estimator in the sense of efficiency. Nevertheless we consider another estimator in the class of minimum power divergence estimators for the normal mean, which is defined by a power parameter to be adaptively selected. The key issue is a selection of estimators, or that of the power parameter, which is contrast with model selection. If the true distribution we consider is not in the normal model but in a normal mixture model with multi modes, then the power loss function for the normal mean has flexibly several local minima for a large power parameter. On the other hand, the log likelihood function always has a unique maximizer, or the sample mean regardless of the true distribution. If we properly select the power parameter, then the set of local minima is shown to be approximately equal to that of component modes in the normal mixture. There is naturally proposed a clustering algorithm based on this local minimization, which leads to automatical detection for the number of clusters by the number of local minima, see Notsu et al. (2014) for detailed discussion. The result is straightforward extended from the normal location model to a general location model. We will show that the expected loss function is convergent to the true density function as the power goes to ∞. The mean integrated square error for the normalized loss function is investigated as a nonparametric density estimation in a usual asymptotic evaluation. As a result we show for the selection for minimum density estimators such that the local minima of the selected minimum density estimators is consistent with the modes of true distribution. This leads to a strong justification for the consistency of the clustering method mentioned above. Furthermore, we discuss the asymptotics for mode estimators depending on the choice for the location model. In this way we expand novel perspectives for SDL beyond usual discussion for robustness and misspecified model.

We study the problem of detecting multiple modes of the true probability density function based on the evidence (points) learning the locality by the minimum divergence method. The paper is organized as follows. Section 1 describes the method of minimum divergence in the class of U-divergence measures. The expected and empirical loss function is formulated given an evidence or data set from the true density function. In Sect. 2 we focus on a power divergence which associates with the loss function with a flexible performance for multi-modality. The flexibility is characterized by the non-convexity with the simple expression for the difference of convex functions. Section 4 discusses the spontaneous property of the power divergence which automatically detects all the modes even when the number of modes is unknown. In a toy example the spontaneous performance is illustrated with a simple algorithm. In Sect. 5 we elucidates a theoretical reason why the power divergence equips with such a spontaneous property in the simplified situation which the power grows to infinity. Finally some concluding remarks on the role and selection of the power are given with future perspectives.

2 Minimum Divergence Estimators

We consider a class of power entropy and divergence including Boltzmann-Shannon entropy and Kullback-Leibler divergence. Let $\mathcal{F}$ be the space of all probability density functions with respect to a base measure ν. Then we call a function D defined on $\mathcal{F} \times \mathcal{F}$ a divergence measure if $D(f, g) \geq 0$ for all f and g of $\mathcal{F}$ with equality if and only if $f(x) = g(x)$ (ν-$a.e.$ x), where ν-$a.e.$ denotes ν-almost everywhere. There is a large class of divergence measures defined by generator functions, cf. Ciszar's f-divergence Csiszr (2008), U-divergence Eguchi and Kano (2001), Eguchi (2008) and the divergence with biduality Zhang (2013). In particular U-divergence has a feasible form based on a data set for statistical estimation, in which the empirical loss function is given by the empirical expectation by the data set. The minimization for the expected loss function is shown to be equivalent to that for the U-divergence of the true density function with the parametric density function. We prepare a set of real-valued functions that are stricly increasing and convex for generating a variety of entropy and divergence measures. Thus we write

$$\mathcal{U} = \{U(s) : U'(s) > 0, U''(s) > 0 \text{ for all s } \in I\}, \tag{1}$$

where I is an open interval of $\mathbb{R}$. We employ U of $\mathcal{U}$ to define a cross entropy C_U, diagonal entropy H_U and divergence D_U as follows:

$$C_U(g, f) = \int \{U(\xi(f(x))) - g(x)\xi(f(x))\} d\nu(x),$$

$H_U(f) = C_U(f, f)$ and $D_U(g, f) = C_U(g, f) - H_U(g)$ for all f and g of $\mathcal{F}$, where ξ is the inverse function of the derivative of U. There is an information inequality:

$$C_U(g, f) \geq H_U(g) \tag{2}$$

with equality if and only if $f(x) = g(x)$ (ν-a.e. x), which guarantees the definition of D_U as a divergence measure. The choice of $U(s) = \exp(s)$ leads to the relative entropy, Boltzmann-Gibbs-Shannon entropy and Kullback-Leibler divergence as for the triple C_U, H_U and D_U, respectively.

Let $t(x)$ be a statistic, or an integrable function of x. Then we focus on a set of all density functions with the expectations are the same, that is,

$$\mathcal{L}_\eta = \left\{ f : \int t(x) f(x) d\nu(x) = \eta \right\},$$

where η is a fixed vector in $\mathbb{R}^d$. We consider a maximum entropy distribution under the constrain $\mathcal{L}_\eta$,

$$f_U = \underset{f \in \mathcal{L}_\eta}{\operatorname{argmax}}\, H_U(f).$$

An argument by Euler's variational calculus gives

$$f_U(x, \theta) = U'(\theta^\top t(x) - \kappa(\theta)), \tag{3}$$

where $\kappa(\theta)$ is the normalizing factor and θ is defined to satisfy the constrain $\mathcal{L}_\eta$, that is, $\int t(x) f_U(x, \theta) d\nu(x) = \eta$. In effect, if f is in $\mathcal{L}_\eta$, then

$$H_U(f_U(\cdot, \theta)) - H_U(f) = D_U(f, f_U(\cdot, \theta))$$

which is always nonnegative, and is 0 only when $f(x) = f_U(x, \theta)$ (ν-a.e. x) because of the information inequality (2). This concludes that $f_U(x, \theta)$ is a maximum entropy density function with respect to the entropy H_U. We will consider a parametric model of maximum entropy density functions

$$\mathcal{M}_U = \{ f_U(\cdot, \theta) : \theta \in \Theta \},$$

which we call U-model, where $\Theta = \{\theta \in \mathbb{R}^d : \kappa(\theta) < \infty\}$. In this way we introduce a class of statistical models generated from $\mathcal{U}$ defined in (1).

Let $\mathcal{M} = \{ f_\theta(x) : \theta \in \Theta \}$ be arbitrarily a fixed statistical model, where Θ is a parameter space. For a true probability density function $g(x)$ the expected loss function based on the U-divergence is given by $C_U(g, f_\theta)$. Let $\{x_i, i = 1, \dots, n\}$ be a data set from the true density function $g(x)$. Then the empirical U-loss function is given by substituting the empirical expectation for the expectation with respect to $g(x)$ as

$$L_U(\theta) = -\frac{1}{n} \sum_{i=1}^{n} \xi(f_\theta(x_i)) + \int U(\xi(f_\theta)) d\nu.$$

and the corresponding estimator for θ is defined as

$$\hat{\theta}_U = \underset{\theta \in \Theta}{\operatorname{argmin}}\, L_U(\theta),$$

called U-estimator for θ. We note that the expectation becomes just the expected U-loss function $C_U(g, f_\theta)$. If the true distribution g is in the parametric model $\mathcal{M}$, that is, there exists θ_0 in Θ such that $g = f_{\theta_0}$ (ν-a.e.), then we confirm that $C_U(g, f_{\theta_0}) \leq$

$C_U(g, f_\theta)$. This is because the difference $C_U(g, f_\theta) - C_U(g, f_{\theta_0})$ is nothing but the U-divergence $D_U(f_{\theta_0}, f_\theta)$, which is always nonnegative and holds equality if and only if $\theta = \theta_0$ from the information inequality (2). This observation shows that $\hat{\theta}_U$ is a consistent estimator for θ for any generator function U of $\mathcal{U}$. Thus there is a variety of consistent estimators for any model $\mathcal{M}$ constructed from a set $\mathcal{U}$ of generators functions as derived above.

A useful choice of U is given by a power exponential function

$$U_\beta(s) = \frac{1}{\beta+1}(1+\beta s)_+^{\frac{\beta+1}{\beta}}$$

with a power parameter β, where A_+ denotes the positive part of A. The derivative function is given by $u_\beta(s) = (1+\beta s)_+^{1/\beta}$; the inverse function $\xi_\beta(t) = (t^\beta - 1)/\beta$. These function $u_\beta(s)$ and $\xi_\beta(t)$ are called power exponential and power log function. The generator U_β leads to β-power cross entropy

$$C_\beta(f, g) = \int \Big\{\frac{g(x)^{\beta+1}}{\beta+1} - \frac{f(x)g(x)^\beta}{\beta}\Big\} d\nu,$$

cf. Basu et al. (1998), Minami and Eguchi (2002). The empirical β-power loss function

$$L_\beta(\theta) = -\frac{1}{\beta n}\sum_{i=1}^n f_\theta(x_i)^{\beta+1} + \frac{1}{\beta+1}\int f_\theta(x)^\beta d\nu(x)$$

provides the robust estimator for $\beta > 0$. The β-power entropy is given by

$$H_\beta(f) = -\frac{1}{\beta(\beta+1)}\int f(x)^{\beta+1} d\nu(x), \tag{4}$$

in which the model of maximum entropy distributions is given by

$$\mathcal{M}_\beta = \{f_\beta(x, \theta) = \{1 - \beta\theta^\top t(x) - \kappa_\beta(\theta)\}_+^{\frac{1}{\beta}} : \theta \in \Theta\}, \tag{5}$$

where $\kappa_\beta(\theta)$ is a normalizing factor and $\Theta = \{\theta \in \mathbb{R}^d : \kappa_\beta(\theta) < \infty\}$. If we put the equal constrain for the mean vector and variance matrix of x, then the family of maximum entropy distributions $f_\beta(x, \theta)$ in (5) for all βs includes t-distributions, normal distribution and Wigner distribution, cf Eguchi et al. (2011). In effect $H_\beta(f)$ is the same as Tsallis q-entropy for $q = \beta + 1$, see Tsallis (1988) for physical understandings. We overview a wide class of U-divergence in which the information inequality (2) plays an important role on the definition.

We have given a class of U-models $\mathcal{M}_\mathcal{U} = \{\mathcal{M}_U : U \in \mathcal{U}\}$ and a class of U-estimators $\mathcal{E}_\mathcal{U} = \{\hat{\theta}_U : U \in \mathcal{U}\}$. We like to focus on the diagonal part of the cross product space $\mathcal{M}_\mathcal{U} \times \mathcal{E}_\mathcal{U}$. Accordingly we consider a statistical behavior of

U-estimator $\hat{\theta}_U$ for θ when the model is taken as U-model $\mathcal{M}_U$. Specifically, if $U_0(s) = \exp(s)$, then $\hat{\theta}_U$ is the maximum likelihood estimator; $\mathcal{M}_U$ is the exponential family with the canonical statistic $t(x)$. On the couple of the estimator and model there has been established a basic concept such as sufficiency, ancillarity, efficiency and so forth. In effect the log likelihood function for data $\{x_i\}$ is written as

$$L_0(\theta) = \theta^\top \bar{t} - \kappa_0(\theta)$$

under the exponential family

$$\mathcal{M}_0 = \{f_0(x, \theta) = \exp(\theta^\top t(x) - \kappa_0(\theta)) : \theta \in \Theta\}, \tag{6}$$

where

$$\bar{t} = \frac{1}{n}\sum_{i=1}^{n} t(x_i), \tag{7}$$

called the canonical statistic. The maximum likelihood estimator for θ is a function of the canonical statistic $\bar{t}$, which is sufficient for θ. In the theory the convex property for the normalizing factor $\kappa(\theta)$ is intrinsic, which leads to the concavity of $L_0(\theta)$ and to the uniqueness for the maximum likelihood estimator. In effect the elegant property holds for the couple of $\hat{\theta}_U$ and $\mathcal{M}_U$. The U-loss function under the U-model is written by

$$L_U(\theta) = -\frac{1}{n}\sum_{i=1}^{n} \xi(f_U(x_i, \theta)) + \int U(\xi(f_U(x, \theta)))d\nu(x),$$

where $f_U(x, \theta)$ is defined in (3). Noting that $\xi(f_U(x, \theta)) = \theta^\top t(x) - \kappa(\theta)$ we get that

$$L_U(\theta) = -\{\theta^\top \bar{t} - c_U(\theta)\}, \tag{8}$$

where

$$c_U(\theta) = \kappa(\theta) - \int U(\theta^\top t(x) - \kappa(\theta))d\nu(x).$$

We refer loss-sufficiency if the loss function depends only on such a statistic $\bar{t}$ as in (8). We observe from (8) that the U-estimator $\hat{\theta}_U$ for θ is a function of $\bar{t}$. We note that, if we are interested in the mean parameter

$$\eta = \int t(x) f_U(x, \theta)d\nu(x),$$

then the U-estimator $\hat{\eta}_U$ for η is exactly equal to $\bar{t}$ since the transform from θ to η is given by the gradient vector of $c_U(\theta)$,

$$\eta = \frac{\partial}{\partial\theta} c_U(\theta).$$

Thus there are a lot of gradient fields for the mean parameter associated with the set $\mathcal{U}$ in (1), in which $c_U(\theta)$ plays a role as a potential function. As a result we conclude that the canonical property for the couple between estimation and model is preserved for the couple of U-estimator and U-model for any generator function U. In particular we find the common property that the U-estimator for the mean parameter η is equal to the canonical statistic $\bar{t}$. This is associated with the convex geometry associated with U-divergence. However the canonical property comes from a specific choice for the couple. For example, take another generator function V to U such that $V(s)$ is in $\mathcal{U}$, in which we can consider V-etimator $\hat{\eta}_V$ for the mean parameter η of U-model, but $\hat{\eta}_V$ is not already $\bar{t}$ whenever $V \neq U$. This aspect will be explored in a subsequent discussion such that the canonical statistic (7) is deformed to a weighted mean.

3 Nonconvexity of the Power Divergence

Let us focus on a class of power divergence with a property of projective invariance. We discuss the parametric estimation under an exponential family $\mathcal{M}_0$ defined in (6). A variant for the β-power cross entropy is defined by

$$C_\gamma(g, f) = -\frac{1}{\gamma(\gamma+1)} \int \left(\frac{f(x)}{\|f\|}\right)^\gamma g(x) d\nu(x)$$

where $\|f\|$ is the Lebesgue p-norm, or $\{\int |f(x)|^p d\nu(x)\}^{1/p}$ with $p = \gamma + 1$, see Fujisawa and Eguchi (2008), Eguchi et al. (2011) for the detailed discussion. The diagonal entropy is defined by $H_\gamma(f) = C_\gamma(f, f)$, which is written as $-\|f\|/(\gamma(\gamma+1))$. Thus $C_\gamma(g, f)$ is ourside the class of U-cross entry, however, $H_\gamma(f)$ is connected with the β-power entropy $H_\beta(f)$ defined in (4) in a one-to-one correspondence if γ equals β, so that the maximum entropy distributions with respect to H_γ and H_β are the same when $\gamma = \beta$. By definition,

$$C_\gamma(f, g) = C_\gamma(f, \lambda g)$$

for all $\lambda > 0$, which is referred to as being projectively invariant. In effect $C_\gamma(f, g)$ is characterized by this projective invariance with some requirements, cf. Fujisawa and Eguchi (2008), Eguchi et al. (2011). The expected γ-power loss function associated with the power divergence under a statistical model $\mathcal{M}$ is given by

$$C_\gamma(g, f_\theta) = c_\gamma(\theta)\int f_\theta(x)^\gamma g(x)d\nu(x)$$

where $c_\gamma(\theta)$ is a normalizing constant defined by $-\|f_\theta\|^{-\gamma}/(\gamma(\gamma+1))$. We remark that if γ is taken a limit to 0, then $C_\gamma(g, f_\theta)+1/(\gamma(\gamma+1))$ is the minus log likelihood function; if γ equals 1, $C_\gamma(g, f_\theta)$ is closely related with the mean integrated squared error, see Basu et al. (1998), Scott (2001), Minami and Eguchi (2002), Murata et al. (2004), Fujisawa and Eguchi (2008), Eguchi (2008), Eguchi and Kato (2010), Eguchi et al. (2011) for the super-robust property, the applications to machine learning and discussion on maximum entropy. The empirical γ-power loss function is given by substituting the empirical expectation for the expectation with respect to $g(x)$ as

$$L_\gamma(\theta) = c_\gamma(\theta)\sum_{i=1}^n f_\theta(x_i)^\gamma.$$

and the corresponding estimator for θ is defined by

$$\hat\theta_\gamma = \operatorname*{argmin}_{\theta\in\Theta} L_\gamma(\theta), \tag{9}$$

which is called γ-power estimator. When $\gamma = 0$, then $\hat\theta_\gamma$ is nothing but the maximum likelihood estimator. If the true distribution is in the parametric model $\mathcal{M}$, that is, $g = f_{\theta_0}$, then we confirm that $C_\gamma(g, f_{\theta_0}) \le C_\gamma(g, f_\theta)$. This is because $C_\gamma(g, f_\theta) - C_\gamma(g, f_{\theta_0})$ is nothing but the γ-power divergence

$$\begin{aligned} D_\gamma(f_{\theta_0}, f_\theta) = &-\frac{1}{\gamma(\gamma+1)}\frac{\int f_\theta(x)^\gamma f_{\theta_0}(x)d\nu(x)}{\left(\int f_\theta(x)^{\gamma+1}d\nu(x)\right)^{\frac{\gamma}{\gamma+1}}} \\ &+\frac{1}{\gamma(\gamma+1)}\left(\int f_{\theta_0}(x)^{\gamma+1}d\nu(x)\right)^{\frac{1}{\gamma+1}}, \end{aligned}$$

which is always nonnegative and holds equality if and only if $\theta = \theta_0$, cf. Eguchi et al. (2011). Thus $\hat\theta_\gamma$ is a consistent estimator for θ for any γ. In general the minimum divergence estimation satisfies Fisher-consistency as explored in the above discussion.

We pay our attentions to a specific case where the model is fixed as the exponential family $\mathcal{M}_0$. The empirical loss function under $\mathcal{M}_0$ is written by

$$L_\gamma(\theta) = -\frac{1}{\gamma(\gamma+1)}\sum_{i=1}^n \exp\left\{\gamma\theta^\top t(x_i) - \frac{\gamma}{\gamma+1}\kappa_0((\gamma+1)\theta))\right\},$$

of which the gradient vector is given by

$$\frac{\partial}{\partial\theta}L_\gamma(\theta) = -\frac{1}{\gamma+1}\sum_{i=1}^n e^{\gamma\theta^\top t(x_i)-\frac{\gamma}{\gamma+1}\kappa_0((\gamma+1)\theta)}\Big\{t(x_i) - \frac{\partial}{\partial\theta}\kappa_0(\theta)\Big|_{(\gamma+1)\theta}\Big\}.$$

Hence, if we take a new parameter defined by a transform

$$\eta_\gamma = \int e^{(\gamma+1)\theta^\top t(x)-\kappa_0((\gamma+1)\theta)}t(x)d\nu(x)$$

then the γ-power estimator for the parameter η_γ is given by

$$\widehat{\eta}_\gamma = \frac{\sum_{i=1}^n f_0(x_i,\hat{\theta}_\gamma)^\gamma t(x_i)}{\sum_{i=1}^n f_0(x_i,\hat{\theta}_\gamma)^\gamma}, \tag{10}$$

where $f_0(x,\theta)$ is defined in (6). The new parameter is expressed by $\eta_\gamma = (\partial/\partial\theta)\kappa_0(\theta)|_{(\gamma+1)\theta}$, which is the mean parameter in the escort model to $\mathcal{M}_0$,

$$\mathcal{M}_0^{\rm esc} = \Big\{\frac{f(x)^{\gamma+1}}{\int f^{\gamma+1}d\nu} : f \in \mathcal{M}_0\Big\}.$$

Thus, the transformation is one-to-one, so that the γ-power estimator for θ is automatically given by the inverse transform from $\widehat{\eta}_\gamma$. This expression (10) directly shows the robustness for the γ-power estimator. Because the weight function

$$\frac{f_0(x_i,\hat{\theta}_\gamma)^\gamma}{\sum_{i=1}^n f_0(x_i,\hat{\theta}_\gamma)^\gamma},$$

for the i-th observation x_i becomes negligibly small in the weighted mean (10) if x_i is an outlier; while it becomes relatively large if x_i is a proper observation. Such a robustness aspect never occurs if we employ the maximum likelihood, which is nothing but the canonical statistic (7) with a uniform weight $1/n$ over observations. Similarly if we assume the γ-power model

$$\mathcal{M}_\gamma = \{f_\gamma(x,\theta) = \{1+\gamma\theta^\top t(x) - \kappa_\gamma(\theta)\}_+^{\frac{1}{\gamma}} : \theta \in \Theta\}$$

in place of the exponential family $\mathcal{M}_0$, then the γ-power estimator for the mean parameter is exactly the canonical statistic. Thus the choice for the couple of model and estimator determines whether there occurs the weighted manner in the estimation or not. Furthermore the empirical γ-power loss is written by

$$-\log\{-\gamma(\gamma+1)L_\gamma(\theta)\} = \frac{\gamma}{\gamma+1}\kappa_0((\gamma+1)\theta) - \log\Big[\sum_{i=1}^n \exp\{\gamma\theta^\top t(x_i)\}\Big], \tag{11}$$

which is an expression as the difference of two convex functions, cf. Yuille and Rangarajan (2003). This suggests that the loss function flexibly learns any nonlinearity according to the choice of γ; while the minus log likelihood function is convex in Θ, so there exists a unique maximizer in Θ regardless of g is in $\mathcal{M}_0$ or not. In effect there is a wide range of the learnability for the γ-power loss function from convexity to high nonconvexity. We remark that the expected γ-power loss function is reduced to a convex function as

$$-\log\{-\gamma(\gamma+1)L_\gamma(\theta)\} = \frac{1}{\gamma+1}\kappa_0((\gamma+1)\theta)$$

if the true distribution satisfies $g(x) = f_\theta(x)$ ν-a.e. x. Such an exact parametric case does associates with not the difference of convex functions but a convex function. If the data set is sampled from a distribution with the density f_θ, then the empirical γ-power loss function (11) becomes convex in a large sample. On the other hand, the true density function is rather away from the model $\mathcal{M}_0$, then (11) becomes quite non-convex with possibly multimodality.

4 Spontaneous Property of the Power Divergence

We consider a location model $\mathcal{M} = \{h(x-\theta) : \theta \in \mathbb{R}^d\}$, where $h(x)$ is a spherically symmetric density function at 0 of $\mathbb{R}^d$. Thus there is a real-valued function $\phi(s)$ such that $h(x) = \phi(x^\top x)$. The empirical γ-power loss function is given by

$$L_\gamma(\theta) = c_\gamma \sum_{i=1}^{n} h(x_i - \theta)^\gamma,$$

where $c_\gamma = -1/(\gamma(\gamma+1))\{\int h(x)^{\gamma+1} d\nu(x)\}^{\frac{\gamma}{\gamma+1}}$. Here and hereafter we fix the base measure ν by the Lebesgue measure on $\mathbb{R}^d$. The γ-power estimator for θ is written by

$$\hat{\theta}_\gamma = \frac{\sum_{i=1}^n w(x_i - \hat{\theta}_\gamma)x_i}{\sum_{i=1}^n w(x_i - \hat{\theta}_\gamma)},$$

where $w(z) = \phi(z^\top z)^{\gamma-1}\phi'(z^\top z)$. This expression suggests a fixed point algorithm to numerically find the γ-power estimator as the update from θ_t to θ_{t+1} defined by

$$\theta_{t+1} = \frac{\sum_{i=1}^n w(x_i - \theta_t)x_i}{\sum_{i=1}^n w(x_i - \theta_t)}.$$

starting from an appropriately chosen value θ_0.

Let us look at a simple situation where the spontaneous data learning works well. Let $\mathcal{M}$ be a d-normal location model, that is

$$f(x,\theta) = (2\pi)^{-d/2} \exp\Big\{-\frac{1}{2}(x-\theta)^\top(x-\theta)\Big\}; \tag{12}$$

the true density function follows a K-normal mixture as $g(x) = \sum_{k=1}^K p_k f(x, \theta_k)$. We observe that

$$E(L_\gamma(\theta)) \propto -\sum_{k=1}^K p_k f(\theta, \theta_k)^{\frac{\gamma}{\gamma+1}}.$$

Therefore the expected loss function $L_\gamma(\theta)$ converges to the minus of the true density function $-g(\theta)$ up to a proportionality constant as γ goes to ∞. We confirm this property in a simple synthetic experiment. The simulated data are given from five-normal mixture with the 5 centers are fixed as $(0,0)$, $(2,2)$, $(2,-2)$, $(-2,2)$, $(-2,-2)$ and the equal mixing proportion 0.2, see for the sample plot in Fig. 1. Then we observe in Fig. 2 that the empirical loss function $-L_\gamma(\theta)$ for $\gamma = 2.0$ efficiently approximates to the true density function of five-normal mixture as in Fig. 1. Such a flexible shape for $L_\gamma(\theta)$ comes from the nonconvexity as expressed as difference of convex functions. This suggests that the set of local minima for $L_\gamma(\theta)$ asymptotically equals to that of modes of the true density for sufficiently large γ. The fact will be extended to a more general setting in the following discussion.

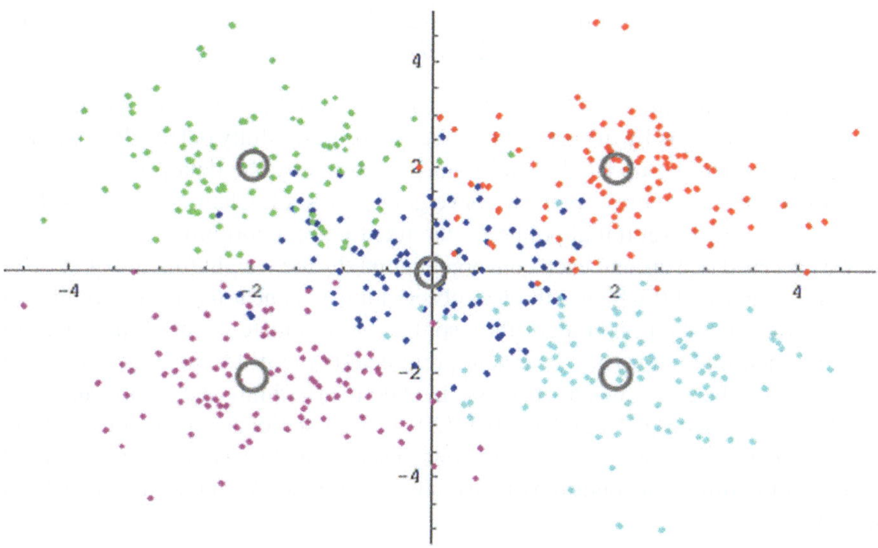

Fig. 1 Scatter plot of the sample

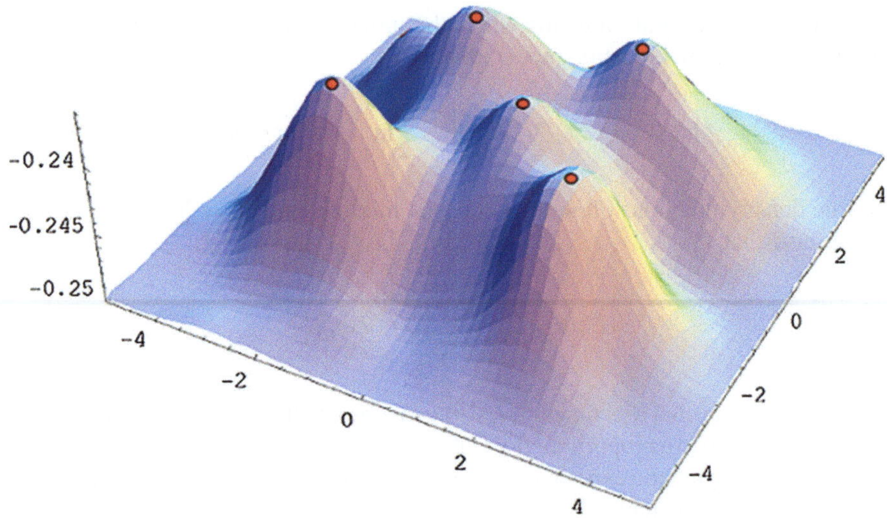

Fig. 2 Scatter plot of the minus empirical loss function $-L_\gamma(\theta)$ against θ

Thus Gamma-clustering proposed in Notsu et al. (2014) is supported a theoretical validation for the consistency for clustering analysis with a selection for γ based on AIC. To find all the local minimizers for $L_\gamma(\theta)$ we repeatedly used for the fixed point algorithm defined by

$$\theta_{t+1} = \frac{\sum_{i=1}^n f(x_i, \theta_t)^\gamma x_i}{\sum_{i=1}^n f(x_i, \theta_t)^\gamma} \tag{13}$$

repeatedly updating the initial point θ_0. Figure 3 shows the process of the detection for five cluster centers by the use of (13). We could successfully detect all five modes for $-L_\gamma(\theta)$ when the updated initial point was taken by the data point that is the remotest from the set of all present convergent points, cf. Chen et al. (2014) for a more greedy way to set initial points for the fixed-point algorithm (13).

We observed that around each convergent point θ_∞ the weight function $f(x_i, \theta_\infty)$ becomes near 1 and 0 whether the i-th observation x_i is near θ_∞ or not. In fact we ran six-times the fixed point algorithm, so that we terminate the algorithm because we confirmed that the sixth convergent point equals the fifth one. The theoretical discussion for the convergence of such a fixed point algorithm is given in Ghassabeh (2015). By the theorem 1 in Ghassabeh (2015) if all stationary points of $L_\gamma(\theta)$ are isolated, then the algorithm converges to one of stationary points. See also Ghassabeh (2015) for the sufficient condition of the convergence when $h(x)$ is a standard normal density.

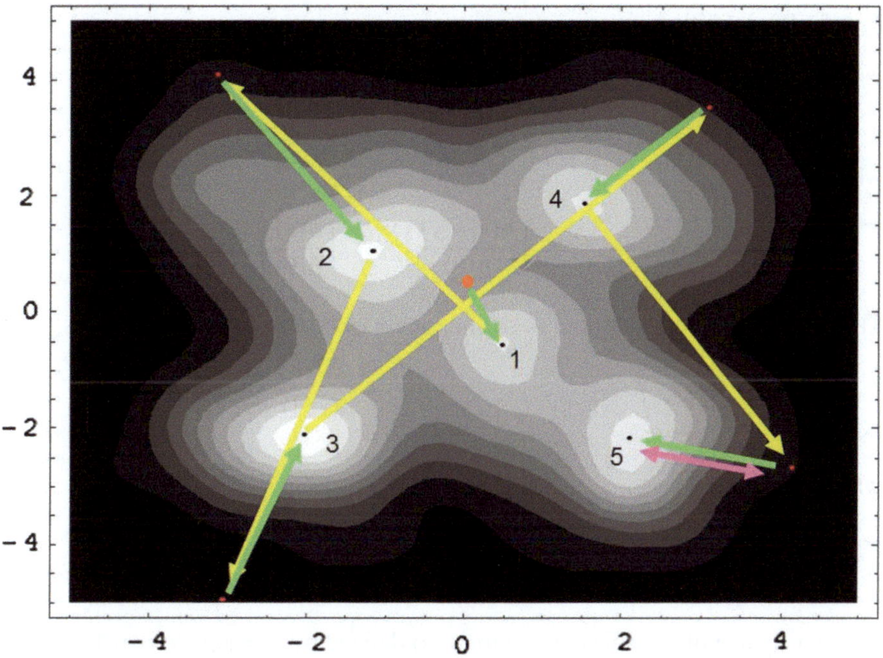

Fig. 3 Contour plot of the empirical loss function

5 Nonparametric Consistency

We discuss the spontaneous learning associated with the γ-power loss function focusing on a specific situation with the normal location model in the preceding section. Such an observation is extended to a more general situation to the location model $\mathcal{M}$. Thus we define a normalized γ-power loss, where $\gamma > 0$, by

$$\hat{L}_\gamma(\theta) = \frac{a_\gamma}{n}\sum_{i=1}^{n} h(x_i - \theta)^\gamma,$$

where $a_\gamma = 1/\int h(x)^\gamma dx$. By definition, if we set $\theta = x$, then $\hat{L}_\gamma(x)$ is viewed as a probability density function on $\mathbb{R}^d$. We show that the normalized γ-power loss function itself is a consistent estimator for the true density function as follows.

Theorem 1 *Let $\mathcal{M} = \{h(x - \theta) : \theta \in \mathbb{R}^d\}$ be a location model assuming that there exists a strictly decreasing positive function $\varphi(s)$ for $s \geq 0$ such that $h(x) = \varphi(\|x\|)$, where $\|x\|$ denotes a norm of x. Then, for a true density function $g(x)$ the normalized γ-power loss function $\hat{L}_\gamma(x)$ almost surely converges to $g(x)$ when the sample size n and the power parameter γ both go to ∞.*

Proof By definition,

$$\mathbb{E}\{\hat{L}_\gamma(x)\} = \int \Psi_\gamma(x-y)g(y)dy,$$

where

$$\Psi_\gamma(x-y) = \Big\{\frac{h(x-y)}{h(0)}\Big\}^\gamma \Big/ \int \Big\{\frac{h(z)}{h(0)}\Big\}^\gamma dz.$$

Thus Ψ_γ satisfies that $\int \Psi(y)dy = 1$. For any $x \neq 0$ in the support of h

$$\int \varphi(\|z\|)^\gamma dz \geq \int_{\{z:\|z\|\leq\|x\|\}} \varphi(\|z\|)^\gamma dz,$$

which is written

$$\int_0^{\|x\|} \varphi(s)^\gamma \Big(\int_{\{u:\|u\|=s\}} |J(u)|du\Big)ds,$$

taking a change of variables of Z into (s,u) defined by $s = \|z\|$, $u = z/\|z\|$, where $J(u)$ is the Jacobian of the transform. The mean-value theorem leads that there exists a s_0, $0 < s_0 < \|x\|$ such that

$$\int_0^{\|x\|} \varphi(s)^\gamma \Big(\int_{\{u:\|u\|=s\}} |J(u)|du\Big)ds = \varphi(s_0)^\gamma \int_0^{\|x\|} \int_{\{u:\|u\|=s\}} |J(u)|duds,$$

which implies that

$$0 \leq \Psi_\gamma(x) \leq \Big(\frac{\varphi(\|x\|)}{\varphi(s_0)}\Big)^\gamma \Big(\int_0^{\|x\|} \int_{\{u:\|u\|=s\}} |J(u)|duds\Big)^{-1}$$

which is convergent to 0 as γ goes to ∞. Because $\varphi(\|x\|) < \varphi(s_0)$ from the assumption for φ. Thus $\Psi_\gamma(x)$ converges to a Dirac delta function $\delta(x)$ as γ goes to ∞, so that the expectation $\mathbb{E}\{\hat{L}_\gamma(x)\}$ converges to $g(x)$. The proof is complete.

This implies that the information about the true density is obtained by a limit of the expected γ-power loss function even if the model is irrelevant to the true density. On the other hand, a kernel density estimator

$$\tilde{g}_\lambda(x) = \frac{1}{n\lambda}\sum_{i=1}^n K\Big(\frac{x-x_i}{\lambda}\Big),$$

is widely employed with an appropriate selection for λ. If we take a standard normal density function for both $h(x)$ and $K(x)$, then $\hat{L}_\gamma(x)$ and $\tilde{g}_\lambda(x)$ eventually coincide with a relation of $\lambda = \gamma^{-1/2}$. However, $\hat{L}_\gamma(x)$ is rather different from any kernel density estimator unless $h(x)$ is a normal density function. An essential difference appears the behaviors for $\psi_\gamma(x-y)$ and $\lambda^{-1}K(\lambda^{-1}(x-y))$ as γ and λ^{-1} go to ∞. The larger the power parameter γ is taken, the more correct information we get in the expected loss function. However, we have to make a more careful selection for γ in the empirical loss function because the behavior of $\hat{L}_\gamma(x)$ becomes unstable and spikey around the data set when $\gamma \gg 0$.

Let us discuss the optimal selection for the power parameter in an asymptotic evaluation for a case where $h(x)$ is a d-variate density function. First, the bias is defined by

$$\text{bias}_\gamma(x) = \frac{\int h(x-y)^\gamma g(y)dy}{\int h(y)^\gamma dy} - g(x), \tag{14}$$

in which taking a scale transform $t = \gamma(x-y)$ for the numerator leads to an approximation

$$\text{bias}_\gamma(x) \approx \frac{1}{2}\frac{1}{\int h(y)^\gamma dy}\text{tr}\Big(V_\gamma \frac{\partial^2 g(x)}{\partial x \partial x^\top}\Big), \tag{15}$$

where tr denotes matrix trace and $V_\gamma = \int ss^\top h(s)^\gamma ds$. See Appendix for the detailed derivation. Secondly, the variance is asymptotically given by

$$\text{var}_\gamma(x) \approx \frac{1}{n}\frac{\int h(s)^{2\gamma}ds}{\big(\int h(y)^\gamma dy\big)^2} g(x), \tag{16}$$

see also Appendix. Hence the optimal γ is given by minimization for the approximate mean integrated square error (MISE)

$$\text{MISE}_\gamma(h,g) \approx \frac{1}{4}\frac{\int \Big\{\text{tr}\Big(V_\gamma \frac{\partial^2 g(x)}{\partial x \partial x^\top}\Big)\Big\}^2 dx}{\big(\int h(y)^\gamma dy\big)^2} + \frac{1}{n}\frac{\int h(y)^{2\gamma}dy}{\big(\int h(y)^\gamma dy\big)^2}. \tag{17}$$

If $h(x)$ is a standard normal density function, then the formula (17) leads to

$$\text{MISE}_\gamma(h,g) \approx \frac{1}{4}\gamma^{-2}\int \Big\{\text{tr}\Big(\frac{\partial^2 g(x)}{\partial x \partial x^\top}\Big)\Big\}^2 dx + \frac{1}{n}2^{-d}\pi^{-\frac{d}{2}}\gamma^{\frac{d}{2}}.$$

Obviously, if h is a univariate standard normal density function, then

$$\text{MISE}_\gamma(\varphi,g) = \frac{1}{4\gamma^2}\int g''(x)^2 dx + \frac{1}{2\sqrt{\pi}n}\gamma^{\frac{1}{2}},$$

which is reduced to the well-known formula for normal kernel density estimator with the bandwidth $\lambda = \gamma^{-1/2}$, cf. Silverman (1986).

Let us consider a case where $h(x)$ is a Wigner semicircle distribution apart from a normalizing factor, that is

$$h(x) = 1_{B_d}(x)(1 - x^\top x)^{\frac{1}{2}},$$

where $B_d = \{x \in \mathbb{R}^d : \|x\|_{L_2} \le 1\}$. The general formula is reduced to

$$\text{MISE}_\gamma(h, g) \approx \frac{d^2}{4}\Big(\frac{\gamma}{e}\Big)^{-2} \int \Big\{\text{tr}\Big(\frac{\partial^2 g(x)}{\partial x \partial x^\top}\Big)\Big\}^2 dx + \frac{1}{n} 2^{-d} \pi^{-\frac{d}{2}} \Big(\frac{\gamma}{e}\Big)^{\frac{d}{2}}. \quad (18)$$

See Appendix for detailed discussion. We remark that the approximate formula (17) for a model density function $h(x)$ is delicate when $h(x)$ has an unbounded support. For example the behavior for the t-distribution case is collapsed in such a large γ asymptotics. In this way the validity for the approximate formula given in (17) depends on the choice of the model density h, while that for kernel density estimator is valid whenever the kernel function has a finite variance. We need more through examination for the power entropy estimator $\hat{L}_\gamma$ as the next project.

6 Concluding Remarks and Discussion

The minimum divergence methods originally focus on the robust properties which have been widely investigated the redescending influence curve, gross error sensitivity and trade between efficiency and robustness, cf Basu et al. (1998), Minami and Eguchi (2002), Fujisawa and Eguchi (2008). In this paper we investigate spontaneous learnability beyond robustness perspectives. If the true density belongs to the model, then the minimum γ-power loss leads to consistent estimation for any γ, and the asymptotic efficiency is attained when $\gamma = 0$, or the maximum likelihood estimation. On the other hand when the true density is away from the model, we found a property of spontaneous data learning with an appropriate selection for γ. This is an approach for selecting the optimal estimator in a bulk of candidate estimators, which is a dualistic analogue to the selection for models.

We gave an approximate mean integrated square error in (17) for a multivariate case, which could be used for the selection for the power γ. Multi-fold cross validation is also a universal tool for the optimal selection for γ. However the implementation is sometimes time-consuming for high-dimensional and massive data, in which a procedure of on-learning may be promising with efficient performance. In fact we proposed an information criterion for the selection, in which AIC is used for the selection based on the approximate normal mixture model, cf Notsu et al. (2014). The method for kernel density estimation is often criticized in a high dimensional case as the curse of dimensionality, cf. Huber (1985). We should carefully check whether is the proposed method trapped into such a difficulty or not.

The choice of the couple of model and estimator is crucial for the aspect of SDL. It is closely related with the maximum entropy model associated with the entropy resulted from the divergence. The γ-power entropy reads to the maximum entropy model indexed by γ including a normal model, t-distribution model and semicircle distribution model. If we select the maximum γ entropy model and the γ' loss function, then the minimum γ divergence estimator is unique irrelevant to data if $\gamma = \gamma'$. The flexibility of the corresponding γ loss function increases as γ and γ' are more different. We have discussed SDL based on the framework of the fixed couple of the exponential model $\mathcal{M}_0$ in (6) and γ-power estimator in (9) Thus the couple does not lead to the canonical statistic but adaptively weighted mean (10) since the model $\mathcal{M}_0$ is totally different the γ-power model defined in (5) with $\beta = \gamma$ unless $\gamma = 0$. The property for SDL is observed if the rigid relation between minimum divergence and maximum entropy is collapsed, cf. Eguchi et al. (2014). For example, if the maximum likelihood is applied to the Laplace location model, then the estimator is not the canonical statistic, or the sample mean but the sample median. Furthermore, if the couple is taken as that of the Cauchy location model and the maximum likelihood, then the property of SDL is observed because the Cauchy model is far from the maximum entropy model, or the normal location model. We like to elucidate this relation associated with SDL in a wider perspective as a future project. The present procedure should be strengthen to tackle with more difficult tasks with adaptive complex and hierarchical system.

Acknowledgements Authors express sincere gratitude to the reviewers for their helpful comments and suggestions for improving the original manuscript, in particular from the viewpoint of information science. SE and OK were supported by Japan Science and Technology Agency (JST), Core Research for Evolutionary Science and Technology (CREST).

Appendix

The Derivation of (15)

If we take a scale transform $t = \gamma(x - y)$, then the integral in (14) is written as

$$\int h(x-y)^\gamma g(y)dy = \frac{1}{\gamma^d}\int h\Big(\frac{t}{\gamma}\Big)^\gamma g\Big(x - \frac{t}{\gamma}\Big)dt$$

which is approximated by

$$\frac{1}{\gamma^d}\int h\Big(\frac{t}{\gamma}\Big)^\gamma \Big\{g(x) - \frac{1}{\gamma}t^\top \frac{\partial g(x)}{\partial x} + \frac{1}{2\gamma^2}t^\top \frac{\partial^2 g(x)}{\partial x \partial x^\top}t\Big\}dt.$$

Therefore we get that

$$\int h(x-y)^\gamma g(y)dy \approx \Big(\int h(s)^\gamma ds\Big)g(x) + \frac{1}{2}\mathrm{tr}\Big\{\Big(\int ss^\top h(s)^\gamma ds\Big)\frac{\partial^2 g(x)}{\partial x \partial x^\top}\Big\},$$

since $\int th(t/\gamma)dt = 0$ from the symmetry assumption for h. This completes the derivation for the formula (15).

The Derivation of (16)

We next show the approximation (16). By definition

$$\mathrm{var}_\gamma(x) = \frac{1}{n}\frac{\int h(x-y)^{2\gamma}g(y)dy}{\big(\int h(y)^\gamma dy\big)^2} - \frac{1}{n}\Big(\frac{\int h(x-y)^\gamma g(y)dy}{\int h(y)^\gamma dy}\Big)^2. \tag{19}$$

An argument similar to the above yields that

$$\int h(x-y)^{2\gamma}g(y)dy \approx \Big(\int h(s)^{2\gamma}ds\Big)g(x)$$

since the second term of (19) is approximated as $-g(x)^2/n$, which can be neglected in the main order. This concludes (16).

The Derivation of (31)

In the case $h(x) = 1_{B_d}(x)(1-x^\top x)^{\frac{1}{2}}$, we have

$$\begin{aligned}\int h(x)^\gamma dx &= \int 1_{B_d}(x)(1-x^\top x)^{\frac{\gamma}{2}}dx \\ &= \int_0^1 (1-t^2)^{\frac{\gamma}{2}} t^{d-1} S^{d-1} dt \\ &= \frac{S^{d-1}}{2}\int_0^1 (1-s)^{\frac{\gamma}{2}} s^{\frac{d-2}{2}} ds \\ &= \frac{S^{d-1}}{2} B\Big(\frac{d}{2}, \frac{\gamma}{2}+1\Big),\end{aligned}$$

where S^{d-1} is the surface of the unit sphere of $d-1$ dimension, and $B(a,b)$ is the beta function such as

$$\begin{aligned} S^{d-1} &= \frac{2\pi^{\frac{d}{2}}}{\Gamma(\frac{d}{2})} \\ B(a,b) &= \int_0^1 t^{a-1}(1-t)^{b-1}dt,\ a>0,\ b>0.\end{aligned}$$

Hence from the Stirling's approximation and $\gamma \gg d$, we have

$$\begin{aligned}\int h(x)^\gamma dx &= \frac{\pi^{\frac{d}{2}}}{\Gamma(\frac{d}{2})}\frac{\Gamma(\frac{d}{2})\Gamma(\frac{\gamma}{2}+1)}{\Gamma(\frac{d}{2}+\frac{\gamma}{2}+1)} \\ &\approx \pi^{\frac{d}{2}}\frac{\sqrt{\pi\gamma}(\frac{\gamma}{2e})^{\frac{\gamma}{2}}}{\sqrt{\pi(d+\gamma)}(\frac{d+\gamma}{2e})^{\frac{d}{2}+\frac{\gamma}{2}}} \\ &\approx \pi^{\frac{d}{2}}\left(\frac{\gamma}{2e}\right)^{-\frac{d}{2}}\end{aligned}$$

Similarly, we have

$$\int h(x)^{2\gamma}dx \approx \pi^{\frac{d}{2}}\left(\frac{\gamma}{e}\right)^{-\frac{d}{2}}$$

Moreover, we have

$$\begin{aligned}\int 1_{B_d}(x)x^\top x(1-x^\top x)^{\frac{\gamma}{2}}dx &= \int_0^1 (1-t^2)^{\frac{\gamma}{2}}t^{d+1}S^{d-1}dt \\ &= \frac{S^{d-1}}{2}B\left(\frac{d}{2}+1,\frac{\gamma}{2}+1\right) \\ &= \frac{\pi^{\frac{d}{2}}}{\Gamma(\frac{d}{2})}\frac{\Gamma(\frac{d}{2}+1)\Gamma(\frac{\gamma}{2}+1)}{\Gamma(\frac{d}{2}+\frac{\gamma}{2}+2)} \\ &\approx \pi^{\frac{d}{2}}\frac{d}{2}\frac{\sqrt{\pi\gamma}(\frac{\gamma}{2e})^{\frac{\gamma}{2}}}{\sqrt{\pi(d+\gamma+2)}(\frac{d+\gamma+2}{2e})^{\frac{d}{2}+\frac{\gamma}{2}+1}} \\ &\approx \pi^{\frac{d}{2}}\frac{d}{2}\left(\frac{\gamma}{2e}\right)^{-\frac{d+2}{2}}\end{aligned}$$

Hence we conclude (31).

In the case that $h(x) = 1_{S_d}(1-\|x\|_{L_1})$, we have

$$\int h(x)^\gamma dx \tag{20}$$

$$= 2^d \int_0^1 \cdots \int_0^{1-x_3-\cdots-x_d}\int_0^{1-x_2-\cdots-x_d}(1-x_1-x_2-\cdots-x_d)^\gamma dx_1dx_2\cdots dx_d \tag{21}$$

$$= \frac{2^d}{(1+\gamma)(2+\gamma)\cdots(d+\gamma)} \tag{22}$$

$$\approx 2^d\gamma^{-d} \tag{23}$$

Similarly, we have

$$\int h(x)^{2\gamma} dx \approx \gamma^{-d}. \tag{24}$$

Here we have

$$\int_0^1 \cdots \int_0^{1-x_3-\cdots-x_d} \int_0^{1-x_2-\cdots-x_d} x_d^2 (1 - x_1 - x_2 - \cdots - x_d)^\gamma dx_1 dx_2 \cdots dx_d \tag{25}$$

$$= \int_0^1 \frac{x_d^2 (1 - x_d)^{d-1+\gamma}}{(1+\gamma)\cdots(d-1+\gamma)} dx_d \tag{26}$$

$$= \frac{2}{(1+\gamma)\cdots(d+2+\gamma)}, \tag{27}$$

where

$$\int_0^1 x_d^2 (1 - x_d)^{d-1+\gamma} dx_d = \frac{2}{(d+\gamma)(d+1+\gamma)(d+2+\gamma)}. \tag{28}$$

Hence from the symmetry regarding $x_1, \ldots, x_d$, we have

$$\int x^\top x h(x)^\gamma dx = \frac{2^{d+1} d}{(1+\gamma)\cdots(d+2+\gamma)} \tag{29}$$

$$\approx 2^{d+1} d \gamma^{-(d+2)} \tag{30}$$

$$\mathrm{MISE}_\gamma(h, g) \approx d^2 \gamma^{-4} \int \left\{ \mathrm{tr}\left(\frac{\partial^2 g(x)}{\partial x \partial x^\top} \right) \right\}^2 dx + \frac{1}{n} 2^{-2d} \gamma^d. \tag{31}$$

References

Amari, S. (1985). *Differential-geometrical methods in statistics*. Lecture notes in statistics (Vol. 28). New York: Springer.

Amari, S., & Nagaoka, H. (2000). *Methods of information geometry*. Oxford: Oxford University Press.

Basu, A., Harris, I. R., Hjort, N. L., & Jones, M. C. (1998). Robust and efficient estimation by minimising a density power divergence. *Biometrika*, *85*, 549–559.

Cichocki, A., & Amari, S. I. (2010). Families of alpha-beta-and gamma-divergences: flexible and robust measures of similarities. *Entropy*, *12*, 1532–1568.

Chen, T. L., Hsieh, D. N., Hung, H., Tu, I. P., Wu, P. S., Wu, Y. M., et al. (2014). γ-SUP: a clustering algorithm for cryo-electron microscopy images of asymmetric particles. *Annals of Applied Statistics*, *8*(1), 259–285.

Csiszár, I. (2008). Axiomatic characterizations of information measures. *Entropy*, *10*(3), 261–273.

Eguchi, S. (1983). Second order efficiency of minimum contrast estimators in a curved exponential family. *The Annals of Statistics*, *11*, 793–803.

Eguchi, S. (1992). Geometry of minimum contrast. *Hiroshima Mathematical Journal*, *22*, 631–647.

Eguchi, S. (2006). Information geometry and statistical pattern recognition. *Sugaku Expositions American Mathematical Society*, *19*, 197–216.

Eguchi, S. (2008). Information divergence geometry and the application to statistical machine learning. In F. Emmert-Streib & M. Dehmer (Eds.), *Information Theory and Statistical Learning* (pp. 309–332). New York: Springer.

Eguchi, S., & Kano, K. (2001). *Robustifying maximum likelihood estimation*. Institute of Statistical Mathematics, Tokyo, Japan: Technical Report.

Eguchi, S., & Kato, S. (2010). Entropy and divergence associated with power function and the statistical application. *Entropy*, *12*, 262–274.

Eguchi, S., Komori, O., & Kato, S. (2011). Projective power entropy and maximum Tsallis entropy distributions. *Entropy*, *13*, 1746–1764.

Eguchi, S., Komori, O., & Ohara, A. (2014). Duality of maximum entropy and minimum divergence. *Entropy*, *16*(7), 3552–3572.

Fisher, R. A. (1912). On an absolute criterion for fitting frequency curves. *Messenger of Mathematics*, *41*, 155–160.

Fisher, R. A. (1922). On the mathematical foundations of theoretical statistics. *Philosophical Transactions of the Royal Society London Series A*, *222*, 309–368.

Fujisawa, H., & Eguchi, S. (2008). Robust parameter estimation with a small bias against heavy contamination. *Journal of Multivariate Analysis*, *99*(9), 2053–2081.

Ghassabeh, A. Y. (2015). A sufficient condition for the convergence of the mean shift algorithm with Gaussian kernel. *Journal of Multivariate Analysis*, *135*, 1–10.

Huber, P. (1985). Projection pursuit. *The Annals of Statistics*, 435–475.

Minami, M., & Eguchi, S. (2002). Robust blind source separation by beta divergence. *Neural Computation*, *14*, 1859–1886.

Murata, N., Takenouchi, T., Kanamori, T., & Eguchi, S. (2004). Information geometry of U-Boost and Bregman divergence. *Neural Computation*, *16*, 1437–1481.

Nielsen, F., & Boltz, S. (2011). The Burbea-Rao and Bhattacharyya centroids. *IEEE Transactions on Information Theory*, *57*(8), 5455–5466.

Nielsen, F., & Nock, R. (2015). Total Jensen divergences: Definition, properties and clustering. In *IEEE International Conference Acoustics, Speech and Signal Processing (ICASSP), 2015*, (pp. 2016–2020).

Notsu, A., Komori, O., & Eguchi, S. (2014). Spontaneous Clustering via Minimum Gamma-divergence. *Neural Computation*, *26*(2), 421–448.

Scott, D. W. (2001). Parametric statistical modeling by minimum integrated square error. *Technometrics*, *43*, 274–285.

Silverman, B. (1986). *Density estimation for statistics and data analysis* (Vol. 26). Florida: CRC press.

Tsallis, C. (1988). Possible generalization of Boltzmann-Gibbs statistics. *Journal of Statistical Physics*, *52*(1–2), 479–487.

Yuille, A. L., & Rangarajan, A. (2003). The concave-convex procedure. *Neural Computation*, *15*, 915–936.

Zhang, J. (2013). Nonparametric information geometry: from divergence function to referential-representational biduality on statistical manifolds. *Entropy*, *15*, 5384–5418.

Extrinsic Projection of Itô SDEs on Submanifolds with Applications to Non-linear Filtering

John Armstrong and Damiano Brigo

AMS classification codes 58J65 · 60H10 · 60J60

1 Introduction

In this paper we consider two notions of projecting a stochastic differential equation (SDE) onto a manifold M. We will call the two approaches Stratonovich projection and the extrinsic Itô projection.

The purpose of these projection methods is to transform an infinite dimensional SDE to a finite dimensional SDE which can then be solved numerically. We will benchmark the performance of these two competing projection techniques using a non-linear filtering problem. We will also compare the performance of our approach to comparable established approaches to non-linear filtering.

To explain the idea, let us first consider projecting an ordinary differential equation (ODE) onto a manifold $M \subseteq \mathbb{R}^r$. *An* ODE can be thought of as defining a vector field in $\mathbb{R}^r$. At every point $x \in M$ we can use the Euclidean metric to project the vector at x onto the tangent space $T_x M$. In this way one obtains a vector field on M which can be thought of as a new ODE on M that approximates the full ODE in $\mathbb{R}^r$.

We now wish to consider projecting stochastic differential equations onto a manifold. One possible answer has been proposed previously which we shall call the Stratonovich projection. The Stratonovich projection is obtained by simply applying the projection operator to the coefficients of the SDE written in Fisk–Stratonovich–

J. Armstrong (✉)
King's College London, WC2R 2LS, London, UK
e-mail: john.1.armstrong@kcl.ac.uk

D. Brigo
Imperial College, SW7 2AZ, London, UK
e-mail: damiano.brigo@imperial.ac.uk

F. Nielsen et al. (eds.), *Computational Information Geometry*,
Signals and Communication Technology, DOI 10.1007/978-3-319-47058-0_5

McShane calculus form (Stratonovich from now on) (Fisk 1963; Stratonovich 1966; McShane 1974). No optimality result has been derived for the Stratonovich projection, it has simply been derived heuristically from the deterministic case. Nevertheless, it appears to be a good approximation in practice and it has been used to find good quality numerical solutions to the non-linear filtering problem (See Brigo et al. 1998, 1999; Armstrong and Brigo 2013, 2016a).

It is obvious to anyone with experience of stochastic differential equations on manifolds that simply applying the projection operator to the coefficients of the SDE written in Itô form will not work. This is because solutions to the projected equation don't stay on the manifold. Nevertheless we will be able to obtain a modification of this idea, which we will call the extrinsic Itô projection, which does give a well defined SDE on the manifold. We will show elsewhere how this extrinsic Itô projection can be derived from an optimality argument and so this new projection is in some sense an optimal approximation of the original SDE on the manifold. The extrinsic Itô projection is described in Sect. 2. We prove directly that it is a well-defined stochastic differential equation. For the benefit of the reader, we include a brief review of stochastic differential equations on manifolds in Sect. 2.1.

Having defined the extrinsic Itô projection, we can apply it to find approximate solutions to difficult stochastic differential equations. In particular we will apply it to the non-linear filtering problem. This application is discussed in Sect. 3. We will derive general projection formulae for the non-linear filtering problem. We will then apply this to the problem of approximating a non-linear filter using a Gaussian distribution. A reader who is unfamiliar with non-linear filtering will want to consult Sect. 3.1 for a brief review.

Gaussian approximations to non-linear filters are widely used in practice (Bain and Crisan 2009). In particular the Extended Kalman Filter is a popular approximation technique. Other Gaussian approximations exist such as Assumed Density Filters and filters derived from the Stratonovich projection. Our theory indicates that all these classical techniques can be improved upon by using the extrinsic Itô projection (at least over small time intervals). We confirm this with a numerical example.

The utility of the projection method is by no means restricted to the filtering problem nor to such simple approximations as Gaussian filters. Our previous work shows how the Stratonovich projection can be used to generate far more sophisticated filters and it is clear that the idea of projection should be widely applicable in the study of SDEs and ODEs. Nevertheless by focussing on Gaussian filters we can examine in detail the idea that there may be many useful ways of approximating an SDE on a submanifold and examine in detail the relative performance of the extrinsic Itô projection. The point we wish to emphasize is that the extrinsic Itô projection is able to tell us something new even about the well-worn topic of approximating the non-linear filtering problem using Gaussian distributions.

Note that the development of the extrinsic Itô projection does not invalidate previous work using the Stratonovich projection, it merely indicates that alternative approximations are possible. In a future paper we will consider in what sense the extrinsic Itô projection is an optimal approximation over a small time horizon. As we will show in that paper, the notion of optimality for a finite dimensional approximation to an SDE is far more subtle than the comparable notion for ODEs. This theoretical analysis will explain why the extrinsic Itô projection gives the excellent results demonstrated numerically in this paper, but will also provide an explanation for why there are occasions when the Stratonovich projection is still a superior approach.

2 Projecting Stochastic Differential Equations

2.1 Itô SDEs on Manifolds

It is well known that one can write SDEs on manifolds in Stratonovich form. However, in our experience there seems to be some confusion about whether one can, or should, write SDEs on manifolds in Itô form. Itô himself (Itô 1950) defined the notion of an SDE on a manfiold using Itô calculus. Nevertheless we believe it may be useful to the non-expert if we explicitly define an Itô SDE on a manifold and explain the motivation behind the definition.

Given a n-dimensional manifold M, we can write down stochastic differential equations in a neighbourhood U of a point x by choosing a chart $\phi : U \to \mathbb{R}^n$ and then writing the stochastic differential equation in local coordinates. The equation written in local coordinates will depend upon the choice of chart ϕ. Thus the data for a stochastic differential equation locally consists of:

1. A vector valued Brownian motion W_t (the theory can also be extended to continuous semi-martingale integrators, but we will use Brownian motion for simplicity)
2. A chart ϕ
3. The coefficient functions a, b of a stochastic differential equation written in local coordinates:

$$dX_t = a(X_t, t)dt + b(X_t, t)dW_t.$$

To define a stochastic differential equation over the entire manifold we will need local data of this form for a complete atlas of charts. Where charts overlap, we will need some compatibility conditions on these local SDEs. The "correct" compatibility condition should be chosen so that the solutions of the SDE in one chart ϕ are mapped to the solutions of the SDE in another chart, Φ by the transition function $\tau = \Phi \circ \phi^{-1}$. Since this requirement is expressed in terms of the *solutions* to a stochastic differential equation it is a mathematically complex requirement. We would prefer to write the requirement in terms of the much simpler data of the *coefficients* of our stochastic differential equation. We can informally calculate the correct compatibility condition on the coefficients using Itô's lemma.

Let (W_t, ϕ, a, b) be the data for the SDE in one chart, and (W_t, Φ, A, B) be the data in another chart. We will suppose that W_t takes valued in $\mathbb{R}^m$ and will write W^α for the components of W_t. Similarly, X^i, a^i, b^i_α are the components of the vectors X, a and the tensor b. We have chosen to label indices such that Roman indices run from 1 through to n (the dimension of our manifold) and Greek indices run from 1 through to m (the dimension of the process W_t).

We can now write out the SDE (3) in full detail in local coordinates as:

$$\mathrm{d}X^i_t = a^i(X_t, t)\mathrm{d}t + \sum_\alpha b^i_\alpha(X_t, t)\mathrm{d}W^\alpha_t.$$

We will write τ^i for the components of τ and will use coordinates x^i for $\mathbb{R}^n$. With this notation in place we can apply Itô's lemma to write out an SDE for $\tau(X_t)$ as follows:

$$\mathrm{d}\tau^i(X_t) = \left(\sum_j \frac{\partial \tau^i}{\partial x^j} a^j(X_t, t) + \sum_{j,k,\alpha,\beta} \frac{1}{2} \frac{\partial^2 \tau^i}{\partial x^j \partial x^k} b^j_\alpha(X_t, t) b^k_\beta(X_t, t) \boldsymbol{g}^{\alpha\beta} \right) \mathrm{d}t$$
$$+ \left(\sum_{j,\alpha} \frac{\partial \tau^i}{\partial x^j} b^j_\alpha(X_t, t) \right) \mathrm{d}W^\alpha_t.$$

Here $\boldsymbol{g}^{\alpha\beta} = [W^\alpha, W^\beta]_t$ denotes the quadratic covariation of W^α and W^β. We use the letter $\boldsymbol{g}$ because this term defines a metric on $\mathbb{R}^m$. Our local coordinate notation is very cumbersome, so we will adopt various conventions taken from differential geometry:

(i) The Einstein summation convention: if an index appears as both an upper index and lower index, one should sum over that index.
(ii) We write ∂_i as an abbrevation for $\frac{\partial}{\partial x^i}$.
(iii) We drop parameters to coefficient functions when it is clear from the context what the parameter values should be. For example, one might abbreviate $a(X_t, t)$ to simply a. Similarly one might write X instead of X_t.

With these conventions in place we can rewrite our SDE for $\tau(X)$ as:

$$\mathrm{d}\tau^i(X) = \left(a^j \partial_j \tau^i + \frac{1}{2} b^j_\alpha b^k_\beta \boldsymbol{g}^{\alpha\beta} \partial_j \partial_k \tau^i \right) \mathrm{d}t + b^j_\alpha \partial_j \tau^i \, \mathrm{d}W^\alpha_t.$$

Note that all the terms on the right are evaluated at X. With all these preliminaries we can now formally define what we mean by a Stochastic differential equation on a manifold M.

Definition 1 *An Itô SDE on a manifold consists of the initial conditions together with an equivalence class of quadruples* (W_t, ϕ, a, b) *under the equivalence relation* $\sim$ *defined by*

$$(W_t, \phi, a, b) \sim (V_t, \Phi, A, B) \text{ if } \begin{cases} W_t = V_t \\ A^j = a^j \partial_j \tau^i + \frac{1}{2} b^j_\alpha b^k_\beta g^{\alpha\beta} \partial_j \partial_k \tau^i \\ B^j = b^j_\alpha \partial_j \tau^i \end{cases}$$

for the transition function $\tau = \Phi \circ \phi^{-1}$.

The first condition could be written in more general terms, but for simplicity we assume pathwise equality between the two Brownian motions.

Rather less formally, one might say that an Itô SDE is an SDE whose coefficients obey Itô's lemma when one changes coordinates.

The definition we have chosen for an Itô SDE is exactly analogous to the common definition of a vector field on a manifold as a set of coordinate functions that transform in a particular way when one changes coordinates.

As is well known, when writing SDEs on $\mathbb{R}^n$ one can choose to write the equation in either Itô or Stratonovich form. We could attempt to define SDEs on manifolds using Stratonovich equations. In this case the local data would be a pair $(W_t, \phi, \overline{a}, b)$ where W_t and ϕ are as before but now $\overline{a}$, b are the coefficients of the Stratonovich equation:

$$\mathrm{d}X_t = \overline{a}\,\mathrm{d}t + b \circ \mathrm{d}W_t.$$

Rather than use Itô's lemma in our informal derivation, we would now use the chain rule. The end result is the following definition:

Definition 2 *A Stratonovich SDE on a manifold consists of the initial conditions together with an equivalence class of quadruples* $(W_t, \phi, \overline{a}, b)$ *under the equivalence relation* $\sim$ *defined by*

$$(W_t, \phi, \overline{a}, b) \sim (V_t, \Phi, \overline{A}, B) \text{ if } \begin{cases} W_t = V_t \\ \overline{A}^j = \overline{a}^j \partial_j \tau^i \\ B^j = b^j_\alpha \partial_j \tau^i \end{cases} . \tag{1}$$

As we know, if the coefficients of the SDEs are smooth enough, Stratonovich SDEs and Itô SDEs on $\mathbb{R}^n$ are essentially equivalent, i.e. an Itô SDE can be transformed to an SDE in Stratonovich form which has the same solutions and vice versa. One sees immediately that Itô SDEs and Stratonovich SDEs on manifolds are essentially equivalent in precisely the same sense.

Since the chain rule is rather simpler than Itô's lemma, the definition of a Stratonovich SDE is rather simpler than that of an Itô SDE. In addition, we can easily replace the complicated index notation with coordinate free notation. We will write $T_x X$ to denote the tangent space at a point x of a manifold X. We will write $f_* : T_x X \to T_{f(x)} Y$ to denote the differential of a smooth map between manifolds $f : X \to Y$. We can then rewrite Eq. (1) more elegantly as:

$$\begin{cases} W_t = V_t, \\ \overline{A} = \tau_*(\bar{a}), \\ \overline{B} = \tau_*(b). \end{cases}$$

Thus the transformation rule for the coefficients in this case is precisely the same as the transformation rule for the coefficients of a vector field. This allows one to devise alternative definitions for stochastic differential equations in terms of vector fields without needing to mention the less attractive details about equivalence relations.

2.2 Projecting SDEs

Let M be a submanifold of $\mathbb{R}^r$ with chart $\psi : U \to \mathbb{R}^n$ for some open neighbourhood in M and inverse $\phi = \psi^{-1}$.

Given an SDE defined on $\mathbb{R}^r$, we would like to approximate solutions in $\mathbb{R}^r$ with solutions to an SDE defined on M.

Definition 3 *Let W_t be an $\mathbb{R}^m$ valued Brownian motion. Given a Stratonovich SDE on $\mathbb{R}^r$*

$$\mathrm{d}X = \overline{a}\,\mathrm{d}t + b_\alpha \circ \mathrm{d}W_t^\alpha$$

and a chart $\psi : U \to \mathbb{R}^n$ for some neighbourhood in $N \subseteq M$ we define the Stratonovich projection *of the SDE to be:*

$$\mathrm{d}\tilde{X} = \overline{A}\,\mathrm{d}t + B_\alpha \circ \mathrm{d}W_t^\alpha$$

where:

$$\begin{aligned} \phi &= \psi^{-1} \\ \overline{A}(\tilde{X}, t) &= (\psi_*)\Pi_{\phi(\tilde{X})}(\overline{a}(\phi(\tilde{X}), t)) \\ B_\alpha(\tilde{X}, t) &= (\psi_*)\Pi_{\phi(\tilde{X})}(b_\alpha(\phi(\tilde{X}), t)) \end{aligned}$$

where Π is the projection of $\mathbb{R}^r$ onto $\phi_*(\mathbb{R}^n)$ defined by the Euclidean metric.

Because we know that the projection of vector fields can be defined similarly, and because we know that the coefficients of Stratonovich SDEs transform like vector fields, we see that the definition above defines a Stratonovich SDE on M. Indeed, if one is willing to accept that projection of vector fields onto a submanifold is well-defined, then one could define the projection of a Stratonovich SDE as the projection of the coefficient functions.

For an Itô SDE one cannot simply apply projection to the coefficient functions because the coefficients of an Itô SDE on a manifold do not transform like vector fields.

The *Stratonovich projection* of an Itô SDE is trivially defined by the recipe:

(i) Rewrite the Itô SDO as a Stratonovich SDE.
(ii) Apply the Stratonovich projection as defined above.
(iii) Rewrite the resulting Stratonovich SDE as an Itô SDE.

We see that the use of Stratonovich calculus is just limited to the projection procedure. After and before that, we wish to use Itô calculus because of the good probabilistic properties of the Itô integral. However, we can try and avoid transiting through Stratonovich calculus and define the *extrinsic Itô projection* as follows:

Definition 4 *Let W_t be an $\mathbb{R}^m$ valued Brownian motion. Given an Itô SDE on $\mathbb{R}^r$*

$$\mathrm{d}X = a\,\mathrm{d}t + b_\alpha\,\mathrm{d}W_t^\alpha$$

and a chart $\psi : U \to \mathbb{R}^n$ for some neighbourhood in $N \subseteq M$ we define the extrinsic Itô projection *of the SDE to be:*

$$\mathrm{d}Y = A\,\mathrm{d}t + B_\alpha \mathrm{d}W_t^\alpha$$

where:

$$\begin{aligned}
\phi &:= \psi^{-1} \\
B_\alpha(Y_t, t) &:= (\psi_*)_{\phi(Y_t)} \Pi_{\phi(Y_t)} b_\alpha(\phi(Y_t), t) \\
A(Y_t, t) &:= (\psi_*)_{\phi(Y_t)} \Pi_{\phi(Y_t)} \left(a(\phi(Y_t), t) - \frac{1}{2}(\nabla_{B_\alpha(\phi(Y_t),t)}\phi_*) B_\beta(\phi(Y_t), t) \boldsymbol{g}^{\alpha\beta} \right)
\end{aligned}$$

where ∇ is the gradient operator defined on $\mathbb{R}^n$.

We will discuss the motivation for this definition in detail in a future paper. For now we will simply remark that it can be derived by searching for the optimal approximation over a small time horizon in the metric defined on $\mathbb{R}^n$. We call it the extrinsic Itô projection because the optimality is defined via the use of the metric on the extrinsic space $\mathbb{R}^n$ rather than using the Riemannian metric of the sub-manifold. As we will show in subsequent papers Armstrong and Brigo (2016b), there is an alternative notion of the intrinsic Itô projection which one may consider. Defining the intrinsic Itô projection is best done using the differential geometric language of 2-jets. Introducing this machinery now would take us too far afield, which is why we have given only this brief motivation for the definition. We will show in our future work that the notion of "optimality" is far more subtle for SDEs than for ODEs. For example, there are occasions where the Stratonovich projection actually out-performs the extrinsic Itô projection.

The Stratonovich projection is manifestly well-defined. We must work harder for the extrinsic Itô projection.

Theorem 1 *The extrinsic Itô projection defines an SDE.*

Proof Let M be a submanifold of $\mathbb{R}^r$ and let $x_1 : U \to \mathbb{R}^k$ and $x_2 : U \to \mathbb{R}^k$ be two charts for some open set U containing a point X_0. Let $\tau = x_1 \circ x_2^{-1}$ be the transition function between the charts.

Write ϕ_i for the inverse of x_i. Write $(P_i)_{x_i} = (x_i)_* \Pi_{\phi_i(x_i)}$ for the projection map associated with the chart x_i. Note that $P_1 = \tau_* P_2$. If the SDE in $\mathbb{R}^n$ has Itô coefficients a and b then the Itô projected SDE w.r.t the coordinates x_i is:

$$\begin{aligned} \mathrm{d}Y =& \left[P_i a(\phi_i(Y), t) - \frac{1}{2} P_i((\nabla_{P_i b(\phi_i(Y))}(\phi_i)_*)(P_i b(\phi_i(Y)))) \right] \mathrm{d}t \\ &+ (P_i b(\phi_i(Y)))\, \mathrm{d}W\,, \\ Y_0 =& x_i(X_0)\,. \end{aligned} \tag{2}$$

What we want to show is that Eq. (2) for x_2 transformed using τ gives the Eq. (2) for x_1. With this in mind we transform equation (2) for x_2 using τ to obtain an equation for $Z = \tau(Y)$:

$$\begin{aligned} \mathrm{d}Z =& \Bigg[\tau_* P_2 a(\phi_1(Z), t) - \frac{1}{2} \tau_* P_2((\nabla_{\tau_* P_2 b(\phi_1(Z),t)}(\phi_2)_*) P_2 b(\phi_1(Z), t)) \\ &+ \frac{1}{2}(\nabla_{\tau_* P_2 b(\phi_1(Z),t)} \tau_*)(P_2 b(\phi_1(Z), t)) \Bigg] \mathrm{d}t \\ &+ \tau_* P_2 b(\phi_1(Z), t) \mathrm{d}W\,, \\ Z_0 =& \, x_1(X_0)\,. \end{aligned} \tag{3}$$

We now simplify this using the following identities:

$$P_1 = \tau_* P_2\,,$$

$$\phi_2 = \phi_1 \circ \tau\,,$$

$$(\phi_2)_* = (\phi_1)_* \circ \tau_*\,.$$

This last identity is the chain rule. So by the product rule we have:

$$\nabla_Y(\phi_2)_* = (\nabla_Y(\phi_1)_*) \circ \tau_* + (\phi_1)_* \circ (\nabla_Y \tau_*) \tag{4}$$

for vectors Y. This allows us to rewrite (3) as follows:

$$\begin{aligned}
\mathrm{d}Z = \Bigg[&(P_1 a(x)) \\
&- \frac{1}{2} P_1((\nabla_{P_1 b(\phi_1(Z),t}\phi^1_*)(\tau_* P_2 b(\phi_1(Z,t)))) \\
&- \frac{1}{2} P_1(\phi^1_*((\nabla_{P_1 b(\phi_1(Z),t)}\tau_*)(P_2 b(\phi_1(Z),t)))) \\
&+ \frac{1}{2}(\nabla_{P_1(b(\phi_1(Z),t))}\tau_*)(P_2 b(\phi_1(Z),t))\Bigg]\mathrm{d}t \\
&+ P_1 b(\phi_1(Z),t)\mathrm{d}W\,, \\
Z_0 = {}& x_1(X_0)\,.]
\end{aligned}$$

We can now use the fact that $P_1(\phi^1_*)$ is the identity and again use the identity $P_1 = \tau_* P_2$ to simplify this. Two unwanted terms cancel leaving use with equation:

$$\begin{aligned}
\mathrm{d}Z = {}& \left[(P_1 a(\phi_1(Z),t)) - \frac{1}{2} P_1((\nabla_{P_1 b(\phi_1(Z),t)}\phi^1_*)(\tau_* P_1 b(\phi_1(Z),t)))\right]\mathrm{d}t \\
&+ P_1 b(\phi_1(Z),t)\mathrm{d}W\,, \\
Z_0 = {}& x_1(X_0)\,.
\end{aligned}$$

This is (2) for x_1 as claimed.

We also want to show that the extrinsic Itô projection is distinct from the Stratonovich projection. We will show this in the next sections by explicitly computing examples. Moreover we will demonstrate numerically that the extrinsic Itô projection gives superior results to the Stratonovich projection when applied to non-linear filtering.

3 Application of the Projection to Non-linear Filtering

3.1 The Kushner Stratonovich Equation

We suppose that the state $X_t \in \mathbb{R}^n$ of a system evolves according to the equation:

$$\mathrm{d}X_t = f(X_t,t)\,\mathrm{d}t + \sigma(X_t,t)\,\mathrm{d}W_t$$

where f and σ are smooth $\mathbb{R}^n$ valued functions and W_t is a Brownian motion.

We suppose that an associated process, the observation process, $Y_t \in \mathbb{R}^d$ evolves according to the equation:

$$\mathrm{d}Y_t = b(X_t, t)\,\mathrm{d}t + \mathrm{d}V_t$$

where b is a smooth $\mathbb{R}^d$ valued function and V_t is a Brownian motion independent of W_t. Note that the filtering problem is often formulated with an additional constant in terms of the observation noise. For simplicity we have assumed that the system is scaled so that this can be omitted.

The filtering problem is to compute the conditional distribution of X_t given a prior distribution for X_0 and the values of Y for all times up to and including t.

Subject to various bounds on the growth of the coefficients of this equation, the assumption that the distribution has a density p_t and suitable bounds on the growth of p_t one can show that p_t satisfies the Kushner–Stratonovich equation:

$$\mathrm{d}p = \mathcal{L}^* p\,\mathrm{d}t + p[b - E_p(b)]^T[\mathrm{d}Y - E_p(b)\mathrm{d}t] \tag{5}$$

where E_p denotes the expectation with respect to the density p and the forward diffusion operator $\mathcal{L}_t^*$ is defined by:

$$\mathcal{L}_t^*\phi = -\sum_{i=1}^{n} \frac{\partial}{\partial x_i}[f_t^i \phi] + \frac{1}{2}\sum_{i,j=1}^{n} \frac{\partial^2}{\partial x_i \partial x_j}[a^{ij}\phi] \tag{6}$$

where $a = \sigma\sigma^T$.

In the event that the coefficient functions f and b are all linear and σ is a deterministic function of time one can show that so long as the prior distribution for X is Gaussian, or deterministic, the density p will be Gaussian at all subsequent times. This allows one to reduce the infinite dimensional equation (5) to a finite dimensional stochastic differential equation for the mean and covariance matrix of this normal distribution. This finite dimensional problem is known as the Kalman filter.

For more general coefficient functions, however, Eq. (5) cannot be reduced to a finite dimensional problem (Hazewinkel et al. 1983). Instead one might seek approximate solutions of (5) that belong to some given statistical family of densities. This is a very general setup and includes, for example, approximating the density using piecewise linear functions to derive a finite difference approximation or approximating the density with Hermite polynomials to derive a spectral method. Other examples include exponential families (considered in Brigo et al. 1999; Brigo 1998) and mixture families (considered in Armstrong and Brigo 2013, 2016a).

Our projection theory tells us how one can find good approximations on a given statistical family with respect to a given metric on the space of distributions. We illustrate this by writing down the extrinsic Itô projection of (5) for the L^2 and Hellinger metrics onto a general manifold.[1]

We will then examine some numerical results regarding the very specific case of seeking approximate solutions using Gaussian distributions. The idea of approxi-

---

[1]Note that it is also possible to consider projecting the Zakai equation. However, as explained in Armstrong and Brigo (2016a), one expects that projecting the Kushner–Stratonovich will lead to smaller error terms.

mating the solution to the filtering problem using a Guassian distribution has been considered by numerous authors who have derived variously, the extended Kalman filter (Pardoux 1991), assumed density filters (Kushner 1967) and Stratonovich projection filters (Brigo 1998). We will be able to derive extrinsic Itô projection filters which outperform all these other filters (assuming performance is measured over small time intervals using the appropriate Hilbert space metric).

3.2 *Itô Projections*

The extrinsic Itô projection filter in the L^2 direct metric Let us suppose that the density p lies in L^2 and so we can use the L^2 norm to measure the accuracy of an approximate solution to Eq. (5). For a discussion on conditions under which a unnormalized version of p is in L^2 (Zakai Equation) see for example Ahmed (1998).

We wish to consider an m-dimensional family of distributions p parameterized by m real valued parameters θ^1, θ^2, …, θ^m. For example we will consider the 2 dimensional Gaussian family:

$$p(x) = \frac{1}{(\theta^2)\sqrt{2\pi}} \exp\left(-\frac{(x-(\theta^1))^2}{2(\theta^2)^2} \right). \tag{7}$$

Note that we have chosen to follow differential geometry convention and use upper indices for the coordinate functions θ^i so we have been careful to distinguish powers from indices using brackets.

More formally, an m-dimensional family is given by a smooth embedding $\phi : \mathbb{R}^m \to L^2(\mathbb{R}^n)$. The tangent vectors $\phi_* \frac{\partial}{\partial \theta^i} \in L^2(\mathbb{R}^n)$ are simply the partial derivatives

$$\frac{\partial p}{\partial \theta^i}.$$

Let us write:

$$g_{ij} = \int_{\mathbb{R}} \frac{\partial p}{\partial \theta^i} \frac{\partial p}{\partial \theta^j} \, \mathrm{d}x.$$

This defines the induced metric tensor on the manifold $\phi(\mathbb{R}^m)$. We will write g^{ij} for the inverse of the matrix g_{ij}. The projection operator $\Pi_{\phi(\theta)}$ is then given by

$$\begin{aligned} \Pi_{\phi(\theta)}(v) &= \sum_{i,j=1}^{m} g^{ij} \left\langle v, \phi_* \frac{\partial}{\partial \theta^i} \right\rangle_{L^2} \phi_* \frac{\partial}{\partial \theta^j} \\ &= \sum_{i,j=1}^{m} g^{ij} \left(\int_{\mathbb{R}^n} v(x) \frac{\partial p}{\partial \theta^i} \, \mathrm{d}x \right) \phi_* \frac{\partial}{\partial \theta^j} . \end{aligned}$$

Thus

$$\phi_*^{-1}\Pi_{\phi(\theta)}(v) = \sum_{i,j=1}^{m} g^{ij} \left(\int_{\mathbb{R}^n} v(x) \frac{\partial p}{\partial \theta^i} \, dx. \right) \frac{\partial}{\partial \theta^j}.$$

We can now write down the extrinsic Itô projection of (5) with respect to the L^2 metric. It is:

$$d\theta^i = A^i \, dt + B^i \, dY_t$$

where:

$$B^i = \sum_{j=1}^{m} g^{ij} \left(\int_{\mathbb{R}} (p(b - E_{p(\theta)}(b)))^T \frac{\partial p}{\partial \theta^j} \, dx. \right)$$

and

$$A^i = \sum_{j=1}^{m} g^{ij} \left(\int_{\mathbb{R}^n} \left(\mathcal{L}^* p - p(b - E_{p(\theta)}(b))^T E_{p(\theta)}(b) - \frac{1}{2} \sum_{k=1}^{m} \frac{\partial^2 p}{\partial \theta^j \partial \theta^k} B^k \right) \frac{\partial p}{\partial \theta^j} \, dx. \right).$$

Example 1 Consider as a test case the 1-dimensional problem with $f(x,t) = 0$, $\sigma(x,t) = 1$ and $b(x,t) = x + \epsilon x^3$ for some small constant ϵ. This problem is a perturbation of a linear filter so one might expect that a Gaussian approximation will perform reasonably well at least for small times. Thus we will use the 2 dimensional manifold of Gaussian distributions given in Eq. (7).

We first calculate the metric tensor g_{ij} which is diagonal in this case:

$$g_{ij} = \frac{1}{4\sqrt{\pi}(\theta^2)^3} \begin{pmatrix} 1 & 0 \\ 0 & \frac{3}{2} \end{pmatrix}.$$

This is easily inverted to compute g^{ij}. We compute the expectation $E_p(b)$:

$$E_p(b) = \frac{\epsilon \left(\sqrt{2\pi}(\theta^1)^3(\theta^2) + 3\sqrt{2\pi}(\theta^1)(\theta^2)^3 \right)}{\sqrt{2\pi}(\theta^2)} + (\theta^1).$$

One can now see that computing the projection equation will simply involve integrating a number of terms of the form a polynomial in x times a Gaussian. The end result is:

$$\begin{aligned} d\theta^1 &= \left(-\frac{1}{4}\theta^1 \left(\theta^2\right)^2 \left(3\epsilon^2 \left(4\left(\theta^1\right)^4 - 4\left(\theta^2\right)^2\left(\theta^1\right)^2 - 3\left(\theta^2\right)^4 \right) + 16\epsilon\left(\theta^1\right)^2 + 4 \right) \right) dt \\ &+ \left(\frac{1}{2}\left(\theta^2\right)^2 \left(3\epsilon\left(2\left(\theta^1\right)^2 + \left(\theta^2\right)^2 \right) + 2 \right) \right) dY_t, \\ d\theta^2 &= \left(-\frac{9\epsilon^2\left(\theta^2\right)^8 + \left(\theta^2\right)^4\left(60\epsilon^2\left(\theta^1\right)^4 + 48\epsilon\left(\theta^1\right)^2 + 4 \right) + 6\epsilon\left(\theta^2\right)^6\left(9\epsilon\left(\theta^1\right)^2 + 2 \right) - 4}{8\theta^2} \right) dt \\ &+ \left(3\epsilon\theta^1\left(\theta^2\right)^3 \right) dY_t. \end{aligned}$$

The extrinsic Itô projection filter in the Hellinger metric The Hellinger metric is a metric on probability measures. In the case of two probability density functions $p(x)$ and $q(x)$ on $\mathbb{R}^n$, that now need only be in L^1, the Hellinger distance is given by the square root of:

$$\frac{1}{2}\int (\sqrt{p(x)} - \sqrt{q(x)})^2 \, \mathrm{d}x.$$

In other words, up to the constant factor of $\frac{1}{2}$ the Hellinger metric corresponds to the L^2 norm on the square root of the density function rather than on the density itself (as in the previous subsection). The Hellinger metric has the important advantage of making the metric independent of the particular background density that is used to express measures as densities. The L^2 direct distance introduced earlier does not satisfy this background independence.

Now, to compute the extrinsic Itô projection with respect to the Hellinger metric we first want to write down an Itô equation for the evolution on $\sqrt{p}$.

Applying Itô's lemma to Eq. (5) we formally obtain:

$$\begin{aligned}
\mathrm{d}\sqrt{p} &= \left(\frac{\mathcal{L}^* p - p(b - E_p(b))^T E_p(b)}{2\sqrt{p}} - \frac{p^2(b - E_p(b))^T (b - E_p(b))}{8p\sqrt{p}} \right) \mathrm{d}t \\
&+ \left(\frac{p(b - E_p(b))^T}{2\sqrt{p}} \right) \mathrm{d}Y_t. \\
&= \left(\frac{\mathcal{L}^* p}{2\sqrt{p}} - \frac{1}{8}\sqrt{p}(b - E_p(b))^T (b + 3E_p(b)) \right) \mathrm{d}t \\
&+ \left(\frac{1}{2}\sqrt{p}(b - E_p(b))^T \right) \mathrm{d}Y_t.
\end{aligned}$$

A family of distributions now corresponds to an embedding ϕ from $\mathbb{R}^m$ to $L^2(\mathbb{R}^n)$ but now $p = \phi(\theta)^2$. The tangent space is spanned by the vectors:

$$\phi_* \frac{\partial}{\partial \theta^i} = \frac{\partial \sqrt{p}}{\partial \theta^i}.$$

We define a metric on the tangent space by:

$$h_{ij} = \int_{\mathbb{R}^n} \frac{\partial \sqrt{p}}{\partial \theta^i} \frac{\partial \sqrt{p}}{\partial \theta^j} \, \mathrm{d}x.$$

We write h^{ij} for the inverse matrix of h_{ij}. The projection operator with respect to the Hellinger metric is:

$$\Pi_{\phi(\theta)}(v) = \sum_{i,j=1}^{m} h^{ij} \left(\int_{\mathbb{R}^n} v(x) \frac{\partial \sqrt{p}}{\partial \theta^i} \, \mathrm{d}x \right) \phi_* \frac{\partial}{\partial \theta^j}.$$

We can now write down the extrinsic Itô projection of (5) with respect to the Hellinger metric. It is:

$$\mathrm{d}\theta^i = A^i\,\mathrm{d}t + B^i\,\mathrm{d}Y_t$$

where:

$$B^i = \sum_{j=1}^{m} h^{ij}\left(\int_{\mathbb{R}} \frac{1}{2}\sqrt{p}(b - E_{p(\theta)}(b))^T \frac{\partial \sqrt{p}}{\partial \theta^j}\,\mathrm{d}x.\right)$$

and

$$A^i = \sum_{j=1}^{m} h^{ij}\left(\int_{\mathbb{R}^n}\left(\frac{\mathcal{L}^* p}{2\sqrt{p}} - \frac{1}{8}\sqrt{p}(b - E_{p(\theta)}(b))^T(b + 3E_{p(\theta)}(b))\right.\right.$$
$$\left.\left. - \frac{1}{2}\sum_{k=1}^{m}\frac{\partial^2\sqrt{p}}{\partial\theta^j\partial\theta^k}B^k\right)\frac{\partial\sqrt{p}}{\partial\theta^j}\,\mathrm{d}x.\right).$$

Example 2 We may repeat Example 1 but projecting using the Hellinger metric. We first calculate the metric tensor h_{ij} which is diagonal also in this case:

$$h_{ij} = \frac{1}{4\theta_2^2}\begin{pmatrix} 1 & 0 \\ 0 & 2 \end{pmatrix}$$

This is easily inverted to compute h^{ij}. We obtain the following SDEs:

$$\mathrm{d}\theta^1 = \left(-\theta^1\left(\theta^2\right)^2\left(3\epsilon^2\left(\left(\theta^1\right)^4 + 4\left(\theta^2\right)^2\left(\theta^1\right)^2 + 6\left(\theta^2\right)^4\right) + \epsilon\left(4\left(\theta^1\right)^2 + 6\left(\theta^2\right)^2\right) + 1\right)\right)\mathrm{d}t$$
$$+ \left(\left(\theta^2\right)^2\left(3\epsilon\left(\left(\theta^1\right)^2 + \left(\theta^2\right)^2\right) + 1\right)\right)\mathrm{d}Y_t,$$
$$\mathrm{d}\theta^2 = \left(-\frac{27\epsilon^2\left(\theta^2\right)^8 + \left(\theta^2\right)^4\left(15\epsilon^2\left(\theta^1\right)^4 + 12\epsilon\left(\theta^1\right)^2 + 1\right) + 9\epsilon\left(\theta^2\right)^6\left(6\epsilon\left(\theta^1\right)^2 + 1\right) - 1}{2\theta^2}\right)\mathrm{d}t$$
$$+ \left(3\epsilon\theta^1\left(\theta^2\right)^3\right)\mathrm{d}Y_t.$$

Note that this satisfies the important check that when ϵ is zero, this reduces to the Kalman filter.

3.3 Other Gaussian Approximate Filters

We will show in a subsequent paper that the filters above are in some sense optimal with respect to the relevant Hilbert space metric. Neverthless many other Gaussian approximate filters have been proposed in the past. We will briefly review a number of different Gaussian approximate filters that can be found in the literature and calculate the relevant stochastic differential equations for our Example 1. We will then compare the performance of these filters numerically.

The Stratonovich projection filter Instead of using the extrinsic Itô projection, one can use the Stratonovich projection.

Example 3 General formulae for performing the Stratonovich L^2 projection are given in Armstrong and Brigo (2016a). In the specfic case of Example 1 the resulting Itô SDEs are:

$$
\begin{aligned}
\mathrm{d}\theta^1 &= \left(-\frac{1}{4}\theta^1\left(\theta^2\right)^2\left(3\epsilon^2\left(4\left(\theta^1\right)^4-4\left(\theta^2\right)^2\left(\theta^1\right)^2-3\left(\theta^2\right)^4\right)+16\epsilon\left(\theta^1\right)^2+4\right)\right)\mathrm{d}t \\
&\quad+\left(\frac{1}{2}\left(\theta^2\right)^2\left(3\epsilon\left(2\left(\theta^1\right)^2+\left(\theta^2\right)^2\right)+2\right)\right)\mathrm{d}Y_t, \\
\mathrm{d}\theta^2 &= \left(-\frac{47\epsilon^2\left(\theta^2\right)^8+\left(\theta^2\right)^4\left(60\epsilon^2\left(\theta^1\right)^4+48\epsilon\left(\theta^1\right)^2+4\right)+2\epsilon\left(\theta^2\right)^6\left(33\epsilon\left(\theta^1\right)^2+8\right)-4}{8\theta^2}\right)\mathrm{d}t \\
&\quad+\left(3\epsilon\theta^1\left(\theta^2\right)^3\right)\mathrm{d}Y_t.
\end{aligned}
$$

Example 4 General formulae for performing the Stratonovich Hellinger projection are given in Brigo et al. (1999). In the specfic case of Example 1 the resulting SDEs are:

$$
\begin{aligned}
\mathrm{d}\theta^1 &= \left(-\theta^1\left(\theta^2\right)^2\left(3\epsilon^2\left(\left(\theta^1\right)^4+4\left(\theta^2\right)^2\left(\theta^1\right)^2+6\left(\theta^2\right)^4\right)+\epsilon\left(4\left(\theta^1\right)^2+6\left(\theta^2\right)^2\right)+1\right)\right)\mathrm{d}t \\
&\quad+\left(\left(\theta^2\right)^2\left(3\epsilon\left(\left(\theta^1\right)^2+\left(\theta^2\right)^2\right)+1\right)\right)\mathrm{d}Y_t, \\
\mathrm{d}\theta^2 &= \left(-\frac{36\epsilon^2\left(\theta^2\right)^8+\left(\theta^2\right)^4\left(15\epsilon^2\left(\theta^1\right)^4+12\epsilon\left(\theta^1\right)^2+1\right)+9\epsilon\left(\theta^2\right)^6\left(6\epsilon\left(\theta^1\right)^2+1\right)-1}{2\theta^2}\right)\mathrm{d}t \\
&\quad+\left(3\epsilon\theta^1\left(\theta^2\right)^3\right)\mathrm{d}Y_t.
\end{aligned}
$$

The Extended Kalman Filter The Extended Kalman Filter (EKF) is a heuristically derived method of finding approximate solutions to the filtering problem based on the idea of linearising the problem and then using the solution to the linear problem. In particular one assumes that the solution can be well approximated by a Gaussian distribution. For the EKF see Jazwinski (1970), Ahmed (1998). A definition and heuristic derivation is given in Bain and Crisan (2009) (which is based, in turn, on the derivation given in Pardoux (1991).

The EKF can be shown to work well on condition that the initial position of the signal is approximated well, the non-linearities of f are small, b is injective and the observation noise is small (Picard 1991). Moreover, the EKF is widely used in practice, see Bain and Crisan (2009) for references to applications.

Example 5 For the example problem $b(x) = x + \epsilon x^3$ the EKF is:

$$\begin{aligned}
\mathrm{d}\theta^1 &= \left(\left(\theta^2\right)^2\left(-\left(3\epsilon\left(\theta^1\right)^2+1\right)\right)\left(\theta^1+\epsilon\left(\theta^1\right)^3\right)\right)\mathrm{d}t \\
&\quad + \left(\left(\theta^2\right)^2\left(3\epsilon\left(\theta^1\right)^2+1\right)\right)\mathrm{d}Y_t, \\
\mathrm{d}\theta^2 &= \left(\frac{1-(\theta^2)^4\left(3\epsilon\left(\theta^1\right)^2+1\right)^2}{2\theta^2}\right)\mathrm{d}t.
\end{aligned}$$

Assumed density filters Assumed density filters (ADFs) provide a finite dimensional method of finding approximate solutions to the filtering problem. They have been considered in, for example, Kushner (1967), Maybeck (1982) and Brigo (1998).

The general setup is to consider a statistical family $\pi(\cdot,\eta)$ of probability measures parameterized by some coordinates $\eta = (\eta^1, \ldots, \eta^m)$. This parameterization is not arbitrary. It must be chosen in such a way that, for elements of the statistical family, the values of η correspond to the expectations of some twice differentiable scalar functions $\{c^1, \ldots, c^m\}$ defined on $\mathbb{R}^n$.

$$\eta^i = E_{\pi(\cdot,\eta)}(c^i) =: E_{\eta_t}(c^i)$$

where for brevity we are using the abbreviation E_{η_t} for $E_{\pi(\cdot,\eta)}$.

For example one might take the statistical family of normal distributions parameterized by its fist and second moments η_1 and η_2, so $c^1(x) = x$, $c^2(x) = (x)^2$.

Given a statistical family parameterized in this way, we define the Itô ADF to be:

$$\mathrm{d}\eta_t^i = E_{\eta_t}(\mathcal{L}_t c^i)\,\mathrm{d}t + \left(E_{\eta_t}(b_t c^i) - E_{\eta_t}(b_t)\eta_t^i\right)^T\left(\mathrm{d}Y_t - E_{\eta_t}(b_t)\,\mathrm{d}t\right).$$

This is motivated by the fact that under the conditions used to derive equation (5), we have that the c_i-moments of π_t, the true solution to the filtering problem, satisfy the Itô equation:

$$\mathrm{d}\pi_t(c_i) = \pi_t(\mathcal{L}_t c^i)\,\mathrm{d}t - \frac{1}{2}(\pi_t(bc^i) - \pi_t(b)\pi_t(c^i))(\mathrm{d}Y - \pi_t(b)\,\mathrm{d}t).$$

Thus if it were true that the true density was a member of our chosen statistical family then the Itô ADF would certainly be satisfied. One just hopes that the Itô ADF will continue to give a reasonable approximation even though we know that the true density isn't a member of the chosen statistical family.

With a similar motivation we define the Stratonovich ADF to be:

$$\begin{aligned}
\mathrm{d}\eta_t^i &= E_{\eta_t}(\mathcal{L}_t c^i)\,\mathrm{d}t - \frac{1}{2}(E_{\eta_t}(|b_t|^2 c^i) - E_{\eta_t}(|b_t|^2)\eta_t^i)\,\mathrm{d}t \\
&\quad + \left(E_{\eta_t}(bc^i) - E(b_t)\eta_t^i\right)^T \circ \mathrm{d}Y_t.
\end{aligned}$$

If it were true that the density was a member of our statistical family then the Itô ADF and the Stratonovich ADF would be equivalent equations. Since we only expect to be able to approximate the true density with our statistical family, we must expect that the Itô ADF and Stratonovich ADF are in fact inequivalent equations. Intuitively, we can say that the local moment matching approximation on which the ADF heuristics are based and the Ito-Stratonovich transformation do not commute.

The justification just given for ADFs is far from convincing. We are relying on little other than hope that these equations will give good approximations. However, it was shown in Brigo (1998) that in fact for exponential families, the Stratonovich projection filter in the Hellinger metric coincides with the Stratonovich ADF, and in Brigo (1995) that for the Gaussian case, this filter approaches the optimal filter under small observation noise.

Example 6 If we calculate the Itô assumed density filter corresponding to Example 1 and the family of normal distributions, and then change coordinates to θ^1 and θ^2 as used in the previous examples, we obtain the SDEs:

$$
\begin{aligned}
\mathrm{d}\theta^1 &= \left(-\theta^1\left(\theta^2\right)^2\left(3\epsilon^2\left(\left(\theta^1\right)^4+4\left(\theta^2\right)^2\left(\theta^1\right)^2+3\left(\theta^2\right)^4\right)+\epsilon\left(4\left(\theta^1\right)^2+6\left(\theta^2\right)^2\right)+1\right)\right)\mathrm{d}t \\
&\quad+\left(\left(\theta^2\right)^2\left(3\epsilon\left(\left(\theta^1\right)^2+\left(\theta^2\right)^2\right)+1\right)\right)\mathrm{d}Y_t, \\
\mathrm{d}\theta^2 &= \left(-\frac{9\epsilon^2\left(\theta^2\right)^8+\left(\theta^2\right)^4\left(15\epsilon^2\left(\theta^1\right)^4+12\epsilon\left(\theta^1\right)^2+1\right)+3\epsilon\left(\theta^2\right)^6\left(15\epsilon\left(\theta^1\right)^2+2\right)-1}{2\theta^2}\right)\mathrm{d}t \\
&\quad+\left(3\epsilon\theta^1\left(\theta^2\right)^3\right)\mathrm{d}Y_t.
\end{aligned}
$$

Example 7 The family of normal distributions is an exponential family, therefore the Stratonovich assumed density filter is equivalent to the Stratonovich projection filter in the Hellinger metric.

3.4 Numerical Results

We simulated the example problem $b(x) = x + \epsilon x^3$ for all of the above approximate filters with $\epsilon = 0.05$. We also computed an "exact" solution using a finite difference method with a fine grid. We define the L^2 residual to be the L^2 distance between the approximate solution and the "exact" solution. We define the Hellinger residual similarly.

In Fig. 1 we see the L^2 residuals for the various methods. As predicted by our theory the extrinsic Itô projection in the L^2 metric results in the lowest residuals. The Stratonovich projection in the L^2 metric comes a close second. The projection methods based on optimizing the Hellinger metric perform the worst.

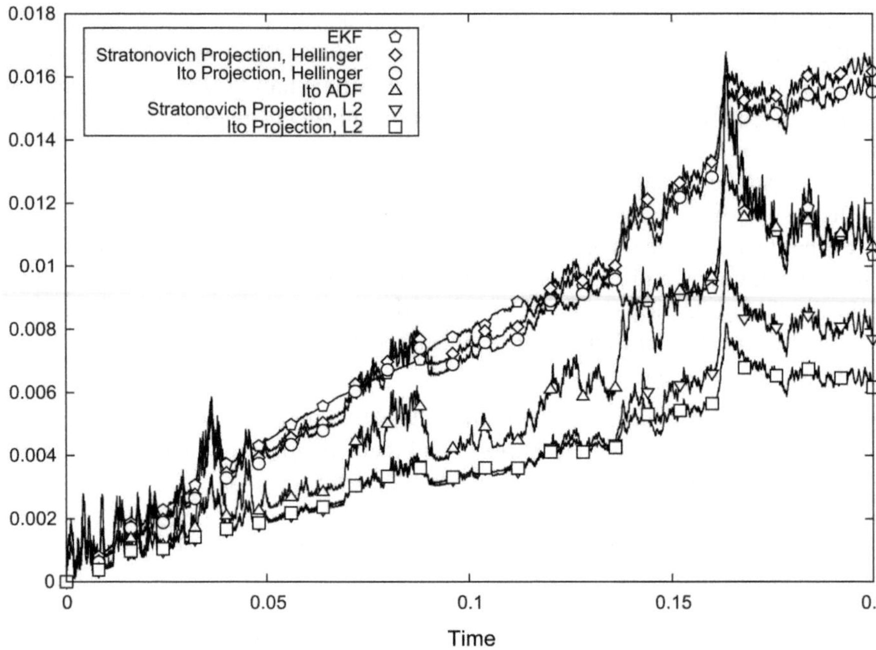

Fig. 1 L^2 residuals for each approximation method

For the Hellinger metric, we have plotted the ratio of the Hellinger residual for each method to the residual of the Itô Hellinger projection method. This is because the residuals themselves are too difficult to distinguish visually. The result is shown in Fig. 2. Over short time periods the expected value of this ratio should be greater than 1 for all the competing methods. This is born out by the numerical experiments. Note that we do not have a theory over which method will perform better in the longer term and so the fact that this relative residual eventually drops below 1 for both the Itô ADF and the Stratonovich projection does not contradict our theoretical results. Having said that, this behaviour does appear to be fairly consistent for our simple example problem. We see therefore that our "greedy" approach of finding the best residual in the short term will not necessarily lead to the best long term result.

3.5 Conclusion

We have defined a new way to approximate a high dimensional SDE with a lower dimensional SDE on a submanifold. This approximation is based on a new notion of projeciton, the extrinsic Itô projection method. We show that this projection leads to an SDE on the submanifold and briefly discuss its optimality property compared to the classic Stratonovich projection used previously in similar contexts.

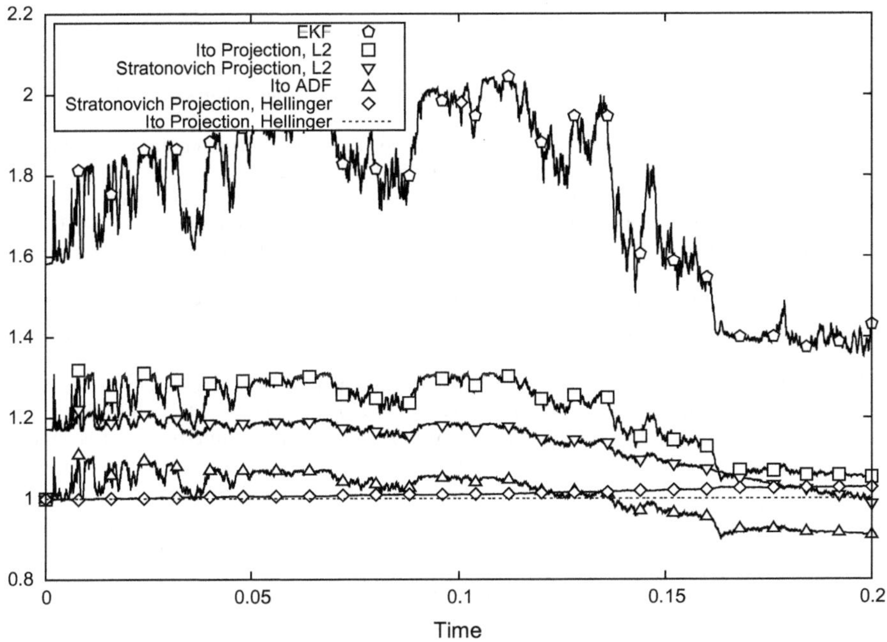

Fig. 2 Hellinger residuals for each approximation method divided by residual for Itô Hellinger projection

We then apply this Itô projection to nonlinear filtering. We project the infinite dimensional stochastic PDE of the optimal filter on a finite dimensional Gaussian family. Our explicit calculations show that the extrinsic Itô projection gives rise to new filters for both the L^2 and Hellinger metrics, and shows in particular that the Itô projection is different from the Stratonovich projection.

Numerical results show that our extrinsic Itô projection filters often out perform existing filters over small time horizons. The difference between the extrinsic Itô projection and the Stratonovich projection approaches is small in practice and will be small whenever the extrinsic Itô projection provides a good approximation. Thus the Stratonovich projection approach can be justified in practice and arguably has the merit of being slightly simpler to calculate.

Importantly, as we will demonstrate in a subsequent paper, unlike the heuristic arguments used to justify existing Gaussian filters, we can show that our filters are in some sense "optimal" for the given Hilbert space metric.

Acknowledgements We would like to thank the organizers of the workshop on computational information geometry for image and signal processing at ICMS in Edinburgh, 21–25 September 2015, and the participants for the lively discussion and feedback.

References

Ahmed, N. U. (1998). *Linear and nonlinear filtering for scientists and engineers*. Singapore: World Scientific.

Armstrong, J., & Brigo, D. (2013). Stochastic filtering via L2 projection on mixture manifolds with computer algorithms and numerical examples. arXiv:1303.6236.

Armstrong, J., & Brigo, D. (2016a). Nonlinear filtering via stochastic PDE projection on mixture manifolds in L^2 direct metric. *Mathematics of Control, Signals and Systems*, *28*(1), 1–33.

Armstrong, J., & Brigo, D. (2016b). Coordinate free stochastic differential equations as jets. arXiv:1602.03931.

Bain, A., & Crisan, D. (2009). *Fundamentals of stochastic filtering*, vol. 3. Springer.

Brigo, D. (1995). On the nice behaviour of the gaussian projection filter with small observation noise. *System and Control Letters*, *26*(5), 363–370.

Brigo, D., Hanzon, B., & Le Gland, F. (1998). A differential geometric approach to nonlinear filtering: the projection filter. *IEEE Transactions on Automatic Control*, *43*(2), 247–252.

Brigo, D., Hanzon, B., & Le Gland, F. (1999). Approximate nonlinear filtering by projection on exponential manifolds of densities. *Bernoulli*, *5*(3), 495–534.

Pardoux, E. (1991). Filtrage non linéaire et équations aux dérivées partielles stochastiques associées. In *Ecole d'Eté de Probabilités de Saint-Flour XIX 1989*, pp. 68–163. Springer.

Fisk, D. (1963). Quasi-martingales and stochastic integrals, Ph.D Dissertation, Michigan State University, Department of Statistics.

Hazewinkel, M., Marcus, S., & Sussmann, H. (1983). Nonexistence of finite-dimensional filters for conditional statistics of the cubic sensor problem. *Systems and Control Letters*, *3*(6), 331–340.

Kushner, H. (1967). Approximations to optimal nonlinear filters. *IEEE Transactions on Automatic Control*, *12*(5), 546–556.

Itô, K. (1950). Stochastic differential equations in a differentiable manifold. *Nagoya Mathematical of Journal*, *1*, 35–47.

Jazwinski, A. H. (1970). *Stochastic processes and filtering theory*. New York: Academic Press.

McShane, E. J. (1974). *Stochastic calculus and stochastic models* (2nd ed.). New York: Academic Press.

Maybeck, P. (1982). *Stochastic models, estimation, and control* (Vol. 3). Cambridge: Academic press.

Picard, J. (1991). Efficiency of the extended kalman filter for nonlinear systems with small noise. *SIAM Journal on Applied Mathematics*, *51*(3), 843–885.

Stratonovich, R. L. (1966). A new representation for stochastic integrals and equations. *SIAM Journal on Control*, *4*(2), 362–371.

Fast $(1 + \epsilon)$-Approximation of the Löwner Extremal Matrices of High-Dimensional Symmetric Matrices

Frank Nielsen and Richard Nock

1 Introduction: Löwner Extremal Matrices and Their Applications

Let $M_d(\mathbb{R})$ denote the space of *square* $d \times d$ matrices with real-valued coefficients, and $\mathrm{Sym}_d(\mathbb{R}) = \{S : S = S^\top\} \subset M_d(\mathbb{R})$ the matrix vector space[1] of *symmetric* matrices. A matrix $P \in M_d(\mathbb{R})$ is said *Symmetric Positive Definite* (Bhatia 2009) (SPD, denoted by $P \succ 0$) iff. $\forall x \neq 0, x^\top P x > 0$ and only *Symmetric Positive Semi-Definite*[2] (SPSD, denoted by $P \succeq 0$) when we relax the strict inequality ($\forall x, x^\top P x \geq 0$). Let $\mathrm{Sym}_d^+(\mathbb{R}) = \{X : X \succeq 0\} \subset \mathrm{Sym}_d(\mathbb{R})$ denote the space of positive semi-definite matrices, and $\mathrm{Sym}_d^{++}(\mathbb{R}) = \{X : X \succ 0\} \subset \mathrm{Sym}_d^+(\mathbb{R})$ denote the space of positive definite matrices. A matrix $S \in \mathrm{Sym}_d(\mathbb{R})$ is defined by $D = \frac{d(d+1)}{2}$ real coefficients, and so is a SPD or a SPSD matrix. Although $\mathrm{Sym}_d(\mathbb{R})$ is a *vector space*, the SPSD matrix space does not have the vector space structure but is rather an abstract *pointed convex cone* with *apex* the zero matrix $0 \in \mathrm{Sym}_d^+(\mathbb{R})$ since $\forall P_1, P_2 \in \mathrm{Sym}_d^+(\mathbb{R}), \forall \lambda \geq 0, \quad P_1 + \lambda P_2 \in \mathrm{Sym}_d^+(\mathbb{R})$. Symmetric matrices can be *partially* ordered using the *Löwner ordering*[3]:

$$P \succeq Q \Leftrightarrow P - Q \succeq 0,$$

and

[1] Although addition preserves the symmetric property, beware that the product of two symmetric matrices may be not symmetric.
[2] Those definitions extend to Hermitian matrices $M_d(\mathbb{C})$.
[3] Also often written Loewner in the literature, e.g., see Siotani (1967).

F. Nielsen (✉)
École Polytechnique, France and Sony Computer Science Laboratories, Tokyo, Japan
e-mail: Frank.Nielsen@acm.org

R. Nock
NICTA & ANU, Sydney, Australia

F. Nielsen et al. (eds.), *Computational Information Geometry*,
Signals and Communication Technology, DOI 10.1007/978-3-319-47058-0_6

$$P \succ Q \Leftrightarrow P - Q \succ 0.$$

When $P \succeq Q$, matrix P is said to *dominate* matrix Q, or equivalently that matrix Q is dominated by matrix P. Note that the difference of two SPSD matrices may not be a SPSD matrix.[4] A non-SPSD symmetric matrix S can be dominated by a SPSD matrix P when $P - S \succ 0$.[5]

The *supremum* operator is defined on n symmetric matrices $S_1, \ldots, S_n$ (not necessarily SPSDs) as follows:

Problem 1 (Löwner maximal matrices)

$$\bar{S} = \inf\{X \in \mathrm{Sym}(\mathbb{R}) \ : \ \forall i \in [n], X \succeq S_i\}, \tag{1}$$

where $[n] = \{1, \ldots, n\}$.

This matrix $\bar{S} = \max(S_1, \ldots, S_n)$ is indeed the "smallest", meaning the *tightest upper bound*, since by definition there does not exist another symmetric matrix X' dominating all the S_i's and dominated by $\bar{S}$. Trivially, when there exists a matrix S_j that dominates all others of a set $S_1, \ldots, S_n$, then the supremum of that set is matrix S_j. Similarly, we define the *minimal/infimum matrix* $\underline{S}$ as the tightest lower bound. Since matrix inversion reverses the Löwner ordering ($A \succ B \Leftrightarrow B^{-1} \succ A^{-1}$), we link those extremal supremum/infimum matrices when considering sets of invertible symmetric matrices as follows:

$$\underline{S} = \left(\max(S_1^{-1}, \ldots, S_n^{-1})\right)^{-1}.$$

Extremal matrices are *rotational invariant*:

$$\max(O^\top S_1 O, \ldots, O^\top S_n O) = O^\top \times \max(S_1, \ldots, S_n) \times O,$$

where O is any orthogonal matrix ($OO^\top = O^\top O = I$). This property is important in DT-MRI processing that should be invariant to the chosen reference frame.

Computing Löwner extremal matrices are useful in many applications: For example, in matrix-valued imaging (Angulo 2013; Burgeth et al. 2007) (morphological operations, filtering, denoising or image pyramid representations), in formal software verification (Allamigeon et al. 2015), in statistical inference with domain constraints (Calvin et al. 1991; Tsai 2007), in structure tensor of computer vision (Förstner 1986) (Förstner-like operators), etc.

This letter is organized as follows: Sect. 2 explains how to transform the extremal matrix problem into an equivalent geometric minimum enclosing ball of balls. Section 3 presents a fast iterative approximation algorithm that scales well in high-dimensions. Section 4 concludes by hinting at further perspectives.

---

[4]For example, consider $P = \mathrm{diag}(1, 2)$ and $Q = \mathrm{diag}(2, 1)$ then $P - Q = \mathrm{diag}(-1, 1)$ and $Q - P = \mathrm{diag}(1, -1)$.

[5]For example, $S = \mathrm{diag}(-1, 1)$ is dominated by $P = \mathrm{diag}(1 = |-1|, 1)$ (by taking the absolute values of the eigenvalues of S).

2 Equivalent Geometric Covering Problems

We build on top of Burgeth et al. (2007) to prove that solving the d-dimensional Löwner maximal matrix amounts to either find (1) the minimal covering Löwner matrix cone (wrt. set containment $\subseteq$) of a corresponding sets of D-dimensional cones (with $D = \frac{d(d+1)}{2}$), or (2) the minimal enclosing ball of a set of corresponding $(D-1)$-dimensional "matrix balls" that we cast into a geometric *vector* ball covering problem for amenable computations.

2.1 Minimal Matrix/Vector Cone Covering Problems

Let $\mathcal{L} = \{X \in \mathrm{Sym}^+(d) \ : \ X \succeq 0\}$ denote the *Löwner ordering cone*, and $\mathcal{L}(S_i)$ the reverted and translated *dominance cone* (termed the penumbra cone in Burgeth et al. (2007)) with apex S_i embedded in the space of symmetric matrices that represents all the symmetric matrices dominated by S_i: $\mathcal{L}(S_i) = \{S \in \mathrm{Sym}_d(\mathbb{R}) \ : \ S_i \succeq S\} = S_i \ominus \mathcal{L}$, where $\ominus$ denotes the *Minkowski set subtraction operator*:

$$\mathcal{A} \ominus \mathcal{B} = \{a - b \ : \ a \in \mathcal{A}, b \in \mathcal{B}\}$$

(hence, $\mathcal{L}(0) = -\mathcal{L}$). A matrix S dominates $S_1, \ldots, S_n$ iff. $\forall i \in [n]$, $\mathcal{L}(S_i) \subseteq \mathcal{L}(S)$. In plain words, S dominates a set of matrices iff. its associated dominance cone $\mathcal{L}(S)$ covers all the dominance cones $\mathcal{L}(S_i)$ for $i \in [n]$. The dominance cones are "abstract" cones defined in the $d \times d$ symmetric matrix space that can be "visualized" as equivalent *vector* cones in dimension $D = \frac{d(d+1)}{2}$ using *half-vectorization*: For a symmetric matrix S, we stack the elements of the lower-triangular matrix part of $S = [s_{i,j}]$ (with $s_{i,j} = s_{j,i}$):

$$\mathrm{vech}(S) = \left[s_{1,1} \ \ldots s_{d,1} \ s_{2,2} \ \ldots \ s_{d,2} \ \ldots \ s_{d,d}\right]^\top \in \mathbb{R}^{\frac{d(d+1)}{2}}.$$

Note that this is not the unique way to half-vectorize symmetric matrices but it is enough for *geometric containment* purposes. Later, we shall enforce that the ℓ_2-norm of vectors $\mathrm{vech}(S)$ matches the Fröbenius matrix norm $\|\cdot\|_F$.

Let $\mathcal{L}_v$ denotes the vectorized matrix Löwner ordering cone: $\mathcal{L}_v = \{\mathrm{vech}(P) \ : \ P \succ 0\}$, and $\mathcal{L}_v(S)$ denote the *vector dominance cone*: $\mathcal{L}_v(S) = \{\mathrm{vech}(X) \ : \ X \in \mathcal{L}(S)\}$. Next, we further transform this minimum D-dimensional matrix/vector cone covering problems as equivalent *Minimum Enclosing Ball* (MEB) problems of $(D-1)$-dimensional matrix/vector balls.

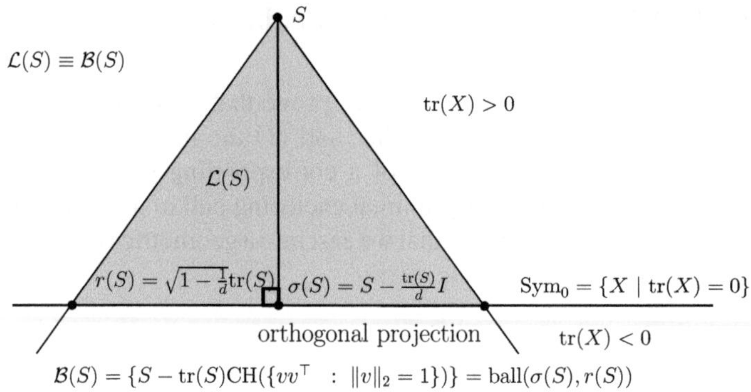

Fig. 1 The dominance *cone* $\mathcal{L}(S)$ associated with matrix S has apex S and base $\mathcal{B}(S) = \mathrm{Ball}(\sigma(S), r(S))$, a ball *centered* at matrix $\sigma(S)$ of radius $r(S)$. The *cone* $\mathcal{L}(S)$ has an equivalent representation $\mathcal{B}(S)$ provided that $\mathrm{tr}(S) \geq 0$

2.2 Minimum Enclosing Ball of Ball Problems

A *basis* $\mathcal{B}$ of a convex cone $\mathcal{C}$ anchored at the origin 0 is a convex subset $\mathcal{B} \subseteq \mathcal{C}$ so that $\forall x \neq 0 \in \mathcal{C}$ there exists a *unique decomposition*: $x = \lambda b$ with $b \in \mathcal{B}$ and $\lambda > 0$. For example, $\mathrm{Sym}_1^+(\mathbb{R}) = \{P \in \mathrm{Sym}^+(\mathbb{R}) : \mathrm{tr}(P) = 1\}$ is a basis of the Löwner cone $\mathcal{L} = \mathrm{Sym}^+(\mathbb{R})$. Informally speaking, a basis of a cone can be interpreted as a *compact* cross-section of the cone. The Löwner cone $\mathcal{L}$ is a smooth convex cone with its interior $\mathrm{Int}(\mathcal{L})$ denoting the space of positive definite matrices $\mathrm{Sym}^{++}(\mathbb{R})$ (full rank matrices), and its border $\partial\mathcal{L} = \mathcal{L}\backslash\mathrm{Int}(\mathcal{L})$ the *rank-deficient* symmetric positive semi-definite matrices (with apex the zero matrix 0 of rank 0). A point x is an *extreme element* of a convex set S iff. $S\backslash\{x\}$ remains convex. It follows from Minkowski theorem that every compact convex set $\mathcal{S}$ in a finite-dimensional vector space can be reconstructed as convex combinations of its extreme points $\mathrm{ext}(\mathcal{S}) \subseteq \partial\mathcal{S}$: That is, the compact convex set is the closed convex hull of its extreme points.

A face $\mathcal{F} \subset \mathcal{C}$ of a closed cone $\mathcal{C}$ is a subcone such that $x + y \in \mathcal{F} \rightarrow x, y \in \mathcal{F}$. The 1-dimensional faces are the *extremal rays* of the cone. The basis of the Löwner ordering cone is Hill and Waters (1987):

$$\mathcal{B}(\mathcal{C}) = \mathrm{CH}(vv^\top : v \in \mathbb{R}^d, \|v\|_2 = 1).$$

Other rank-deficient or full rank matrices can be constructed by convex combinations of these rank-1 matrices, the extremal rays.

For any square matrix $X = [x_{i,j}]$, the *trace operator* is defined by $\mathrm{tr}(X) = \sum_{i=1}^d x_{i,i}$, the sum of the diagonal elements of the matrix. The trace also amounts to the sum of the eigenvalues $\lambda_i(X)$ of matrix X: $\mathrm{tr}(X) = \sum_{i=1}^d \lambda_i(X)$. The basis $\mathcal{B}_i$ of a dominance cone $\mathcal{L}(S_i)$ is:

$$\mathcal{B}_i = \{S_i - \mathrm{tr}(S_i) \times \mathcal{B}(\mathcal{L})\}.$$

Note that all the basis of the dominance cones lie in the *subspace* H_0 of symmetric matrices with zero trace. Let $\langle X, Y\rangle_F = \mathrm{tr}(X^\top Y)$ denote the *matrix inner product* and $\|M\|_F = \sqrt{\langle M, M\rangle_F} = \sqrt{\sum_{i,j} m_{i,j}^2}$ the matrix *Fröbenius norm*. Two matrices X and Y are orthogonal (or perpendicular) iff. $\langle X, Y\rangle_F = 0$. It can be checked that the identity matrix I is perpendicular to any zero-trace matrix X since $\langle X, I\rangle_F = \mathrm{tr}(X) = 0$. The center of the ball basis of the dominance cone $\mathcal{L} = \mathcal{L}(S)$ is obtained as the *orthogonal projection* of S onto the zero-trace subspace H_0: $\sigma(S) = S - \frac{\mathrm{tr}(S)}{d} I$. The dominance cone basis is a *matrix ball* since for any rank-1 matrix $E = vv^\top$ with $\|v\|_2 = 1$ (an extreme point), we have the radius:

$$r(S) = \|S - \mathrm{tr}(S)vv^\top - \sigma(S)\|_F = \mathrm{tr}(S)\sqrt{1 - \frac{1}{d}}, \tag{2}$$

that is non-negative since we assumed that $\mathrm{tr}(S) \geq 0$. Reciprocally, to a basis ball $B = \mathrm{Ball}(\sigma, r)$, we can associate the apex of its corresponding dominance cone $\mathcal{L}(B)$:

$$\sigma + \frac{r}{d}\frac{I}{\sqrt{1-\frac{1}{d}}}.$$

Figure 1 illustrates the notations and the representation of a cone by its corresponding basis and apex. Thus we associate to each dominance cone $\mathcal{L}(S_i)$ its corresponding ball basis $B_i = \mathrm{Ball}(\sigma(S_i), r_i)$ on the subspace H_0 of zero trace matrices:

$$\sigma_i = \sigma(S_i) = S_i - \frac{\mathrm{tr}(S_i)}{d} I,$$

$$r_i = r(S_i) = \mathrm{tr}(S_i)\sqrt{1 - \frac{1}{d}}.$$

We have the following containment relationships:

$$P \succ Q \Leftrightarrow \mathcal{L}(P) \supset \mathcal{L}(Q) \Leftrightarrow B(P) \supset B(Q),$$

and

$$P \succeq Q \Leftrightarrow \mathcal{L}(P) \supseteq \mathcal{L}(Q) \Leftrightarrow B(P) \supseteq B(Q).$$

Finally, we transform this minimum enclosing *matrix* ball problem into a minimum enclosing *vector* ball problem using a half-vectorization that preserves the notion of distances, i.e., using an isomorphism between the space of symmetric matrices and the space of half-vectorized matrices. The ℓ_2-norm of the vectorized matrix should match the matrix Fröbenius norm: $\|s\|_2 = \|\mathrm{vec}^+(S)\|_2 = \|S\|_F$. Since

$$\|S\|_F = \sqrt{\sum_{i=1}^{d}\sum_{j=1}^{d} s_{i,j}^2} = \sqrt{\sum_{i=1}^{d} s_{i,i}^2 + 2\sum_{i=1}^{d-1}\sum_{j=i+1}^{d} s_{i,j}^2} = \|s\|_2,$$

it follows that

$$s = \|\mathrm{vec}^+(S)\|_2 = \left[s_{1,1} \ \ldots \ s_{d,d} \ \sqrt{2}s_{1,2} \ \sqrt{2}s_{1,d} \ldots \sqrt{2}s_{d-1,d}\right]^\top \in \mathbb{R}^{\frac{d(d+1)}{2}}.$$

We can convert back a vector $v \in \mathbb{R}^D$ into a corresponding symmetric matrix.

Since we have considered all dominance cones with basis rooted on H_0^+ : $\mathrm{tr}(X) \geq 0$ in order to compute the ball basis as orthogonal projections, we need to *pre-process* the symmetric matrices to ensure that property as follows: Let $t = \min\{\mathrm{tr}(S_1), \ldots, \mathrm{tr}(S_n)\}$ denote the minimal trace of the input set of symmetric matrices $S_1, \ldots, S_n$, and define $S_i' = S_i - tI$ for $i \in [n]$ where I denotes the identity matrix. Recall that $\mathrm{tr}(X_1 + \lambda X_2) = \mathrm{tr}(X_1) + \lambda\mathrm{tr}(X_2)$. By construction, the transformed input set satisfies $\mathrm{tr}(S_i') \geq 0$, $\forall i \in [n]$. Furthermore, observe that $S \succeq S_i$ iff. $S' \succeq S_i'$ where $S' = S - tI$, so that $\max(S_1, \ldots, S_n) = \max(S_1', \ldots, S_n') + tI$.

As a side note, let us point out that the reverse basis-sphere-to-cone mapping has been used to compute the convex hull of d-dimensional spheres (convex homothets) from the convex hull of $(d+1)$-dimensional equivalent points (Boissonnat et al. 1996; Boissonnat and Karavelas 2003).

Finally, let us notice that there are severals ways to majorize/minorize matrices: For example, once can seek extremal matrices that are invariant up to an *invertible transformation* (Allamigeon et al. 2015), a stronger requirement than the invariance by orthogonal transformation. In the latter case, it amounts to geometrically compute the Minimum Volume Enclosing Ellipsoid of Ellipsoids (MVEEE) (Allamigeon et al. 2015; Jambawalikar and Kumar 2008).

2.3 *Defining* $(1+\epsilon)$*-Approximations of* $\bar{S}$

First, let us summarize the algorithm for computing the Löwner maximal matrix of a set of n symmetric matrices $S_1, \ldots, S_n$ as follows:

1. Normalize matrices so that they have all non-negative traces:

$$S_i' = S_i - tI, \quad t = \min\{\mathrm{tr}(S_1), \ldots, \mathrm{tr}(S_n)\}.$$

2. Compute the vector ball representations of the dominance cones:

$$B_i = \mathrm{Ball}\,(\sigma_i, r_i)$$

with

$$\sigma_i = \mathrm{vec}^+\left(S_i' - \frac{\mathrm{tr}(S_i')}{d} I\right)$$

and

$$r_i = \mathrm{tr}(S_i')\sqrt{1-\frac{1}{d}}$$

3. Compute the small(est) enclosing ball $B' = \mathrm{Ball}(\sigma', r')$ of basis balls (either exactly or an approximation):

$$B' = \mathrm{Small(est)EnclosingBall}(B_1, \ldots, B_n)$$

4. Convert back the small(est) enclosing ball B' to the dominance cone, and recover its apex S':

$$\bar{S}' = \sigma' + \frac{r'}{d}\frac{I}{\sqrt{1-\frac{1}{d}}}.$$

5. Adjust back the matrix trace:

$$\bar{S} = \bar{S}' + tI, \quad t = \min\{\mathrm{tr}(S_1), \ldots, \mathrm{tr}(S_n)\}.$$

Computing *exactly* the extremal Löwner matrices suffer from the *curse of dimensionality* of computing MEBs (Fischer et al. 2003). In Burgeth et al. (2007), proceed by discretizing the basis spheres by sampling[6] the extreme x points $vv^\top$ for $\|v\|_2 = 1$. This yields an approximation term, requires more computation, and even worse the method does not scale (Fischer and Gärtner 2004) in high-dimensions. Thus in order to handle high-dimensional matrices met in software formal verification (Allamigeon et al. 2015) or in computer vision (structure tensor (Förstner 1986)), we consider $(1+\epsilon)$-approximation of the extremal Löwner matrices. The notion of tightness of approximation of $\bar{S}$ (the epsilon) is imported straightforwardly from the definition of the tightness of the geometric covering problems. A $(1+\epsilon)$-approximation $\tilde{S}$ of $\bar{S}$ is a matrix $\tilde{S} \succ \bar{S}$ such that: $r(\tilde{S}) \leq (1+\epsilon)r(\bar{S})$. It follows from Eq. 2 that a $(1+\epsilon)$-approximation satisfies $\mathrm{tr}(\tilde{S}) \leq (1+\epsilon)\mathrm{tr}(\bar{S})$.

We present a fast guaranteed approximation algorithm for approximating the minimum enclosing ball of a set of balls (or more generally, for sets of compact geometric objects).

3 Approximating the Minimum Enclosing Ball of Objects and Balls

We extend the incremental algorithm of Bădoiu and Clarkson (2008) (BC) designed for finite point sets to *ball sets* or *compact object sets* that work in large dimensions. Let $B_1 = \mathrm{Ball}(c_1, r_1)$, ..., $B_n = \mathrm{Ball}(c_n, r_n)$ denote a set of n balls. For an object $\mathcal{O}$

[6]In 2D, we sample $v = [\cos\theta, \sin\theta]^\top$ for $\theta \in [0, 2\pi[$. In 3D, we use spherical coordinates $v = [\sin\theta\cos\phi, \sin\theta\sin\phi, \cos\theta]^\top$ for $\theta \in [0, 2\pi[$ and $\phi \in [0, \pi[$.

and a query point q, denote by $D^f(q, \mathcal{O})$ the *farthest* distance from q to $\mathcal{O}$:

$$D^f(q, \mathcal{O}) = \max_{o \in \mathcal{O}} \|q - o\|,$$

and let $F(q, \mathcal{O})$ denote the farthest point of $\mathcal{O}$ from q. The generalized BC (Bădoiu and Clarkson 2008) algorithm for approximating the circumcenter of the minimum volume enclosing ball of n objects (MVBO) $\mathcal{O}_1, \ldots, \mathcal{O}_n$ is summarized as follows:

- Let $e_1 \leftarrow x \in \mathcal{O}_1$ and $i \leftarrow 1$.
- Repeat l times:
  - Find the farthest object $\mathcal{O}_f$ to current center:

$$f = \arg\max_{j \in [n]} D^f(e_i, \mathcal{O}_j)$$

  - Update the circumcenter:

$$e_{i+1} = \frac{i}{i+1} e_i + \frac{1}{i+1}(F(e_i, \mathcal{O}_f) - e_i)$$

  - $i \leftarrow i + 1$.

When considering balls as objects, the farthest distance of a point x to a ball $B_j = \mathrm{Ball}(c_j, r_j)$ is

$$D^f(e_i, B_j) = \|c_j - e_i\| + r_j,$$

and the circumcenter updating rule is:

$$e_{i+1} = \frac{i}{i+1} e_i + \frac{1}{i+1}(c_f - e_i)\left(1 + \frac{r_f}{\|c_f - e_i\|}\right).$$

See Fig. 2 and online video[7] for an illustration. (MVBO can also be used to approximate the MEB of ellipsoids.) It is proved in Bădoiu and Clarkson (2003) that at iteration i, we have

$$\|e_i - e^*\| \leq \frac{r^*}{\sqrt{i}},$$

where $B^* = \mathrm{Ball}(e^*, r^*)$ is the unique smallest enclosing ball. Hence the radius of the ball centered at e_i is bounded by $(1 + \frac{1}{\sqrt{i}})r^*$. To get a $(1 + \epsilon)$-approximation, we need $\frac{1}{\epsilon^2}$ iterations. It follows that a $(1 + \epsilon)$-approximation of the smallest enclosing ball of n D-dimensional balls can be computed in $O(\frac{D}{n}\epsilon^2)$-time (Bădoiu and Clarkson 2003), and since $D = O(d^2)$ we get:

[7] https://www.youtube.com/watch?v=w1ULgGAK6vc.

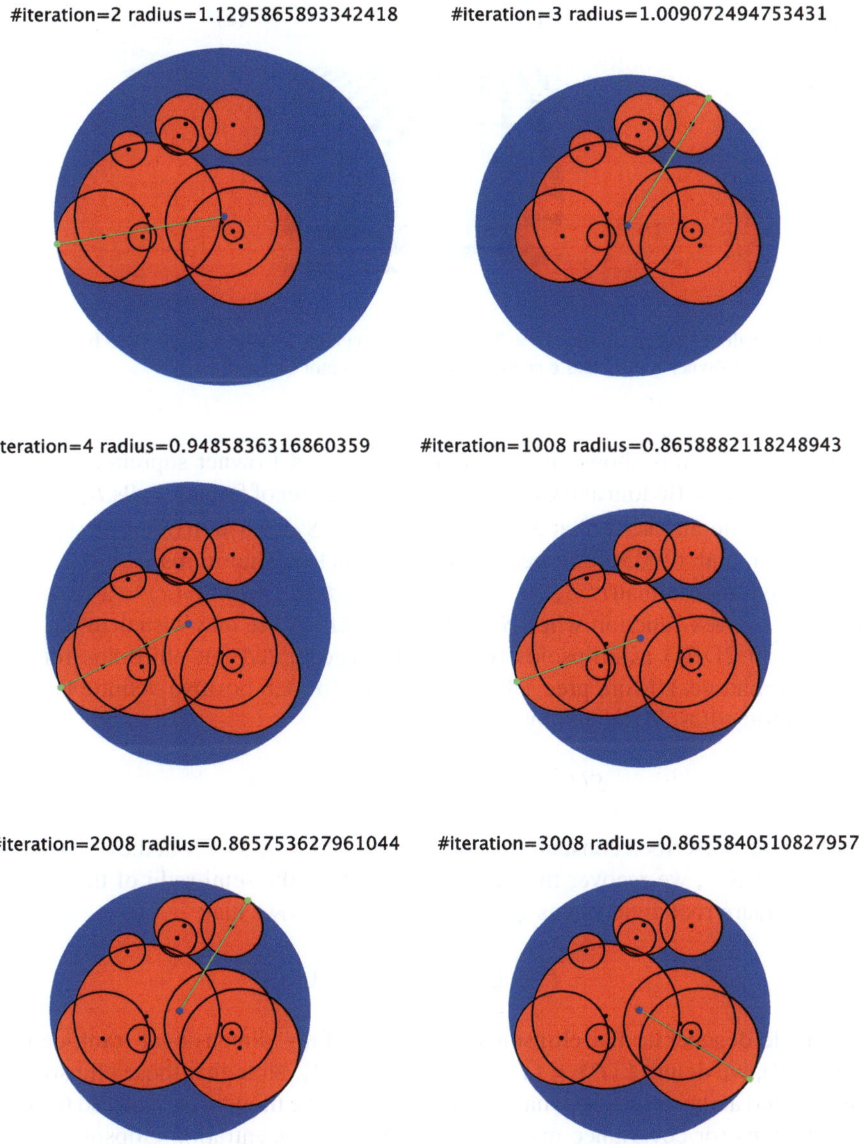

Fig. 2 Approximating the minimum enclosing ball of balls iteratively: Snapshots at iterations 1, 2, 3,1008, 2008 and 3008

Theorem 1 *The Löwner maximal matrix* $\bar{S}$ *of a set of* n d*-dimensional symmetric matrices can be approximated by a matrix* $\tilde{S} \succ \bar{S}$ *such that* $\mathrm{tr}(\tilde{S}) \leq (1+\epsilon)\mathrm{tr}(\bar{S})$ *in* $O(\frac{d^2}{n}\epsilon^2)$*-time.*

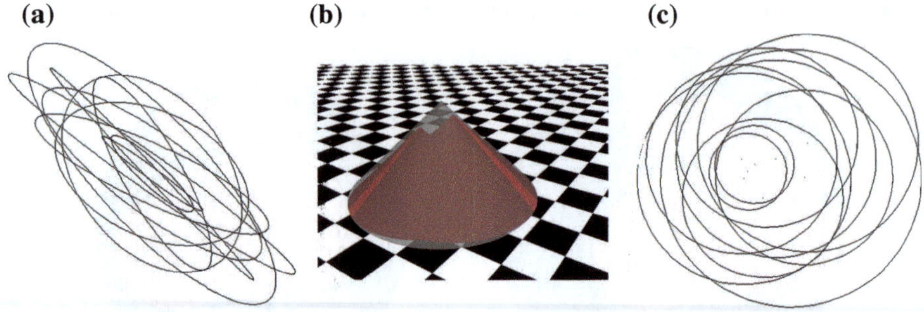

Fig. 3 Equivalent visualizations: **a** 2×2 PSD matrices visualized as *ellipsoids*, with **b** corresponding 3D vector Löwner *cones*, and **c** corresponding *cone* vector ball basis

Interestingly, this shows that the approximation of Löwner supremum matrices admits core-sets (Bădoiu and Clarkson 2003), the subset of farthest balls $B_{f(i)}$ chosen during the l iterations, so that $\tilde{S} = \max(S_{f(1)}, \ldots, S_{f(l)})$ with $\text{tr}(\tilde{S}) \leq (1+\epsilon)\text{tr}(\bar{S})$. See Kumar et al. (2003) for other MEB approximation algorithms.

To a symmetric matrix S, we associate a *quadratic form* $q_S(x) = x^\top S x$ that is a strictly convex function when S is PSD. Therefore, we may visualize the SPSD matrices in 2D/3D as ellipsoids (potentially degenerated flat ellipsoids for rank-deficient matrices). More precisely, we associate to each positive definite matrix S, a geometric ellipsoid defined by

$$\mathcal{E}(S) = \{x \in \mathbb{R}^d \ : \ x^\top S^{-1} x = \rho\},$$

where ρ is a prescribed constant (usually set to $\rho = 1$, Fig. 3). From the SVD decomposition of S^{-1}, we recover the rotation matrix, and the semi-radii of the ellipsoid are the square root eigenvalues $\sqrt{\lambda_1}, \ldots, \sqrt{\lambda_d}$. It follows that:

$$P \succeq Q \Leftrightarrow \mathcal{E}(P) \supseteq \mathcal{E}(Q).$$

To handle degenerate flat ellipsoids that are not fully dimensional (rank-deficient matrix P), we define $\mathcal{E}(P) = \{x \in \mathbb{R}^d \ : \ xx\top \preceq P\}$. Note that those ellipsoids are all centered at the origin, and may also conceptually be thought as centered Gaussian distributions (or covariance matrices denoting the concentration ellipsoids of estimators (Siotani 1967) in statistics). We can also visualize the Löwner ordering cone and dominance cones for 2×2 matrices embedded in the vectorized 3D space of symmetric matrices (Fig. 3), and the corresponding half-vectorized ball basis (Fig. 3).

4 Concluding Remarks

Our novel extremal matrix approximation method allows one to leverage further related results related to core-sets (Bădoiu and Clarkson 2008) for dealing with high-dimensional extremal matrices. For example, we may consider clustering PSD matrices with respect to Löwner order and use the k-center clustering technique with guaranteed approximation (Mihelic and Robic 2003; Chen 2009). A Java™ code of our method is available for reproducible research.

Acknowledgements This work was carried out during the Matrix Information Geometry (MIG) workshop (Nielsen and Bhatia 2013), organized at École Polytechnique, France in February 2011 (https://www.sonycsl.co.jp/person/nielsen/infogeo/MIG/). Frank Nielsen dedicates this work to the memory of his late father Gudmund Liebach Nielsen who passed away during the last day of the workshop.

References

Allamigeon, X., Gaubert, S., Goubault, E., Putot, S., & Stott, N. (1980). A scalable algebraic method to infer quadratic invariants of switched systems. In *2015 International Conference on Embedded Software (EMSOFT)* (pp. 75–84), October 2015.

Angulo, J. (2013). Matrix Information Geometry. In F. Nielsen & R. Bhatia (Eds.), *Supremum/infimum and nonlinear averaging of positive definite symmetric matrices* (pp. 3–33). Heidelberg: Springer.

Bădoiu, M., & Clarkson, K. L. (2003). Smaller core-sets for balls. In *Proceedings of the Fourteenth Annual ACM-SIAM Symposium on Discrete Algorithms*, SODA '03, Philadelphia, PA, USA, 2003 (pp 801–802). Society for Industrial and Applied Mathematics.

Bădoiu, M., & Clarkson, K. L. (2008). Optimal core-sets for balls. *Computational Geometry*, *40*(1), 14–22.

Bhatia, R. (2009). *Positive definite matrices*. Princeton: Princeton university press.

Boissonnat, J.-D., Cérézo, A., Devillers, O., Duquesne, J., & Yvinec, M. (1996). An algorithm for constructing the convex hull of a set of spheres in dimension d. *Computational Geometry*, *6*(2), 123–130.

Boissonnat, J.-D., & Karavelas, M. I. (2003). On the combinatorial complexity of euclidean Voronoi cells and convex hulls of d-dimensional spheres. In *Proceedings of the Fourteenth Annual ACM-SIAM Symposium on Discrete Algorithms* (pp. 305–312). Society for Industrial and Applied Mathematics.

Burgeth, B., Bruhn, A., Didas, S., Weickert, J., & Welk, M. (2007). Morphology for matrix data: Ordering versus PDE-based approach. *Image and Vision Computing*, *25*(4), 496–511.

Burgeth, B., Bruhn, A., Papenberg, N., Welk, M., & Weickert, J. (2007). Mathematical morphology for matrix fields induced by the Loewner ordering in higher dimensions. *Signal Processing*, *87*, 277–290.

Calvin, J. A., & Dykstra, R. L. (1991). Maximum likelihood estimation of a set of covariance matrices under Löwner order restrictions with applications to balanced multivariate variance components models. *The Annals of Statistics*, *19*, 850–869.

Chen, K. (2009). On coresets for k-median and k-means clustering in metric and euclidean spaces and their applications. *SIAM Journal on Computing*, *39*(3), 923–947.

Fischer, K., & Gärtner, B. (2004). The smallest enclosing ball of balls: Combinatorial structure and algorithms. *International Journal of Computational Geometry & Applications*, *14*(04n05), 341–378.

Fischer, K., Gärtner, B., & Kutz, M. (2003). Fast smallest-enclosing-ball computation in high dimensions. In G. Di Battista & U. Zwick (Eds.), *Algorithms-ESA 2003* (pp. 630–641). Heidelberg: Springer.

Förstner, W. (1986). A feature based correspondence algorithm for image matching. *International Archives of Photogrammetry and Remote Sensing*, *26*(3), 150–166.

Hill, R. D., & Waters, S. R. (1987). On the cone of positive semidefinite matrices. *Linear Algebra and its Applications*, *90*, 81–88.

Jambawalikar, S., & Kumar, P. (2008). A note on approximate minimum volume enclosing ellipsoid of ellipsoids. In *International Conference on Computational Sciences and Its Applications, 2008. ICCSA'08* (pp. 478–487). IEEE.

Kumar, P., Mitchell, J. S. B., & Yildirim, E. A. (2003). Approximate minimum enclosing balls in high dimensions using core-sets. *Journal of Experimental Algorithmics (JEA)*, *8*, Article No. 1.1.

Mihelic, J., & Robic, B. (2003). Approximation algorithms for the k-center problem: An experimental evaluation. In *Selected papers of the International Conference on Operations Research (SOR 2002)* (p. 371) Heidelberg:Springer.

Nielsen, F., & Bhatia, R. (2013). *Matrix Information Geometry*. Heidelberg: Springer Publishing Company, Incorporated.

Siotani, M. (1967). Some applications of Loewner's ordering on symmetric matrices. *Annals of the Institute of Statistical Mathematics*, *19*(1), 245–259.

Tsai, M.-T. (2007). Maximum likelihood estimation of Wishart mean matrices under Löwner order restrictions. *Journal of Multivariate Analysis*, *98*(5), 932–944.

Dimensionality Reduction for Information Geometric Characterization of Surface Topographies

C.T.J. Dodson, M. Mettänen and W.W. Sampson

1 Introduction

Stochastic textures with features spanning many length scales arise in a range of contexts in physical and natural sciences. Whereas the features of interest may differ when considering cosmological scale data for galactic density distributions, from those at the global scale representation of oceanographic temperatures or nanoscale features such as the surface topography of synthetic bone, the common format for the data is as a two-dimensional array, which is typically rendered as an image. In general, the challenge is the extraction of the features of interest which may be obscured within an inherently noisy data set. Since paper is made as a web from the continuous filtration of a stochastic dispersion of cellulose fibres, there is a standard reference structure which can be used: a planar isotropic Poisson process of the given fibres, for which the structure is known (Dodson 1971).

Here, we use information geometry and dimensionality reduction to bypass the extraction of features from textures and instead make a direct assessment of whether they are different or not. We illustrate our approach using the example of two-dimensional stochastic textures arising from measurements of the surface topography of different grades of paper. Whereas paper represents a convenient source of data available with a wide range of surface topographies, it turns out that there is genuine

C.T.J. Dodson
School of Mathematics, University of Manchester, Manchester M13 9PL, UK

M. Mettänen
Department of Automation Science & Engineering, Tampere University of Technology, PL 692, FI-33101 Tampere, Finland

W.W. Sampson (✉)
School of Materials, University of Manchester, Manchester M13 9PL, UK
e-mail: william.sampson@manchester.ac.uk

F. Nielsen et al. (eds.), *Computational Information Geometry*,
Signals and Communication Technology, DOI 10.1007/978-3-319-47058-0_7

interest in the papermaking industry in characterizing this structural feature of the material and its influence on product performance (Mettänen and Hirn 2015).

Information geometry uses the Fisher information metric on smoothly parametrized families of probability density functions to provide a natural distance structure. Gaussians parametrized by mean and standard deviation yield a 2-dimensional curved surface and bivariate Gaussians yield a 5-dimensional curved space, *cf.* Amari (2016). Thus, the information metric gives an arc length function along any curve between two probability density functions in the given family. The geometry of commonly occurring families of probability density functions is well-known, see Arwini and Dodson (2008) for relevant examples. The technical algorithmic difficulty is that, in the curved space of probability density functions, the true information distance between two points is the infimum of arc length taken over all curves joining the points. This infimum is the length of the geodesic between the points.

Materials scientists study the interdependence of the structure and properties of materials and how these may be influenced by manufacturing processes. Typically, the properties of the material are the product specifications for end-use and employed for quality control; examples include mechanical behaviour, thermal or electrical conductivity, permeability, etc. Our focus here is identification of differences in structure that may be difficult to identify using conventional data handling methodologies.

We illustrate our approach using measurements of the surface topography of paper. This is a particularly convenient material to study: almost all grades of paper have principally the same chemical structure – they consist of natural cellulosic fibres with length of order a millimeter or two and width a few tens of micrometers; sheets may be filled or coated with minerals such as clay or calcium carbonate. Structural variability in paper is observed at scales corresponding to the fibre dimensions and above and, importantly, depends on the fibre dimensions and the manufacturing processes employed; these dependencies are discussed in detail in, e.g. Deng and Dodson (1994), Sampson (2009b). Papermakers control global average structural properties to influence the sheet properties for a given application, so, for example, other than weight per unit area, the principle difference between newsprint and bathroom tissue is the density of the sheet. Local average variability in such properties is far more difficult to characterize and control because of the underlying stochastic variability (Deng and Dodson 1994) arising from the finite length fibres and influencing the autocovariance function of the planar ensemble.

The stochastic variability of mass of paper, i.e.the distribution of local averages of areal density in the plane of the sheet is a fundamental structural property of paper and characterizes the extent to which fibres are clustered. Recently, through analysis of simulated textures representing the distribution of mass, we demonstrated that information geometry could be used to discriminate variability arising from the size and intensity of clusters (Dodson and Sampson 2013, 2014).

In what follows we illustrate the differences of features in given data sets obtained from measurements of the surface topography of different paper samples. For each sample, our source information is an array of surface heights, which we process to yield a 3×3 covariance matrix Σ and mean vector μ arising from pixels and their first and second neighbours. We proceed to use dimensionality reduction to extract the three most significant features from the set of samples so that all samples can be displayed graphically in a 3-dimensional plot. The aim is to reveal groupings of data points that distinguish among and within grades of paper. The method depends on extracting the three largest eigenvalues and their eigenvectors from a matrix of pairwise information distances among distributions representing the samples in the data set. The number of samples in the data set is unimportant, except for the computation time in finding eigenvalues.

2 Data Sets

Data was acquired as local height values from the surfaces of paper samples using a photometric stereo device; details of the measurement technique are provided in Mettänen and Hirn (2015). Data were acquired at different times under subtly different optical conditions; though in all cases we handled arrays of at least 2400×2400 pixels with spatial resolution between 4 and 7 μm per pixel, which is smaller than the expected width of the constituent fibres. All measurements were made on industrially manufactured paper samples. Measurements were made on 3 groups of samples:

Group 1: Packaging and printing grades. Ten samples: coated packaging paper and cardboard; uncoated packaging paper and cardboard; uncoated wrapping grades. Measurements made on both sides of each sample.

Group 2: Tissue. Five samples of two-ply bathroom tissue. Measurements made on one side only.

Group 3: Printing, writing and sack grades. Five samples: one high quality coated grade and three utility grades for printing and writing; one grade for making paper sacks. Measurements made on one side only; two measurements made of each sample.

Graphical representations of three examples of the surface height distribution are provided in Fig. 1. These show three very different surfaces: a coated board surface, an uncoated packaging paper surface and the surface of a bathroom tissue. In the figure, dark regions correspond to low height and vice versa; each image represents a square of side 1500 pixels.

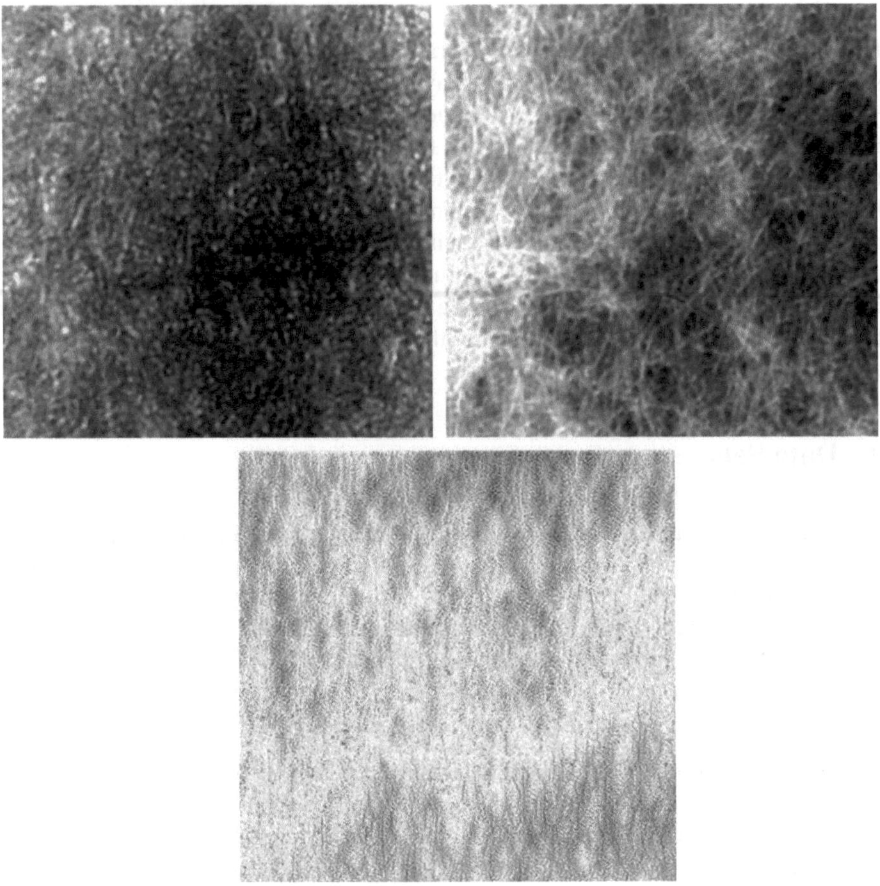

Fig. 1 Graphical representations of sample height data. *Top left* coated board; *top right* uncoated packaging paper; *bottom* bathroom tissue. Each image represent a square of side 1500 pixels with approximate resolution 5 μm per pixel

3 Information Geometry Model

Each of our source data sets consists of a two-dimensional array of local average height values $\tilde{h}_i$. From each of these, we generate two numbers: the average height of the 8 first-neighbour pixels, $\tilde{h}_{1,i}$ and the average height of the 16 second-neighbour pixels, $\tilde{h}_{2,i}$. Thus, we have a trivariate distribution of the random variables $(\tilde{h}_i, \tilde{h}_{1,i}, \tilde{h}_{2,i})$ with $\bar{h}_2 = \bar{h}_1 = \bar{h}$ and the marginal distributions of $\tilde{h}_i$, $\tilde{h}_{1,i}$ and $\tilde{h}_{2,i}$ are well approximated by Gaussian distributions.

The geodesic distance between two multivariate Gaussians, A, B, with probability density functions f^A, f^B mean vectors μ^A, μ^B and covariance matrices Σ^A, Σ^B of the *same* number n of variables is known analytically in two particular cases (Atkinson and Mitchell 1981):

Common covariance matrix, different mean vectors:
$\mu^A \neq \mu^B, \Sigma^A = \Sigma^B = \Sigma; f^A = (n, \mu^A, \Sigma), f^B = (n, \mu^B, \Sigma)$

$$D_\mu(f^A, f^B) = \sqrt{\left(\mu^A - \mu^B\right)^T \cdot \Sigma^{-1} \cdot \left(\mu^A - \mu^B\right)}. \tag{1}$$

Common mean vector, different covariance matrices:
$\mu^A = \mu^B = \mu, \Sigma^A \neq \Sigma^B : f^A = (n, \mu, \Sigma^A), f^B = (n, \mu, \Sigma^B)$

$$D_\Sigma(f^A, f^B) = \sqrt{\frac{1}{2}\sum_{j=1}^{n} \log^2(\lambda_j)}, \quad \text{with } \{\lambda_j\} = \mathrm{Eig}\left((\Sigma^A)^{-\frac{1}{2}} \cdot \Sigma^B \cdot (\Sigma^A)^{-\frac{1}{2}}\right). \tag{2}$$

Here we shall take the simplest choice and sum the two components (1) and (2) to give a net measure of distance between two arbitrary n-variate Gaussians f^A, f^B

$$D(f^A, f^B) = \frac{1}{2}\left(D_\mu(f^A, f^B) + D_\mu(f^B, f^A)\right) + D_\Sigma(f^A, f^B) \tag{3}$$

where we have to take the average of (1) using Σ^A and Σ^B so (3) gives an upper bound on the true distance.

4 Dimensionality Reduction

Now, our family of 35 data sets gives us a 35×35 symmetric matrix of pairwise information distances between pairs of samples, each sample represented by a trivariate Gaussian distribution. Graphically, we can comprehend a 3-dimensional representation of features so we need a method to reduce the feature representation in our data set of 35 to fit into a 3-dimensional image. Accordingly, we follow the methods described by Carter et al. (2007, 2009) to reduce the dimensionality of our data sets and hence identify clustering of data sets with similar topographies through 3-dimensional rendering of the resultant plots. Briefly, we follow a series of computational steps:

1. Obtain pairwise 'information distances' $D(i, j)$ among the members of the dataset of textures $X_1, X_2, \ldots, X_N$ characterised by pixel arrays representing height values.
2. The array of $N \times N$ distances $D(i, j)$ is a symmetric matrix with diagonal zero. This is centralized by subtracting row and column means and then adding back the grand mean to give *CD(i,j)*.

3. The centralized matrix *CD(i,j)* is again symmetric with diagonal zero. We obtain its N eigenvalues *ECD(i)*, which are necessarily real, and the N corresponding N-dimensional eigenvectors *VCD(i)*.
4. Make a 3×3 diagonal matrix A of the first three eigenvalues of largest absolute magnitude and a $3 \times N$ matrix B of the corresponding eigenvectors. The matrix product $A \cdot B$ yields a $3 \times N$ matrix and its transpose is an $N \times 3$ matrix T, which gives us N coordinate values (x_i, y_i, z_i) to embed the N samples in 3-space.

Of course, any pairwise divergence matrix could be used in this situation and might yield different numerical values. However, the qualitative effect will be the same due to a one-to-one monotonic relationship.

5 Results

We illustrate the effectiveness of the approach for the surface textures of the three groups of samples described in Sect. 2. We consider first an application of the theory from samples in cases when both μ and σ are known and then when μ is disregarded. We proceed to show the reproducibility of the approach for discrimination among samples and their position when embedded in 3-space. Finally, we examine the influence of applying the algorithm to data subjected to a simple high-pass filter, as applied in conventional image processing of such data.

5.1 Sensitivity to Mean Vector μ

The top row of Fig. 2 shows the plot of $D(f^A, f^B)$ from Eq. (3) as a cubic-smoothed surface (left), and as a contour plot (right), for trivariate Gaussian information distances among our 35 data sets.[1] On first inspection it is clear that there is structure in the assembled information and the three groups of data can be readily identified from these graphics. The resultant 3-dimensional embedding is shown on the bottom row of Fig. 2; here we observe that the data from Groups 1 and 3 occupy a different region from those for Group 2 which is consistent with the observed surface texture of tissue being manifestly different from those of printing, writing and packaging grades of paper. The first 10 eigenvalues arising from the dimensionality reduction are plotted as a bar chart in Fig. 3, showing clearly that the majority of the spectral information is captured by the 3-largest eigenvalues, in this case approximately 75 %.

Figure 4 shows graphics corresponding to those in Fig. 2 but computed using only the covariances to estimate distances $D_\Sigma(f^A, f^B)$ from Eq. (2) among samples. We see that for this data set the influence on information distance of the changes in mean are rather minor compared with those of the covariances.

---

[1]The small positive values visible in the diagonal in these and subsequent contour plots are an artefact arising from the cubic interpolation.

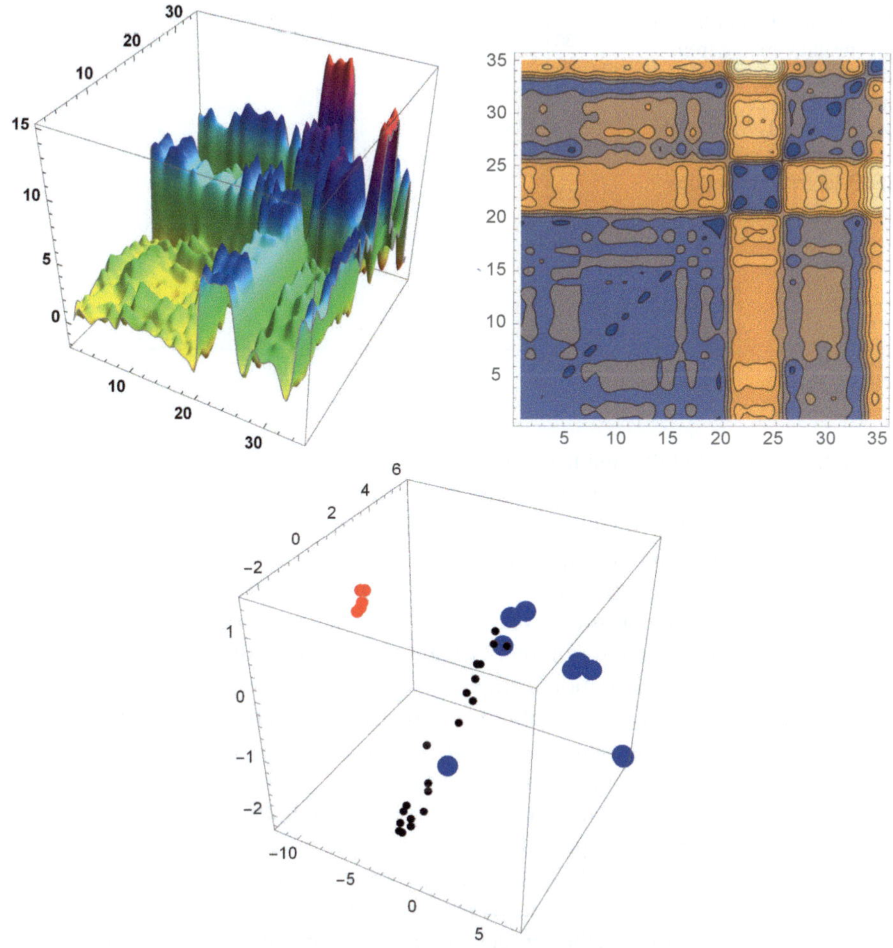

Fig. 2 *Top row* Plot of $D(f^A, f^B)$ from Eq. (3) as a cubic-smoothed surface (*left*), and as a contour plot (*right*), for trivariate Gaussian information distances among 35 datasets of surface heights capturing all three data groups. Axes numbering corresponds to data sets: 1–20, Group 1; 21–25, Group 2; 26–35, Group 3. *Bottom row* Dimensionality reduction embedding of the same data. Group 1 (*small black*), Group 2 (*medium red*), Group 3 (*large blue*)

5.2 Reproducibility

A potential application of the methods we present is the on-line monitoring of change in manufacturing processes. For such applications, repeated sampling and computation of the information distance will yield a surface representing the operating region of the process. Through qualitative and quantitative calibration processes, we might anticipate that data sampled when the process is manufacturing on-specification product would yield embedded data that populate a well-defined region that surface,

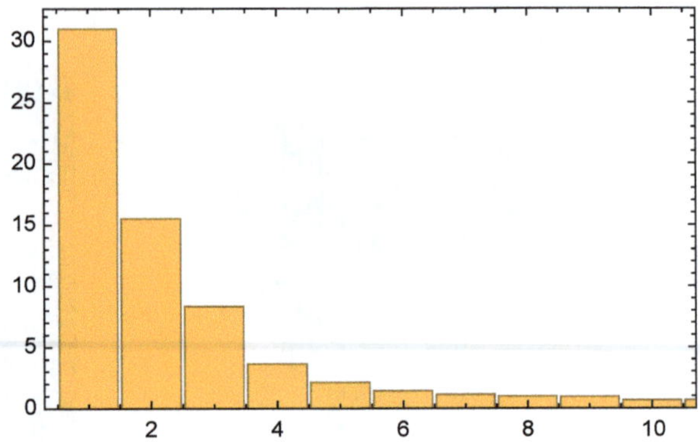

Fig. 3 Bar chart of first 10 eigenvalues arising from dimensionality reduction shown in Fig. 2. Approximately 75 % of the information is captured by the 3-largest eigenvalues

such that when data fall outside this region, operators may be alerted that the process may have altered to give out-of-specification product.

Recall that the data in Group 3 consist of duplicated measurements made on the same side of five different paper specimens. Further, since paper is an inherently stochastic material, we expect some variability from region to region when sampling its surface textures. Accordingly, we use the paired data within Group 3 to investigate the reproducibility of the measurements made on nominally identical samples, which is a prerequisite for on-line monitoring processes of the type proposed.

The plot of $D(f^A, f^B)$ as a cubic-smoothed surface (left), and as a contour plot (right), for trivariate Gaussian information distances among the 10 data sets representing duplicate measurements from five samples is given in the top row of Fig. 5; the resultant 3-dimensional embedding is shown at the bottom of the figure. Again, the first three eigenvalues captured about 75 % of the information

Discrimination among different paper samples within the embedded space is clear and it is noteworthy that the three utility printing and writing grades occupy a different region of the plot from the other grades, which are themselves clearly differentiated. Note that the surface uniformity of these grades are also very different and this manifests itself in the reproducibility of the paired data: for the utility printing and writing grades the two points representing each pair are close; for the high quality coated grade the surface is very smooth and the pair of red points are almost coincident; finally, the structure of sack grade packaging paper is highly non-uniform due to the long fibres used to achieve high mechanical strength and in this case the data points exhibit the greatest separation, though still occupy a manifestly different region from the those representing the other grades.

5.3 Filtered Data

It is common in conventional image processing of stochastic data to apply a high-pass filter to two-dimensional data prior to analysis to characterise features of interest, in

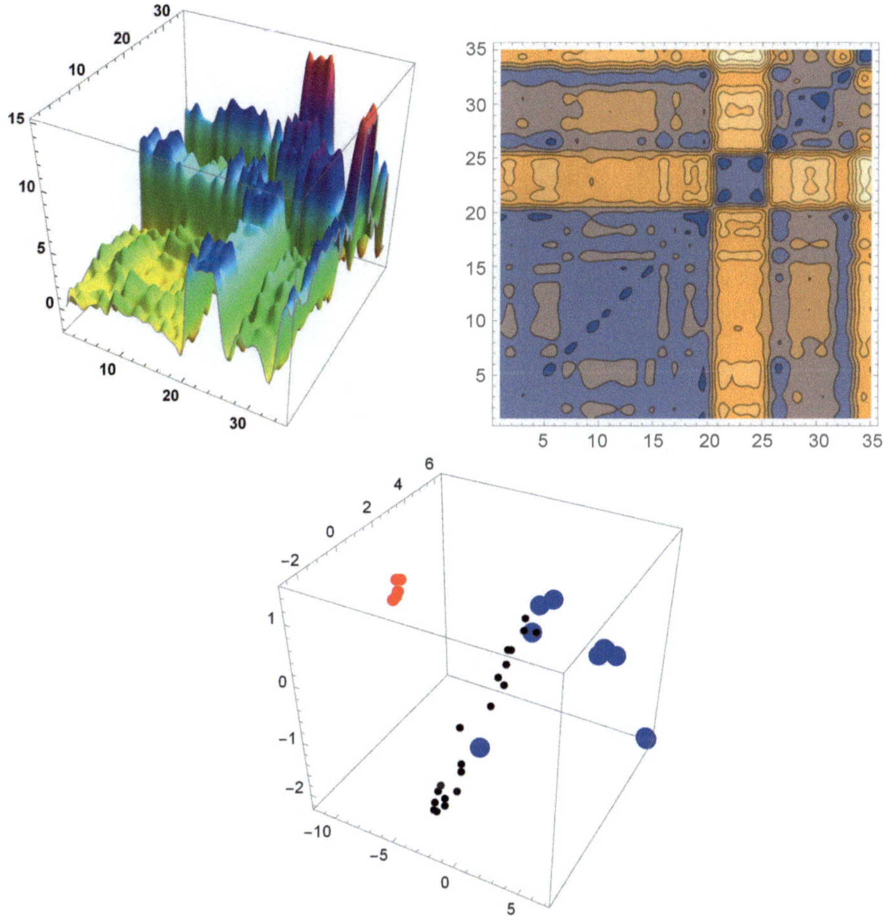

Fig. 4 Plot of $D_\Sigma(f^A, f^B)$ from Eq. (2) as a cubic-smoothed surface (*left*), and as a contour plot (*right*), for trivariate Gaussian covariance only information distances among 35 datasets of surface heights capturing all three data groups. Axes numbering corresponds to data sets: 1–20, Group 1; 21–25, Group 2; 26–35, Group 3. *Bottom row* Dimensionality reduction embedding of the same data. Group 1 (*small black*), Group 2 (*medium red*), Group 3 (*large blue*)

this case the surface roughness relevant to, e.g. printing. Indeed, such processing was applied to the textures described in Mettänen and Hirn (2015) which were similar to those analysed here. Accordingly, we have applied our treatment to the data in Group 3 after application of a high-pass filter.

Plots corresponding to those in Fig. 5 are shown in Fig. 6 and we observe a similar quality of discrimination among samples and a similar level of reproducibility, with similar eigenvalue distribution.

Figure 7 shows the effect of using filtered data in comparison to unfiltered data. On the top, the plot combines the embeddings shown at the bottom of Figs. 5 and 6; on the

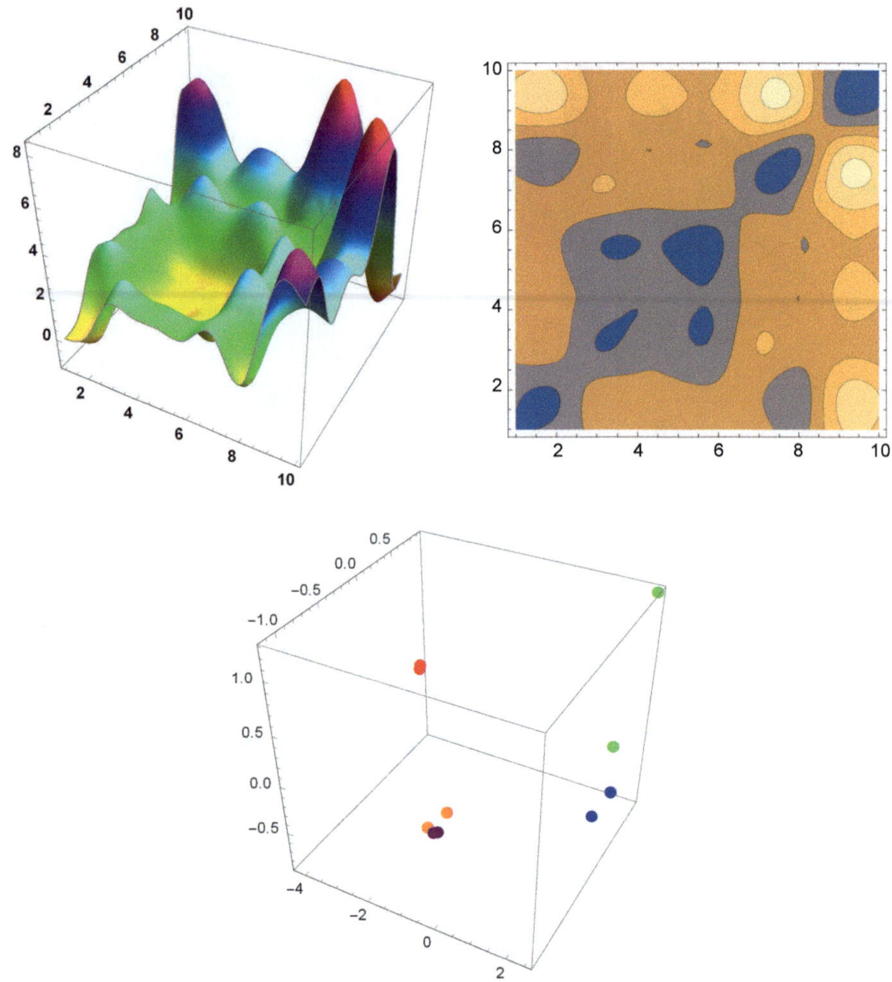

Fig. 5 *Top row* plot of $D(f^A, f^B)$ from Eq. (3) as a cubic-smoothed surface (*left*), and as a contour plot (*right*), for trivariate Gaussian information distances among 10 datasets of surface heights arising from duplicated measurements of the five samples in Group 3. *Bottom* Dimensionality reduction embedding of the same data. High quality coated paper (*red*), sack paper (*green*), three utility printing and writing grades (*orange*, *purple*, *blue*)

bottom the figure shows the embedding obtained by combining the filtered and unfiltered data sets for all sampled to yield a group of 20 arrays (2 filter-states $\times$ 2 repeats $\times$ 5 samples) and computing the information distance $D(f^A, f^B)$ from Eq. (3). Note that although the different processes yield different embeddings, each discriminates well between samples and yields good reproducibility, indicating excellent potential for the use of raw, unfiltered and noisy data in on-line monitoring by application of the approach described.

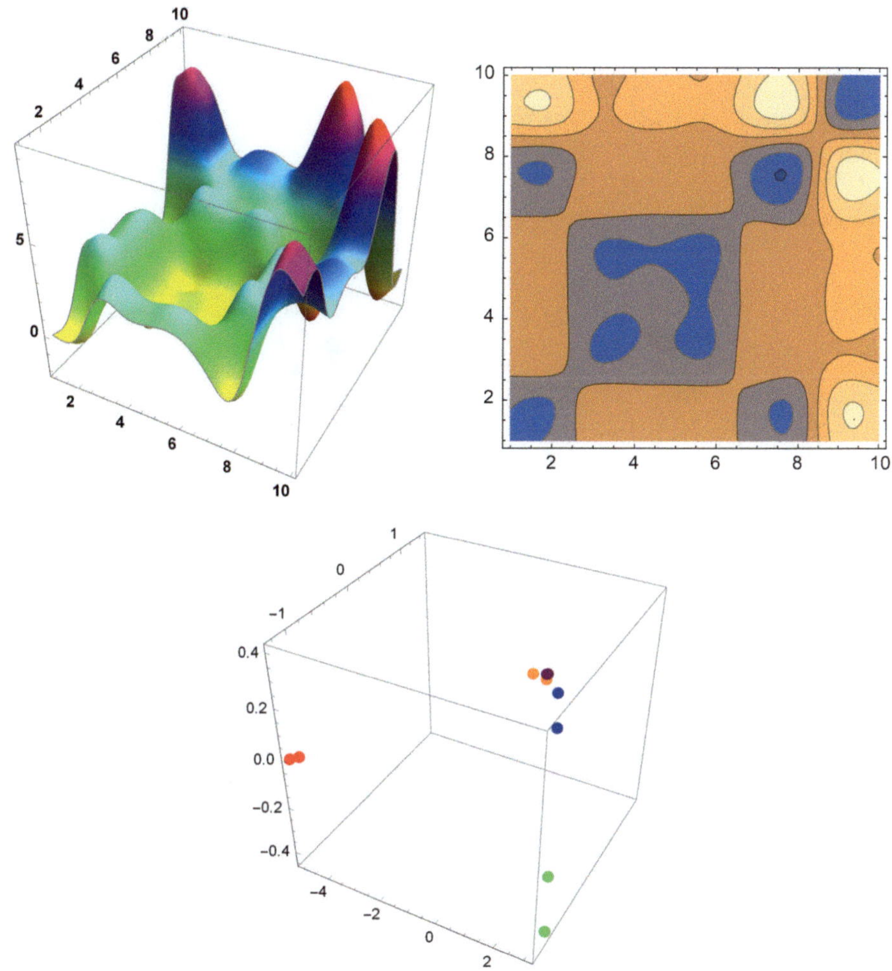

Fig. 6 *Top row* plot of $D(f^A, f^B)$ from Eq. (3) as a cubic-smoothed surface (*left*), and as a contour plot (*right*), for trivariate Gaussian information distances among 10 datasets of surface heights arising from duplicated measurements of the five samples in Group 3 subjected to a high-pass filter. *Bottom* Dimensionality reduction embedding of the same data. High quality coated paper (*red*), sack paper (*green*), three utility printing and writing grades (*orange*, *purple*, *blue*)

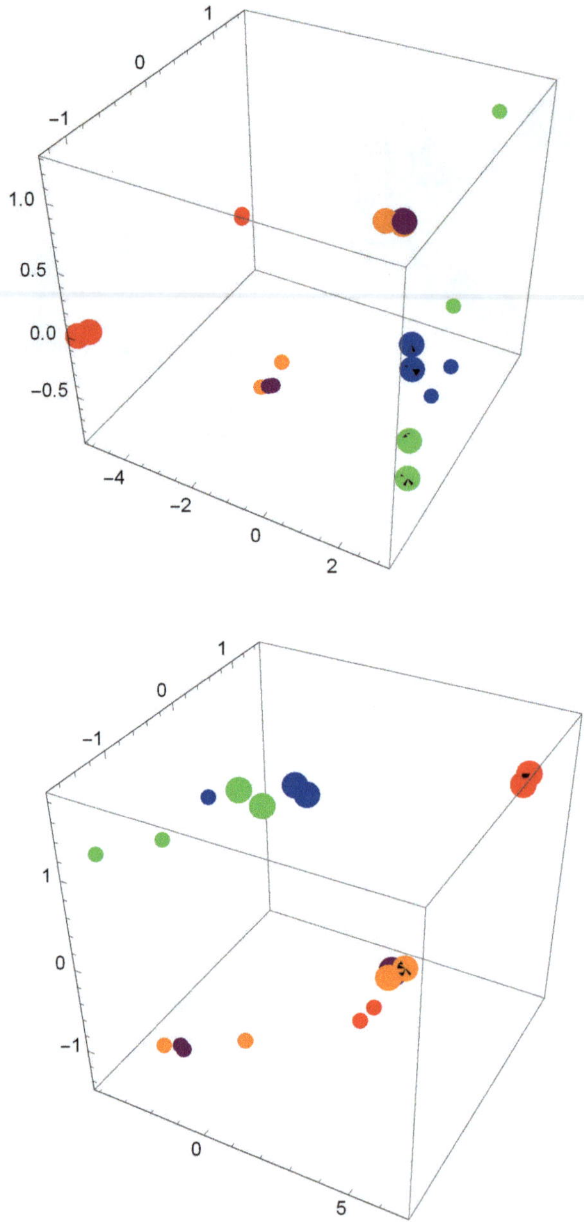

Fig. 7 Combined dimensionality reduction embedding using $D(f^A, f^B)$ from Eq. (3) for unfiltered data (*small circles*) and data subjected to a high-pass filter (*large circles*). High quality coated paper (*red*), sack paper (*green*), three utility printing and writing grades (*orange*, *purple*, *blue*). *Top* embedding shown in Fig. 5 superimposed on that shown in Fig. 6; *bottom* information distances and embedding calculated for combined data set of filtered and unfiltered data from Group 3 ($2 \times 2 \times 5$)

6 Random Fibre Networks

A natural choice of reference structure for the surface of heterogeneous fibrous web-like materials such as paper is a thin network of fibres with uniform orientation and with centres distributed according to a planar Poisson point process (Dodson 1971; Deng and Dodson 1994; Sampson 2009a). In Dodson and Sampson (2014) Sect. 3 we outlined for such structures the analytic derivation of the spatial variance function for local averages $\tilde{c}$ of the coverage by fibres all of length λ and width ω, which tends to a Gaussian random variable. For sampling of the process using, say square inspection pixels of side length x, the variance of their density $\tilde{c}(x)$ is the expectation of the point autocorrelation function α

$$Var(\tilde{c}(x)) = Var(c(0)) \int_0^{\sqrt{2}x} \alpha(r, \omega, \lambda)\, b(r)\, dr \tag{4}$$

where b is the probability density function for the distance r between two points chosen independently and at random in the given type of pixel; it was derived by Ghosh (1951).

For practical variance computations we usually have the case of sampling using large square pixels of side mx say, which themselves consist of exactly m^2 small square pixels of side x. The variance $Var(\tilde{c}(m\,x))$ is related to $Var(\tilde{c}(x))$ through the covariance $Cov(x, mx)$ of x-pixels in mx-pixels (Dodson 1971):

$$Var(\tilde{c}(m\,x)) = \frac{1}{m^2} Var(\tilde{c}(x)) + \frac{m^2 - 1}{m^2} Cov(x, m\,x). \tag{5}$$

As $m \to \infty$, the small pixels tend towards points, $\frac{1}{m^2} Var(\tilde{c}(x)) \to 0$ so $Var(\tilde{c}(m\,x))$ admits interpretation as $Cov(0, m\,x)$, the covariance among points inside $m\,x$-pixels, the intra-pixel covariance, precisely $Var(\tilde{c}(m\,x))$ from Eq. (4).

Then by rearranging Eq. (5) the fractional between pixel variance for x-pixels is

$$\tilde{\rho}(x) = \frac{Cov(0, x)}{Var(c(0))} = \frac{Var(\tilde{c}(x))}{Var(c(0))} \tag{6}$$

which increases monotonically with fibre length λ and with fibre width ω but decreases monotonically with mx, see Deng and Dodson (1994) for more details. In fact, for a Poisson process of such rectangles the variance of coverage at points is precisely the mean coverage, $Var(c(0)) = \bar{c}$, so if we agree to measure coverage as a fraction of the mean coverage then Eq. (4) reduces to the integral

$$\frac{Var(\tilde{c}(x))}{\bar{c}} = \int_0^{\sqrt{2}x} \alpha(r, \omega, \lambda)\, b(r)\, dr = \tilde{\rho}(x). \tag{7}$$

The covariance among points inside $m\,x$-pixels, $Cov(0, m\,x)$, is the expectation of the covariance between pairs of points separated by distance r, taken over the possible values for r in an $m\,x$-pixel; that amounts to the integral in Eq. (4). By this means we have continuous families of 2×2 covariance matrices for $x \in \mathbb{R}^+$ and $2 < m \in \mathbb{Z}^+$ given by

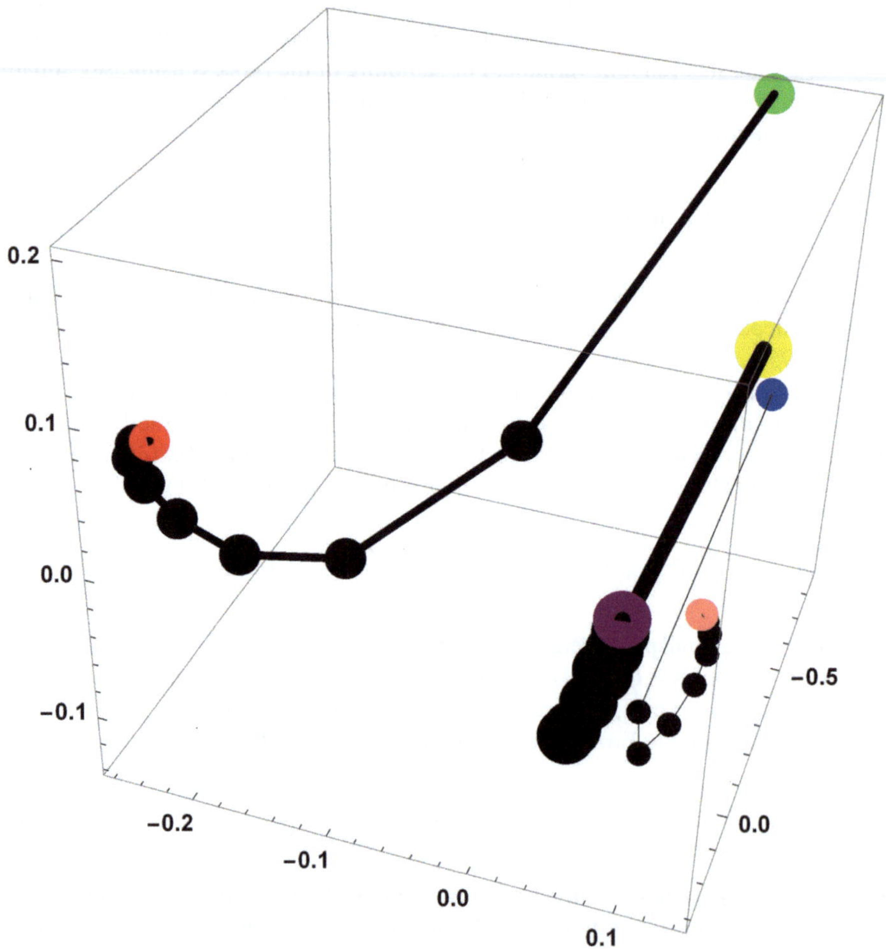

Fig. 8 Dimensionality reduction embedding for coverage autocovariances for planar Poisson processes of fibres of lengths $\lambda = 1.0, 1.5, 2.0$ with width $\omega = 0.1$, for sampling with square pixels of side length $x = 0.1, 0.2, \ldots, 1.0$. For the three increasing fibre lengths, the embeddings have respectively the endpoints, *blue* to *pink*, *green* to *red*, and *yellow* to *purple*, with *points* and *line* thicknesses in increasing size

$$\Sigma^{x,m} = \begin{pmatrix} \sigma_{11} & \sigma_{12} \\ \sigma_{12} & \sigma_{22} \end{pmatrix} - \begin{pmatrix} Var(\tilde{c}(x)) & Cov(x, m\,x) \\ Cov(x, m\,x) & Var(\tilde{c}(x)) \end{pmatrix}$$
$$= \begin{pmatrix} \tilde{\rho}(x) & \tilde{\rho}(m\,x) \\ \tilde{\rho}(m\,x) & \tilde{\rho}(x) \end{pmatrix}. \quad (8)$$

which encodes information about the spatial structure formed from the Poisson process of fibres, for each choice of fibre dimensions $\omega \le \lambda \in \mathbb{R}^+$.

The embedding generated by applying Eq. (2) to compute D_Σ from analytic autocovariance matrices from Eq. (8) for a planar Poisson process of fibres having width $\omega = 0.1$, lengths $\lambda = 0.5, 1.0, 1.5$, for square pixels of side length $x = 0.1, 0.2, \ldots, 1.0$ is shown in Fig. 8. We see that fibre length separates the three sets, and in each set the decreasing covariance with increasing pixel size separates the points.

References

Amari, S. (2016). Information geometry and its applications. *Applied mathematical sciences* (Vol. 194). Japan: Springer.

Arwini, K., & Dodson, C. T. J. (2008). *Information geometry near randomness and near independence*. Lecture notes in mathematics. New York: Springer.

Atkinson, C., & Mitchell, A. F. S. (1981). Rao's distance measure. *Sankhya: Indian Journal of Statistics*, *48A*(3), 345–365.

Carter, K.M. (2009). Dimensionality reduction on statistical manifolds. PhD thesis, University of Michigan. http://tbayes.eecs.umich.edu/kmcarter/thesis.

Carter, K. M., Raich, R., & Hero, A. O. (2007). Learning on statistical manifolds for clustering and visualization. In *45th Allerton Conference on Communication, Control, and Computing, Monticello, Illinois, 2007*. https://wiki.eecs.umich.edu/global/data/hero/images/c/c6/Kmcarter-learnstatman.pdf.

Deng, M., & Dodson, C. T. J. (1994). *Paper: An engineered stochastic structure*. Atlanta: Tappi Press.

Dodson, C. T. J. (1971). Spatial variability and the theory of sampling in random fibrous networks. *Journal of the Royal Statistical Society: Series B*, *33*(1), 88–94.

Dodson, C. T. J., & Sampson, W. W. (2013). Information geometry and dimensionality reduction for statistical structural features of paper. In S. J. I'Anson (Ed.), *Advances in Pulp and Paper Research, Cambridge, 2013. Transactions on XVth Fund. Res. Symposium* (pp. 55–69). Manchester: FRC.

Dodson, C. T. J., & Sampson, W. W. (2014). Dimensionality reduction for classification of stochastic texture images. In F. Nielsen (Ed.), *Geometric Theory of Information*. Signals and Communication Technology Series. Switzerland: Springer International Publishing.

Ghosh, B. (1951). Random distances within a rectangle and between two rectangles. *Calcutta Mathematical Society*, *43*(1), 17–24.

Mettänen, M., & Hirn, U. (2015). A comparison of five optical surface topography measurement methods. *Tappi Journal*, *14*(1), 27–37.

Sampson, W. W. (2009a). *Modelling stochastic fibre materials with Mathematica*. New York: Springer.

Sampson, W. W. (2009b). Materials properties of paper as influenced by its fibrous architecture. *International Materials Reviews*, *54*(3), 134–156.

On Clustering Financial Time Series: A Need for Distances Between Dependent Random Variables

Gautier Marti, Frank Nielsen, Philippe Donnat and Sébastien Andler

1 Clustering for Financial Risk Modelling

In financial applications, the variance-covariance matrix is an essential tool to assess the risk of a portfolio. Assuming that assets' returns are following a Gaussian multivariate distribution, the variance-covariance matrix captures both their joint behaviour (in this case, their Pearson correlation) and the specific risk of each asset which corresponds to its returns' standard deviation (also named volatility in finance). However, using an empirical variance-covariance matrix suffers from at least two shortcomings:

(i) if the assets' returns are following another multivariate distribution, then the variance-covariance matrix only measures a mixed information of linear dependence perturbed by the (possibly heavy-tailed) marginals. In this case, the variance-covariance matrix is not a relevant tool to quantify the risk between financial assets from their past returns time series;

(ii) estimating the empirical variance-covariance matrix from data is a problem in itself (Laloux et al. 2000). For N assets, one has to estimate $N(N-1)/2$ coefficients from N time series of length T. If T is small compared to N, the coefficients will be noisy and the matrix to some extent random.

Shortcoming (ii) has been adressed in the literature by several approaches. One of them leverages results from the Random Matrix Theory (RMT) and can be found under the terms "noise dressing" in the econophysics literature (Laloux et al. 1999,

G. Marti (✉) · P. Donnat
Hellebore Capital Limited, Michelin House, 81 Fulham Road, London SW3 6RD, UK
e-mail: gautier.marti@polytechnique.edu

G. Marti · F. Nielsen
Ecole Polytechnique, LIX - UMR 7161, Bâtiment Alan Turing, 1 rue Honoré d'Estienne d'Orves, 91120 Palaiseau, France

S. Andler
Ecole Normale Supérieure de Lyon, 46 Allée d'Italie, 69007 Lyon, France

F. Nielsen et al. (eds.), *Computational Information Geometry*,
Signals and Communication Technology, DOI 10.1007/978-3-319-47058-0_8

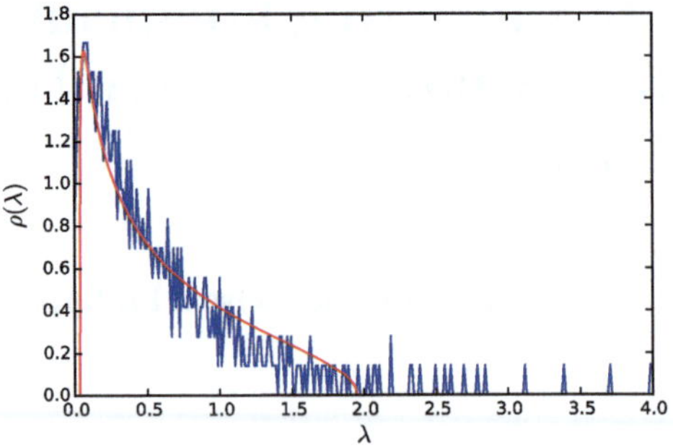

Fig. 1 Theoretical eigenvalues density for a purely random correlation matrix (*red*) versus empirical density (*blue*) of the correlation matrix eigenvalues

2000; Plerou et al. 2002; Potters et al. 2005; Allez et al. 2014; Bun et al. 2015). For example, authors in Laloux et al. (1999) compare the distribution of the empirical correlation eigenvalues to the known theoretical distribution given by RMT, and find that 94 % of the total number of eigenvalues falls in the support of the theoretical distribution. This experiment was led on stock market data, more precisely using $N = 406$ assets of the S&P500 during the years 1991–1996. We can observe that this stylized fact about correlation between stocks also applies to different markets and different periods. For example, we illustrate this empirical property on the credit default swaps (CDS) market. Let X be the matrix storing the standardized daily returns of $N = 560$ credit default swaps (5-year maturity) during the years 2006–2015 ($T \approx 2500$ values for each time series). Then, the empirical correlation matrix of the returns is $C = \frac{1}{T} X X^\top$. We can compute the empirical density of its eigenvalues $\rho(\lambda) = \frac{1}{N}\frac{dn(\lambda)}{d\lambda}$, where $n(\lambda)$ counts the number of eigenvalues of C less than λ. From random matrix theory, the limit distribution as $N \to \infty$, $T \to \infty$ and T/N fixed reads:

$$\rho(\lambda) = \frac{T/N}{2\pi} \frac{\sqrt{(\lambda_{\max} - \lambda)(\lambda - \lambda_{\min})}}{\lambda}, \tag{1}$$

where $\lambda_{\min}^{\max} = 1 + N/T \pm 2\sqrt{N/T}$, and $\lambda \in [\lambda_{\min}, \lambda_{\max}]$. We can observe in Fig. 1 that the theoretical distribution fits well the empirical one meaning that most of the information contained in the empirical correlation matrix can be considered noise. Only 26 eigenvalues are greater than $\lambda_{\max}$, i.e. 95 % of the total number of eigenvalues falls in the support of the theoretical distribution.

These results are important to take into account: for example, they have "interesting potential applications to risk management and portfolio optimisation. It is clear [...] that Markowitz's portfolio optimisation scheme based on a purely historical determination of the correlation matrix is not adequate, since its lowest eigenvalues (corresponding to the smallest risk portfolios) are dominated by noise" (Laloux et al. 2000). It motivates the need for filtering procedures of correlation matrices. Besides the RMT approach, several other methods have been proposed and compared (Tumminello et al. 2007; Pantaleo et al. 2011). From these papers it stems that hierarchical clustering yields better results (Tola et al. 2008) then other estimators such as shrinkage or RMT-based estimators for correlation matrices of financial time

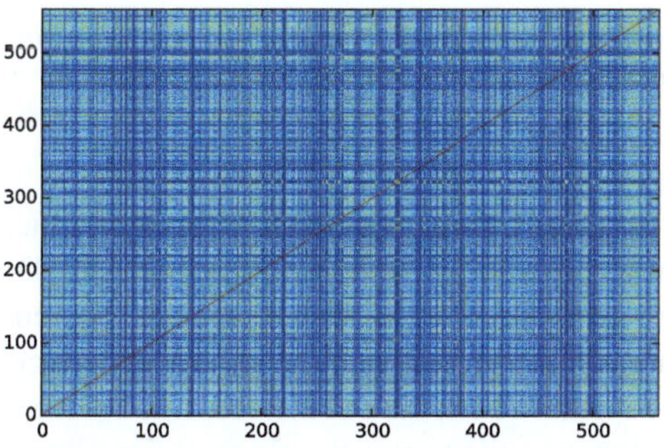

Fig. 2 An empirical and noisy correlation matrix computed on the log-returns of $N = 560$ credit default swap time series of length $T \approx 2500$

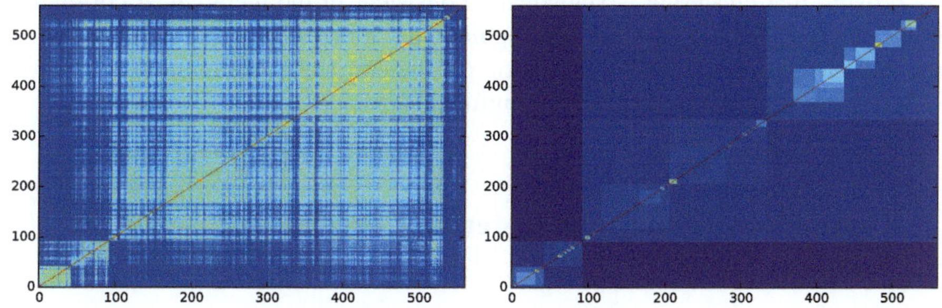

Fig. 3 The same noisy correlation matrix re-ordered by a hierarchical clustering algorithm; one can notice its noisy hierarchical correlation structure (*left*); The filtered correlation matrix resulting from the method described in Mantegna and Stanley (1999) (*right*)

series. The hierarchical clustering filtering procedure first described in Mantegna and Stanley (1999) is illustrated in Figs. 2 and 3. In Fig. 2, we display the empirical correlation matrix as estimated on our CDS dataset of $N = 560$ time series of length $T \approx 2500$. Then, we run a hierarchical clustering algorithm (such as average linkage for example) which gives a re-ordering of the time series, and thus a seriation of the correlation matrix. The re-ordered correlation matrix is displayed in Fig. 3 (left). We can now notice its noisy hierarchical correlation structure. According to the hierarchical clustering computed, we can finally filter the correlation coefficients to obtain the correlation matrix displayed in Fig. 3 (right).

Mantegna (1999) and many following papers insist on the hierarchical correlation pattern present in financial time series. This intrinsic structure may be an explanation to the efficiency of the hierarchical clustering filtering procedure. Taking into account other known empirical properties of daily asset returns in liquid financial markets which are well documented in Cont (2001), we do not consider vector autoregression (VAR) modelling and the frequency domain approaches:

> Mandelbrot expressed this property by stating that arbitrage tends to whiten the spectrum of price changes. This property implies that traditional tools of signal processing which are based on second-order properties, in the time domain - autocovariance analysis, ARMA

modelling - or in the spectral domain - Fourier analysis, linear filtering - cannot distinguish between asset returns and white noise. This points out the need for nonlinear measures of dependence in order to characterize the dependence properties of asset returns. *Excerpt from* Cont (2001)

Now, assuming that data follow this underlying hierarchical correlation model, we may wonder if these clustering procedures are consistent. Do they always recover the underlying model provided that the time series are long enough? If yes, another interesting point for the practitioner is knowing the convergence rate. How much data is enough for the result to be reliable? Indeed, since these time series may not be stationary, the practitioner wants to use the shortest time interval possible provided that the results are still relevant. In the following section, we justify the validity of the clustering approach for the analysis of correlation between financial time series by proving that clustering is statistically consistent in the hierarchical correlation block model. We also provide some guidelines to select a good combination of the clustering algorithm, the correlation coefficient, and the minimum number of observations required to obtain meaningful clusters.

2 On the Consistency of Clustering Correlated Random Variables

We show that clustering correlated random variables from their observations is statistically consistent. More precisely, when the underlying clusters of correlated random variables satisfy a strong enough separation condition and when there are enough observations, we prove that many of the celebrated clustering algorithms recover these cluster structures with high probability. We corroborate our theoretical results with an empirical study of the convergence rates.

Clustering consistency has been widely studied, starting from Hartigan's proof of Single Linkage (Hartigan 1981) and Pollard's proof of k-means consistency (Pollard et al. 1981) to recent work such as the consistency of spectral clustering Von Luxburg et al. (2008), or modified k-means (Terada 2013, 2014). However, these papers assume that N data points are independently sampled from an underlying probability distribution in dimension T fixed. They show that in the large sample limit, $N \to \infty$, the clustering structures constructed by the given algorithm converge to a clustering of the whole underlying space. Much less work has been done to prove consistency of clustering in the *Time Series Asymptotics*, i.e. $(N \to \infty, T \to \infty, T/N \to \infty)$ and (N fixed, $T \to \infty$). We should mention (Borysov et al. 2014) which shows the asymptotic behavior of three hierarchical clustering algorithms, namely Single, Average and Ward Linkage, and their consistency on the task of clustering $N = n + m$ observations from a mixture of two T dimensional Gaussian distributions $\mathcal{N}(\mu_1, \sigma_1^2 I_T)$ and $\mathcal{N}(\mu_2, \sigma_2^2 I_T)$ and Ryabko (2010a), Khaleghi et al. (2012) who prove the consistency of k-means for clustering processes according only to their distribution. In this work, we show the consistency of clustering N random variables from their T observations according to their observed correlations. The consistency results presented hold for several well-known clustering algorithms, and unlike (Borysov et al. 2014), we do not assume Gaussian distribution for the random variables, but data assumptions are adjusted to the natural scope of the correlation coefficients (e.g. Gaussian for Pearson correlation, elliptical copula for Kendall tau rank correlation).

Notations

- $X_1, \ldots, X_N$ univariate random variables
- X_i^t is the tth observation of variable X_i
- $X_i^{(t)}$ is the tth sorted observation of X_i
- F_X is the cumulative distribution function of X
- $\rho_{ij} = \rho(X_i, X_j)$ correlation between X_i, X_j
- $d_{ij} = d(X_i, X_j)$ distance between X_i, X_j
- $D_{ij} = D(C_i, C_j)$ distance between clusters C_i, C_j
- $P_k = \{C_1^{(k)}, \ldots, C_{l_k}^{(k)}\}$ is a partition of $X_1, \ldots, X_N$
- $C^{(k)}(X_i)$ denotes the cluster of X_i in partition P_k
- $\|\Sigma\|_\infty = \max_{ij} \Sigma_{ij}$
- $X = O_p(k)$ means X/k is stochastically bounded, i.e. $\forall \varepsilon > 0, \exists M > 0, P(|X/k| > M) < \varepsilon$.

2.1 Correlations

The most common correlation coefficient is the Pearson correlation coefficient defined by

$$\rho(X, Y) = \frac{\mathbb{E}[XY] - \mathbb{E}[X]\mathbb{E}[Y]}{\sqrt{\mathbb{E}[X^2] - \mathbb{E}[X]^2}\sqrt{\mathbb{E}[Y^2] - \mathbb{E}[Y]^2}} \quad (2)$$

which can be estimated by

$$\hat{\rho}(X, Y) = \frac{\sum_{t=1}^{T}(X^t - \overline{X})(Y^t - \overline{Y})}{\sqrt{\sum_{t=1}^{T}\left(X^t - \overline{X}\right)^2}\sqrt{\sum_{t=1}^{T}\left(Y^t - \overline{Y}\right)^2}} \quad (3)$$

where $\overline{X} = \frac{1}{T}\sum_{t=1}^{T} X^t$ is the empirical mean of X. This coefficient suffers from several drawbacks: it only measures linear relationship between two variables; it is not robust to noise and may be undefined if the distribution of one of these variables have infinite second moment. More robust correlation coefficients are copula-based dependence measures such as Kendall's tau

$$\tau(X,Y) = 4\int_0^1\int_0^1 C(u,v)dC(u,v) - 1 \tag{4}$$

$$= \mathbb{E}\left[\text{sign}\left((X-\tilde{X})(Y-\tilde{Y})\right)\right] \tag{5}$$

where $\tilde{X}$ is an independent copy of X, C is a copula, and its statistical estimate

$$\hat{\tau}(X,Y) = \frac{\sum_{1\le i<j\le T} \text{sign}\left(\left(X^i - X^j\right)\left(Y^i - Y^j\right)\right)}{\binom{T}{2}} \tag{6}$$

and Spearman's rho

$$\rho_S(X,Y) = 12\int_0^1\int_0^1 C(u,v)dudv - 3 \tag{7}$$

$$= 12\,\mathbb{E}\left[F_X(X), F_Y(Y)\right] - 3 \tag{8}$$

$$= \rho\left(F_X(X), F_Y(Y)\right) \tag{9}$$

and its statistical estimate

$$\hat{\rho}_S(X,Y) = 1 - \frac{6}{T(T^2-1)}\sum_{t=1}^{T}\left(X^{(t)} - Y^{(t)}\right)^2. \tag{10}$$

These correlation coefficients are robust to noise (since rank statistics normalize outliers) and invariant to monotonous transformations of the random variables (since copula-based measures benefit from the probability integral transform $F_X(X) \sim \mathcal{U}[0,1]$).

2.2 *Clustering of Correlations: The Hierarchical Correlation Block Model*

We assume that the N univariate random variables $X_1, \ldots, X_N$ follow a Hierarchical Correlation Block Model (HCBM). This model consists in correlation matrices having a hierarchical block structure (Balakrishnan et al. 2011; Krishnamurthy et al. 2012). Each block corresponds to a correlation cluster that we want to recover with a clustering algorithm. In Fig. 4, we display a correlation matrix from the HCBM. Notice that in practice one does not observe the hierarchical block diagonal structure displayed in the left picture, but a correlation matrix similar to the one displayed in the right picture which is identical to the left one up to a permutation of the data. The HCBM defines a set of nested partitions $\mathcal{P} = \{P_0 \supseteq P_1 \supseteq \ldots \supseteq P_h\}$ for some $h \in [1, N]$, where P_0 is the trivial partition, the partitions $P_k = \{C_1^{(k)}, \ldots, C_{l_k}^{(k)}\}$, and $\sqcup_{i=1}^{l_k} C_i^{(k)} = \{X_1, \ldots, X_N\}$. For all $1 \le k \le h$,

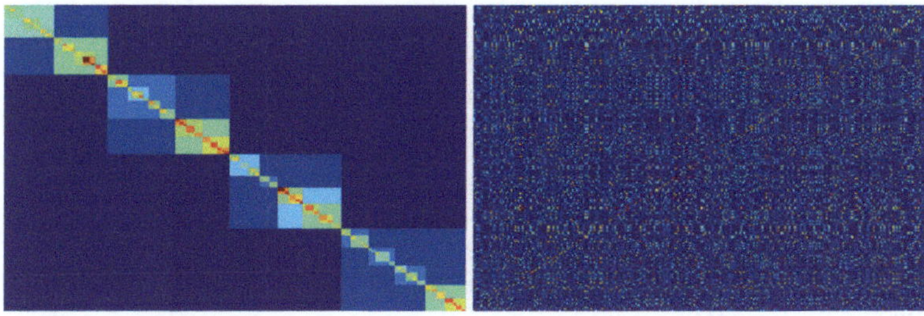

Fig. 4 (*left*) hierarchical correlation block model; (*right*) observed correlation matrix (following the HCBM) identical to the *left one up* to a permutation of the data

we define $\underline{\rho}_k$ and $\overline{\rho}_k$ such that for all $1 \leq i, j \leq N$, we have $\underline{\rho}_k \leq \rho_{ij} \leq \overline{\rho}_k$ when $C^{(k)}(X_i) = C^{(k)}(X_j)$ and $C^{(k+1)}(X_i) \neq C^{(k+1)}(X_j)$, i.e. $\underline{\rho}_k$ and $\overline{\rho}_k$ are the minimum and maximum correlation respectively within all the clusters $C_i^{(k)}$ in the partition P_k at depth k. In order to have a proper nested correlation hierarchy, we must have $\overline{\rho}_k < \underline{\rho}_{k+1}$ for all k.

Without loss of generality and for ease of demonstration we will consider the one-level HCBM with K blocks of size $n_1, \ldots, n_K$ such that $\sum_{i=1}^{K} n_i = N$. We explain later how to extend the results to the general HCBM. Since clustering methods usually require a distance matrix as input, we also consider the corresponding distance matrix with coefficients $d_{ij} = \frac{1-\rho_{ij}}{2}$, where $0 < \rho_{ij} < 1$ is a correlation coefficient (Pearson, Spearman, Kendall).

2.3 Clustering Methods

Many paradigms exist in the literature for clustering data. We consider in this work only hard (in opposition to soft) clustering methods, i.e. algorithms producing partitions of the data (in opposition to methods assigning several clusters to a given data point). Within the hard clustering family, we can classify for instance these algorithms in hierarchical clustering methods (yielding nested partitions of the data) and flat clustering methods (yielding a single partition) such as k-means.

We will consider the infinite Lance-Williams family which further subdivides the hierarchical clustering since many of the popular algorithms such as Single Linkage, Complete Linkage, Average Linkage (UPGMA), McQuitty's Linkage (WPGMA), Median Linkage (WPGMC), Centroid Linkage (UPGMC), and Ward's method are members of this family (cf. Table 1). It will allow us a more concise and unified treatment of the consistency proofs for these algorithms. Interesting and recently designed hierarchical agglomerative clustering algorithms such as Hausdorff Linkage (Basalto et al. 2007) and Minimax Linkage (Ao et al. 2005) do not belong to this family (Bien and Tibshirani 2011), but their linkage functions share a convenient property for cluster separability.

Table 1 Many well-known hierarchical agglomerative clustering algorithms are members of the Lance-Williams family, i.e. the distance between clusters can be written: $D(C_i \cup C_j, C_k) = \alpha_i D_{ik} + \alpha_j D_{jk} + \beta D_{ij} + \gamma |D_{ik} - D_{jk}|$ (Murtagh and Contreras 2011)

	α_i	β	γ
Single	1/2	0	−1/2
Complete	1/2	0	1/2
Average	$\frac{\lvert C_i\rvert}{\lvert C_i\rvert+\lvert C_j\rvert}$	0	0
McQuitty	1/2	0	0
Median	1/2	−1/4	0
Centroid	$\frac{\lvert C_i\rvert}{\lvert C_i\rvert+\lvert C_j\rvert}$	$-\frac{\lvert C_i\rvert\lvert C_j\rvert}{(\lvert C_i\rvert+\lvert C_j\rvert)^2}$	0
Ward	$\frac{\lvert C_i\rvert+\lvert C_k\rvert}{\lvert C_i\rvert+\lvert C_j\rvert+\lvert C_k\rvert}$	$-\frac{\lvert C_k\rvert}{C_i\rvert+\lvert C_j\rvert+\lvert C_k\rvert}$	0

2.4 Separability Conditions for Clustering

In our context the distances between the points we want to cluster are random and defined by the estimated correlations. However by definition of the HCBM, each point X_i belongs to exactly one cluster $C^{(k)}(X_i)$ at a given depth k, and we want to know under which condition on the distance matrix we will find the correct clusters defined by P_k. We call these conditions the separability conditions. A separability condition for the points $X_1, \ldots, X_N$ is a condition on the distance matrix of these points such that if we apply a clustering procedure whose input is the distance matrix, then the algorithm yields the correct clustering $P_k = \{C_1^{(k)}, \ldots, C_{l_k}^{(k)}\}$, for all k. For example, for $\{X_1, X_2, X_3\}$ if we have $C(X_1) = C(X_2) \neq C(X_3)$ in the one-level two-block HCBM, then a separability condition is $d_{1,2} < d_{1,3}$ and $d_{1,2} < d_{2,3}$.

Separability conditions are deterministic and depend on the algorithm used for clustering. They are generic in the sense that for any sets of points that satisfy the condition the algorithm will separate them in the correct clusters. In the Lance-Williams algorithm framework (Chen and Van Ness 1996), they are closely related to "space conserving" properties of the algorithm and in particular on the way the distances between clusters change during the clustering process.

In Chen and Van Ness (1996), the authors define what they call a semi-space-conserving algorithm.

Semi-space-conserving algorithms (Chen and Van Ness 1996)
An algorithm is semi-space-conserving if for all clusters C_i, C_j, and C_k,

$$D(C_i \cup C_j, C_k) \in \left[\min(D_{ik}, D_{jk}), \max(D_{ik}, D_{jk})\right]$$

Among the Lance-Williams algorithms we study here, Single, Complete, Average and McQuitty algorithms are semi-space-conserving. Although Chen and Van Ness only considered Lance-Williams algorithms the definition of a space conserving algorithm is useful for any agglomerative hierarchical algorithm. An alternative formulation of the semi-space-conserving property is:

Space-conserving algorithms. A linkage agglomerative hierarchical algorithm is space-conserving if $D_{ij} \in \left[\min_{x \in C_i, y \in C_j} d(x, y), \max_{x \in C_i, y \in C_j} d(x, y)\right]$.

Such an algorithm does not "distort" the space when points are clustered which makes the sufficient separability condition easier to get. For these algorithms the separability condition does not depend on the size of the clusters.

The following two propositions are easy to verify.

Proposition. The semi-space-conserving Lance-Williams algorithms are space-conserving.

Proposition. Minimax linkage and Hausdorff linkage are space-conserving.

For space-conserving algorithms we can now state a sufficient separability condition on the distance matrix.

Proposition. The following condition is a separability condition for space-conserving algorithms:

$$\max_{\substack{1 \leq i,j \leq N \\ C(i)=C(j)}} d(X_i, X_j) < \min_{\substack{1 \leq i,j \leq N \\ C(i)\neq C(j)}} d(X_i, X_j) \tag{S1}$$

The maximum distance is taken over any two points in a same cluster (intra) and the minimum over any two points in different clusters (inter).

Proof Consider the set $\{d_{ij}^s\}$ of distances between clusters after s steps of the clustering algorithm (therefore $\{d_{ij}^0\}$ is the initial set of distances between the points). Denote $\{d_{inter}^s\}$ (resp. $\{d_{intra}^s\}$) the sets of distances between subclusters belonging to different clusters (resp. the same cluster) at step s. If the separability condition is satisfied then we have the following inequalities:

$$\min d_{intra}^0 \leq \max d_{intra}^0 < \min d_{inter}^0 \leq \max d_{inter}^0 \tag{S2}$$

Then the separability condition implies that the separability condition S2 is verified for all step s because after each step the updated intra distances are in the convex hull of the intra distances of the previous step and the same is true for the inter distances. Moreover since S2 is verified after each step, the algorithm never links points from different clusters and the proposition entails. □

2.5 Concentration Bounds for the Correlation Matrix

We have determined configurations of points such that the clustering algorithm will find the right partition. The proof of the consistency now relies on showing that these configurations are likely. In fact the probability that our points fall in these configurations goes to 1 as $T \to \infty$.

The precise definition of what we mean by consistency of an algorithm is the following:

Consistency of a clustering algorithm. Let $(X_1^t, \ldots, X_N^t)$, $t = 1, \ldots, T$, be N univariate random variables observed T times. A clustering algorithm $\mathcal{A}$ is consistent with respect to the Hierarchical Correlation Block Model (HCBM) defining a set of nested partitions $\mathcal{P}$ if the probability that the algorithm $\mathcal{A}$ recovers all the partitions in $\mathcal{P}$ converges to 1 when $T \to \infty$.

We now get explicit lower bounds on the probability of finding the right clusters with the clustering algorithms using concentration bounds on the empirical correlation matrix.

As we have seen in the previous section the correct clustering can be ensured if the estimated correlation matrix verifies some separability condition. This condition can be guaranteed by requiring the error on each entry of the matrix $\hat{R}_T$ to be smaller than the contrast, i.e. $\frac{\underline{\rho}_1 - \overline{\rho}_0}{2}$, on the theoretical matrix R. In general the error on the matrix $\hat{R}_T$ is of the order $\|R - \hat{R}_T\|_\infty = O_P\left(\sqrt{\frac{\log N}{T}}\right)$ and thus, if $T \gg \log(N)$ then the clustering will find the correct partition.

Results and proofs are the object of an upcoming publication. Below we only give the results for the Kendall's tau coefficient, but Spearman's bound is similar.

Concentration bound on the Kendall's tau correlation matrix *U*

Let X^t, $t = 1, \ldots, T$, be T independent realizations of a N-dimensional distribution having elliptical copula and any margins. We have

$$\mathbb{P}\left(\|\hat{U}_T - U\|_\infty \leq \epsilon\right) \geq 1 - 2N^2 e^{-\frac{T}{8}\epsilon^2}. \tag{11}$$

The lower bound on the probability of success now follows by requiring that the error on the estimated correlation matrix is small enough. Moreover ρ is taken to be a generic correlation and Σ the corresponding generic correlation matrix.

Space-conserving algorithms

The separability condition is satisfied if $\|\Sigma - \hat{\Sigma}\|_\infty < \frac{\underline{\rho}_1 - \overline{\rho}_0}{2}$. Therefore with probability at least

$$1 - 2N^2 e^{-\frac{T\left(\underline{\rho}_1 - \overline{\rho}_0\right)^2}{32}} \tag{12}$$

for Kendall correlation, the algorithm finds the correct partition.

Therefore we obtain consistency for the presented algorithms with respect to the one-level HCBM.

2.6 From the One-Level to the General HCBM

To go from the one-level HCBM to the general case we need to get a separability condition on the nested partition model. For space-conserving algorithms, this is done by requiring the corresponding separability condition for each level of the hierarchy.

For all $1 \leq k \leq h$, we define $\underline{d}_k$ and $\overline{d}_k$ such that for all $1 \leq i, j \leq N$, we have $\underline{d}_k \leq d_{ij} \leq \overline{d}_k$ when $C^{(k)}(X_i) = C^{(k)}(X_j)$ and $C^{(k+1)}(X_i) \neq C^{(k+1)}(X_j)$. Notice that $\underline{d}_k = (1 - \overline{\rho}_k)/2$ and $\overline{d}_k = (1 - \underline{\rho}_k)/2$.

Separability condition for space-conserving algorithms in the case of nested partitions. The separability condition reads:

$$\overline{d}_h < \underline{d}_{h-1} < \ldots < \overline{d}_{k+1} < \underline{d}_k < \ldots < \underline{d}_1.$$

This condition can be guaranteed by requiring the error on each entry of the matrix $\hat{\Sigma}$ to be smaller than the lowest contrast. Therefore the maximum error we can have for space-conserving algorithms on the correlation matrix is

$$\|\Sigma - \hat{\Sigma}\|_\infty < \min_k \frac{\underline{\rho}_{k+1} - \overline{\rho}_k}{2}.$$

We finally obtain consistency for the presented algorithms with respect to the HCBM from the previous concentration bounds.

2.7 *Empirical Rates of Convergence*

Researchers have used from 30 days to several years of daily returns as source data for clustering financial time series based on their correlations. How long should the time series be? If too short, the clusters found can be spurious; if too long, dynamics can be smoothed out.

For illustration purpose, we consider the simple case where we have two correlation blocks C_1 and C_2. The correlation within the block C_1 is ρ and within the block C_2 is 2ρ and both blocks are independent. C_2 counts for 70 % of the N points. The underlying correlation matrix is thus of the form:

$$\begin{pmatrix}
1 & 2\rho & \cdots & 2\rho & 0 & \cdots & \cdots & \cdots & \cdots & 0 \\
2\rho & \ddots & \ddots & \vdots & \vdots & \ddots & \ddots & \ddots & \ddots & \vdots \\
\vdots & \ddots & \ddots & 2\rho & \vdots & \ddots & \ddots & \ddots & \ddots & \vdots \\
2\rho & \cdots & 2\rho & 1 & 0 & \cdots & \cdots & \cdots & \cdots & 0 \\
0 & \cdots & \cdots & 0 & 1 & \rho & \cdots & \cdots & \cdots & \rho \\
\vdots & \ddots & \ddots & \vdots & \rho & \ddots & \ddots & \ddots & \ddots & \vdots \\
\vdots & \ddots & \ddots & \vdots & \vdots & \ddots & \ddots & \ddots & \ddots & \vdots \\
\vdots & \ddots & \ddots & \vdots & \vdots & \ddots & \ddots & \ddots & \ddots & \vdots \\
\vdots & \ddots & \ddots & \vdots & \vdots & \ddots & \ddots & \ddots & \ddots & \rho \\
0 & \cdots & \cdots & 0 & \rho & \cdots & \cdots & \cdots & \rho & 1
\end{pmatrix}$$

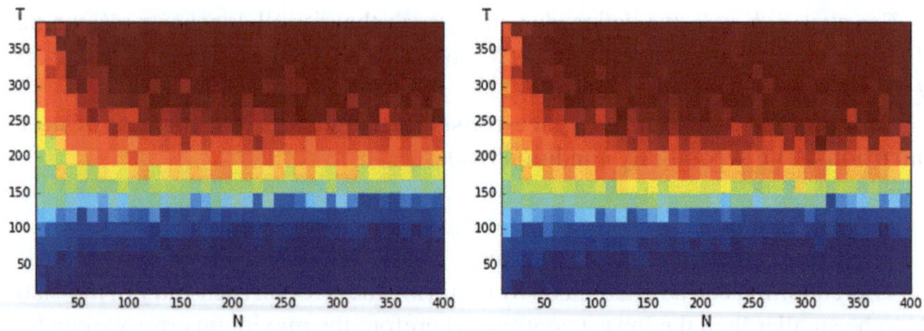

Fig. 5 Single Linkage applied on (*left*) Spearman dissimilarity, (*right*) Pearson dissimilarity; the x axis is $N = 10 \dots 400$, the y axis is $T = 10 \dots 390$

We then simulate Gaussian and Student (with $\nu = 3$ degrees of freedom, i.e. heavy-tailed) random vectors, create the different correlation matrices and cluster with these matrices using the Ward, Single, Complete and Average Linkage algorithms. We then count the number of success of these clustering procedures, i.e. finding the correct partition, over 100 trials. This experiment has been done for the two sets of parameters (N, T) and (ρ, T). We produce the heat maps (relative to the number of successes) for these different experiments.

(N, T) experiments. In this first experiment ρ is fixed at 0.1 and we do the clustering procedure for different values of N and T.

As one can see in Fig. 5, there is a "transition" area between zones with probability almost 1 and almost 0 of finding the right clusters. The absolute level of this transition zone depends on the clustering algorithm. What we can see in these examples is that the dependence in T is much quicker than in N and that in fact in our sample for $N > 100$ there is little dependence in N.

For moderately sized group of points, typically $100 \leq N \leq 400$, we can deduce that for $T \geq 250$ all of the clustering algorithms find the correct partition in the HCBM model with very high probability (cf. Table 2).

(ρ, T) experiments. For the (ρ, T) experiments, we made two different sets of experiments both with the Spearman correlation matrix and the Pearson correlation matrix. One with Gaussian random variables and the other with multivariate Student variables (with $\nu = 3$ degrees of freedom) which exhibit fatter tails.

Table 2 Number of success out of 100 trials for $T = 250$ and $N = 400$

	Single	Average	Complete
Pearson	98	98	99
Spearman	95	99	100

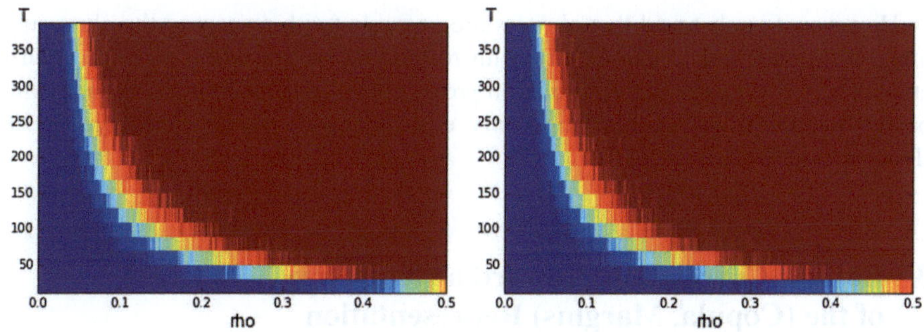

Fig. 6 Gaussian case for Spearman (*left*) and Pearson (*right*) and for the average Linkage. The x axis is $\rho = 0 \ldots 0.5$ and the y axis is $T = 10 \ldots 390$

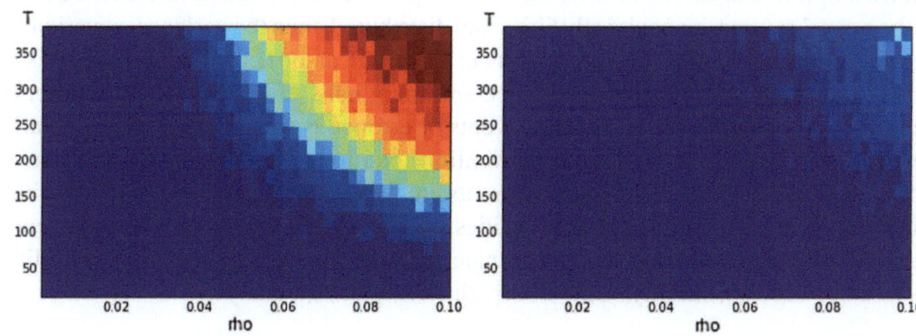

Fig. 7 Student case for Spearman (*left*) and Pearson (*right*) and for the average Linkage. The x axis is $\rho = 0 \ldots 0.1$ and the y axis is $T = 10 \ldots 390$

As expected with the Student distribution, the Pearson correlation coefficient is not robust to fatter tails and the clustering rate of success is much lower than in the Gaussian case (Fig. 6) as it can be seen in Fig. 7.

Concretely, our results suggest that for properly clustering $N \simeq 400$ correlated financial time series, the practitioner should need $T \geq 250$, i.e. at least a year of daily prices. We also advise to measure correlation with the Kendall coefficient since

- more generic: Kendall can be used with any elliptical copula and any margins,
- unbiased (unlike Spearman),
- faster convergence rate (than Spearman corrected from the bias),
- can be computed efficiently in $O(T \log T)$ versus $O(T \log T)$ for Spearman and $O(T)$ for Pearson.

We notice that there are isoquants of clustering accuracy for many sets of parameters, e.g. (N, T), (ρ, T). Such isoquants are displayed in Fig. 6. Further work may aim at characterizing these curves. We can also observe in Fig. 6 that for $\rho \leq 0.08$, the critical value for T explodes. It would be interesting to determine this asymptotics as ρ tends to 0.

However it is observed that clusters are unstable (with respect to the clustering method (Lemieux et al. 2014), and with respect to the clustering distance (Marti et al. 2015). It suggests that information present in the financial time series may not be summarized by cross-correlation only, even under the random walk hypothesis (Donnat et al. 2016).

3 Beyond Correlation: Toward a Geometry of the (Copula, Margins) Representation

In this section, we provide avenues for tackling shortcoming (i) when clustering, i.e. the assumption that assets' returns are following a Gaussian multivariate distribution. If the assets' returns are not jointly Gaussian distributed, then the variance-covariance matrix does not capture their dependence: linear (Pearson) correlation measures a mixed information of linear dependence and marginals' effect on it. Few 'outliers' returns of some assets due to specific events or erroneous values in the data (i.e. tail-realizations from an heavy-tailed distribution) can lower drastically the measured correlation making one to believe that assets are weakly correlated and that investing in them is a diversified investment. Besides, even if several assets are perfectly 'correlated', one may still want to discriminate between assets that have high volatility from those of low volatility while doing clustering or risk analysis.

3.1 A First Approach with N Univariate Time Series

A naive but often used distance between random variables to measure similarity and to perform clustering is the L_2 distance $\mathbb{E}[(X-Y)^2]$. Yet, this distance is not suited to our task.

Example 1 (*Distance L_2 between two Gaussians*) Let (X, Y) be a bivariate Gaussian vector, with $X \sim \mathcal{N}(\mu_X, \sigma_X^2)$, $Y \sim \mathcal{N}(\mu_Y, \sigma_Y^2)$ and whose correlation is $\rho(X, Y) \in [-1, 1]$. We obtain $\mathbb{E}[(X-Y)^2] = (\mu_X - \mu_Y)^2 + (\sigma_X - \sigma_Y)^2 + 2\sigma_X\sigma_Y (1 - \rho(X, Y))$. Now, consider the following values for correlation:

- $\rho(X, Y) = 0$, so $\mathbb{E}[(X-Y)^2] = (\mu_X - \mu_Y)^2 + \sigma_X^2 + \sigma_Y^2$. The two variables are independent (since uncorrelated and jointly normally distributed), thus we must discriminate on distribution information. Assume $\mu_X = \mu_Y$ and $\sigma_X = \sigma_Y$. For $\sigma_X = \sigma_Y \gg 1$, we obtain $\mathbb{E}[(X-Y)^2] \gg 1$ instead of the distance 0, expected from comparing two equal Gaussians.
- $\rho(X, Y) = 1$, so $\mathbb{E}[(X-Y)^2] = (\mu_X - \mu_Y)^2 + (\sigma_X - \sigma_Y)^2$. Since the variables are perfectly correlated, we must discriminate on distributions. We actually compare them with a L_2 metric on the mean $\times$ standard deviation half-plane. However, this is not an appropriate geometry for comparing two Gaussians Costa et al.

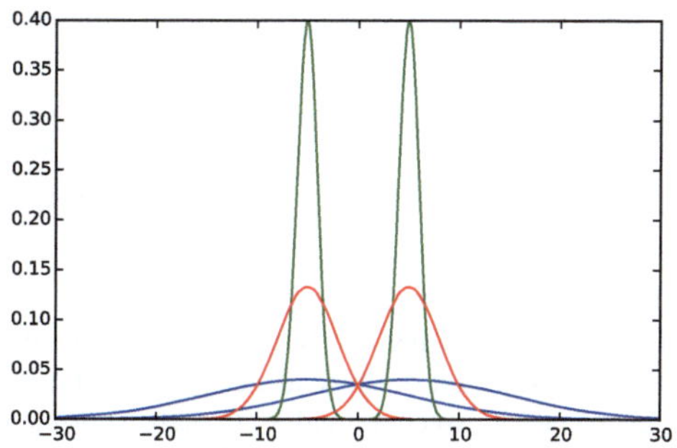

Fig. 8 Probability density functions of Gaussians $\mathcal{N}(-5, 1)$ and $\mathcal{N}(5, 1)$ (in *green*), Gaussians $\mathcal{N}(-5, 3)$ and $\mathcal{N}(5, 3)$ (in *red*), and Gaussians $\mathcal{N}(-5, 10)$ and $\mathcal{N}(5, 10)$ (in *blue*). *Green, red* and *blue* Gaussians are equidistant using L_2 geometry on the parameter space (μ, σ)

(2014). For instance, if $\sigma_X = \sigma_Y = \sigma$, we find $\mathbb{E}[(X - Y)^2] = (\mu_X - \mu_Y)^2$ for any values of σ. As σ grows, probability attached by the two Gaussians to a given interval grows similar (cf. Fig. 8), yet this increasing similarity is not taken into account by this L_2 distance.

$\mathbb{E}[(X - Y)^2]$ considers both dependence and distribution information of the random variables, but not in a relevant way with respect to our task. Our purpose is to introduce a new data representation and a suitable distance which takes into account both distributional proximities and joint behaviours.

Let $(\Omega, \mathcal{F}, \mathbb{P})$ be a probability space. Ω is the sample space, $\mathcal{F}$ is the σ-algebra of events, and $\mathbb{P}$ is the probability measure. Let $\mathcal{V}$ be the space of all continuous real-valued random variables defined on $(\Omega, \mathcal{F}, \mathbb{P})$. Let $\mathcal{U}$ be the space of random variables following a uniform distribution on $[0, 1]$ and $\mathcal{G}$ be the space of absolutely continuous cumulative distribution functions (cdf).

The copula transform Let $X = (X_1, \ldots, X_N) \in \mathcal{V}^N$ be a random vector with cdfs $G_X = (G_{X_1}, \ldots, G_{X_N}) \in \mathcal{G}^N$. The random vector $G_X(X) = (G_{X_1}(X_1), \ldots, G_{X_N}(X_N)) \in \mathcal{U}^N$ is known as the copula transform.

Uniform margins of the copula transform $G_{X_i}(X_i)$, $1 \leq i \leq N$, are uniformly distributed on $[0, 1]$.

Proof $x = G_{X_i}(G_{X_i}^{-1}(x)) = \mathbb{P}(X_i \leq G_{X_i}^{-1}(x)) = \mathbb{P}(G_{X_i}(X_i) \leq x)$.

We define the following representation of random vectors that actually splits the joint behaviours of the marginal variables from their distributional information.

Dependence $\oplus$ distribution space projection. Let $\mathcal{T}$ be a mapping which transforms $X = (X_1, \ldots, X_N)$ into its generic representation, an element of $\mathcal{U}^N \times \mathcal{G}^N$ representing X, defined as follow

$$\begin{aligned} \mathcal{T} : \mathcal{V}^N &\to \mathcal{U}^N \times \mathcal{G}^N \\ X &\mapsto (G_X(X), G_X). \end{aligned} \tag{13}$$

$\mathcal{T}$ is a bijection.

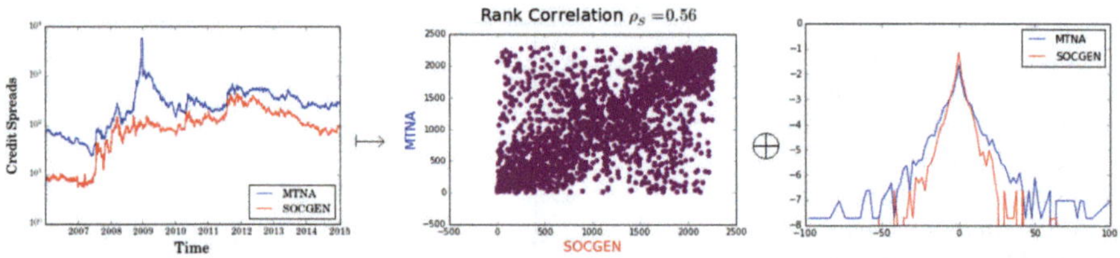

Fig. 9 ArcelorMittal and Société générale prices (T observations $(X_1^t, X_2^t)_{t=1}^T$ from $(X_1, X_2) \in \mathcal{V}^2$) are projected on dependence $\oplus$ distribution space; $(G_{X_1}(X_1), G_{X_2}(X_2)) \in \mathcal{U}^2$ encode the dependence between X_1 and X_2 (a perfect correlation would be represented by a sharp diagonal on the scatterplot); (G_{X_1}, G_{X_2}) are the margins (their log-densities are displayed above), notice their heavy-tailed exponential distribution (especially for ArcelorMittal)

Proof $\mathcal{T}$ is surjective as any element $(U, G) \in \mathcal{U}^N \times \mathcal{G}^N$ has the fiber $G^{-1}(U)$. $\mathcal{T}$ is injective as $(U_1, G_1) = (U_2, G_2)$ *a.s.* in $\mathcal{U}^N \times \mathcal{G}^N$ implies that they have the same cdf $G = G_1 = G_2$ and since $U_1 = U_2$ *a.s.*, it follows that $G^{-1}(U_1) = G^{-1}(U_2)$ *a.s.*

This result replicates the seminal result of copula theory, namely Sklar's theorem (Sklar 1959), which asserts one can split the dependency and distribution apart without losing any information. Figure 9 illustrates this projection for $N = 2$.

We leverage the propounded representation to build a suitable yet simple distance between random variables which is invariant under diffeomorphism.

Distance d_θ between two random variables Let $\theta \in [0, 1]$. Let $(X, Y) \in \mathcal{V}^2$. Let $G = (G_X, G_Y)$, where G_X and G_Y are respectively X and Y marginal cdfs. We define the following distance

$$d_\theta^2(X, Y) = \theta d_1^2(G_X(X), G_Y(Y)) + (1 - \theta) d_0^2(G_X, G_Y), \tag{14}$$

where

$$d_1^2(G_X(X), G_Y(Y)) = 3\mathbb{E}[|G_X(X) - G_Y(Y)|^2], \tag{15}$$

and

$$d_0^2(G_X, G_Y) = \frac{1}{2} \int_{\mathbf{R}} \left(\sqrt{\frac{dG_X}{d\lambda}} - \sqrt{\frac{dG_Y}{d\lambda}} \right)^2 \mathrm{d}\lambda. \tag{16}$$

In particular, $d_0 = \sqrt{1 - BC}$ is the Hellinger distance related to the Bhattacharyya (1/2-Chernoff) coefficient BC upper bounding the Bayes' classification error. To quantify distribution dissimilarity, d_0 is used rather than the more general α-Chernoff divergences since it satisfies the invariance to a monotonous transform of the variables (significant for practitioners as it ensures to be insensitive to scaling (e.g. choice of units) or measurement scheme (e.g. device, mathematical modelling) of the underlying phenomenon). In addition, d_θ can thus be efficiently implemented as a scalar product. $d_1 = \sqrt{(1-\rho_S)/2}$ is a distance correlation measuring statistical dependence between two random variables, where ρ_S is the Spearman's correlation between X and Y. Notice that d_1 can be expressed by using the copula $C : [0, 1]^2 \rightarrow [0, 1]$ implicitly defined by the relation $G(X, Y) = C(G_X(X), G_Y(Y))$ since $\rho_S(X, Y) = 12 \int_0^1 \int_0^1 C(u, v)\, du\, dv - 3$ (Fredricks and Nelsen 2007).

Example 2 (*Distance d_θ between two Gaussians*) Let (X, Y) be a bivariate Gaussian vector, with $X \sim \mathcal{N}(\mu_X, \sigma_X^2)$, $Y \sim \mathcal{N}(\mu_Y, \sigma_Y^2)$ and $\rho(X, Y) = \rho$. We obtain,

$$d_\theta^2(X, Y) = \theta \frac{1-\rho_S}{2} + (1-\theta)\left(1 - \sqrt{\frac{2\sigma_X\sigma_Y}{\sigma_X^2+\sigma_Y^2}} e^{-\frac{1}{4}\frac{(\mu_X-\mu_Y)^2}{\sigma_X^2+\sigma_Y^2}}\right).$$

Remember that for perfectly correlated Gaussians ($\rho = \rho_S = 1$), we want to discriminate on their distributions. We can observe that

- for $\sigma_X, \sigma_Y \rightarrow +\infty$, then $d_0(X, Y) \rightarrow 0$, it alleviates a main shortcoming of the basic L_2 distance which is diverging to $+\infty$ in this case;
- if $\mu_X \neq \mu_Y$, for $\sigma_X, \sigma_Y \rightarrow 0$, then $d_0(X, Y) \rightarrow 1$, its maximum value, i.e. it means that two Gaussians cannot be more remote from each other than two different Dirac delta functions.

This distance is a fast and good proxy for distance d_θ when the first two moments μ and σ predominate. Nonetheless, for datasets which contain heavy-tailed distributions, it fails to capture this information.

To apply the propounded distance d_θ on sampled data without parametric assumptions, we have to define its statistical estimate $\tilde{d}_\theta$ working on realizations of the i.i.d. random variables. Distance d_1 working with continuous uniform distributions can be approximated by normalized rank statistics yielding to discrete uniform distributions, in fact coordinates of the multivariate empirical copula (Deheuvels 1979) which is a non-parametric estimate converging uniformly toward the underlying copula (Deheuvels 1981). Distance d_0 working with densities can be approximated by using its discrete form working on histogram density estimates.

The empirical copula transform. Let $X^T = (X_1^t, \ldots, X_N^t)$, $t = 1, \ldots, T$, be T observations from a random vector $X = (X_1, \ldots, X_N)$ with continuous margins $G_X = (G_{X_1}(X_1), \ldots, G_{X_N}(X_N))$. Since one cannot directly obtain the corresponding copula observations $(G_{X_1}(X_1^t), \ldots, G_{X_N}(X_N^t))$ without knowing a priori G_X, one can instead estimate the N empirical margins $G_{X_i}^T(x) = \frac{1}{T}\sum_{t=1}^T \mathbf{1}(X_i^t \leq x)$ to obtain T empirical observations $(G_{X_1}^T(X_1^t), \ldots, G_{X_N}^T(X_N^t))$ which are thus related to normalized rank statistics as $G_{X_i}^T(X_i^t) = X_i^{(t)}/T$, where $X_i^{(t)}$ denotes the rank of observation X_i^t.

Empirical distance. Let $(X^t)_{t=1}^T$ and $(Y^t)_{t=1}^T$ be T realizations of real-valued random variables $X, Y \in \mathcal{V}$ respectively. An empirical distance between realizations of random variables can be defined by

$$\tilde{d}_\theta^2\left((X^t)_{t=1}^T, (Y^t)_{t=1}^T\right) \stackrel{a.s.}{=} \theta \tilde{d}_1^2 + (1-\theta)\tilde{d}_0^2, \tag{17}$$

where

$$\tilde{d}_1^2 = \frac{3}{T(T^2-1)} \sum_{t=1}^{T} \left(X^{(t)} - Y^{(t)}\right)^2 \tag{18}$$

and

$$\tilde{d}_0^2 = \frac{1}{2} \sum_{k=-\infty}^{+\infty} \left(\sqrt{g_X^h(hk)} - \sqrt{g_Y^h(hk)}\right)^2, \tag{19}$$

h being here a suitable bandwidth, and $g_X^h(x) = \frac{1}{T}\sum_{t=1}^{T} \mathbf{1}(\lfloor \frac{x}{h} \rfloor h \leq X^t < (\lfloor \frac{x}{h} \rfloor + 1)h)$ being a density histogram estimating pdf g_X from $(X^t)_{t=1}^T$, T realizations of random variable $X \in \mathcal{V}$.

To use effectively d_θ and its statistical estimate, it boils down to select a particular value for θ. We suggest here an exploratory approach where one can test (i) distribution information ($\theta = 0$), (ii) dependence information ($\theta = 1$), and (iii) a mix of both information ($\theta = 0.5$). Ideally, θ should reflect the balance of dependence and distribution information in the data. In a supervised setting, one could select an estimate $\hat{\theta}$ of the right balance $\theta^\star$ optimizing some loss function by techniques such as cross-validation. Yet, the lack of a clear loss function makes the estimation of $\theta^\star$ difficult in an unsupervised setting. For clustering, many authors Lange et al. (2004), Shamir and Tishby (2007), Shamir and Tishby (2008), Meinshausen and Bühlmann (2010) suggest stability as a tool for parameter selection.

3.2 How to Extend the Approach to N Multivariate Time Series?

We are now interested in clustering N assets which are described by more than one time series. Though a stock is usually described by a single time series, its market price, other assets such as credit default swaps can be described by several maturities, their term structure. In practice, a CDS term structure time series is a 5-variate time series. At each time t, it consists in $d = 5$ prices for the different traded maturities: $1, 3, 5, 7, 10$ years. In our opinion, the case where each object is described by several time series has not been thoroughly explored in the machine learning literature (Yang and Shahabi 2004; Singhal and Seborg 2002; Dasu et al. 2005). We suggest ways to develop a geometry based methodology to address this clustering problem. At least three avenues of research can be explored:

- distances from Information Geometry theory,
- distances from Optimal Transport theory,
- distances from kernel embedding of distributions Smola et al. (2007).

Intra-dependence and margins. We suppose that the d time series describing a given asset follow a d-variate distribution of density $f(x) := f(x_1, \ldots, x_d)$. According to Sklar's Theorem (Sklar 1959), we have

$$f(x_1, \ldots, x_d) = c(F_1(x_1), \ldots, F_d(x_d)) \prod_{i=1}^{d} f_i(x_i), \tag{20}$$

where c is the copula density, F_i are the marginal cumulative distribution functions and f_i their densities.

Assuming a parametric modelling, we can derive the Fisher-Rao geodesic distance between two assets represented by their parametric multivariate densities $f(x_1, \ldots, x_d; \theta_1)$ and $f(x_1, \ldots, x_d; \theta_2)$ respectively. Since the copula density c has its own set of parameters θ_c and the margins f_i also have their own parameters θ_{m_i}, we have $f(x_1, \ldots, x_d; \theta) = f(x_1, \ldots, x_d; \theta_c, \theta_m)$ which is equal to $c(F_1(x_1; \theta_{m_1}), \ldots, F_d(x_d; \theta_{m_d}); \theta_c) \prod_{i=1}^{d} f_i(x_i; \theta_{m_i})$. To compute the Fisher-Rao geodesic distance D between $f(x; \theta_1)$ and $f(x; \theta_2)$:

$$D(f(x; \theta_1), f(x; \theta_2)) = \int_{\theta_1}^{\theta_2} ds = \int_0^1 \sqrt{\sum_{i,j} g_{ij}(\theta(t)) \frac{d\theta^i}{dt} \frac{d\theta^j}{dt}} dt, \tag{21}$$

we first compute the Fisher information matrix $g_{ij}(\theta)$:

$$\begin{aligned} g_{ij}(\theta) = & -\mathbb{E}_X \left[\frac{\partial^2}{\partial\theta^i \partial\theta^j} \log c(F_1(x_1; \theta_{m_1}), \ldots, F_d(x_d; \theta_{m_d}); \theta_c) \right] & (22) \\ & -\mathbb{E}_X \left[\frac{\partial^2}{\partial\theta^i \partial\theta^j} \log \prod_{k=1}^{d} f_k(x_k; \theta_{m_k}) \right] & (23) \\ = & -\mathbb{E}_X \left[\frac{\partial^2}{\partial\theta^i \partial\theta^j} \log c(F_1(x_1; \theta_{m_1}), \ldots, F_d(x_d; \theta_{m_d}); \theta_c) \right] & (24) \\ & -\sum_{k=1}^{d} \mathbb{E}_X \left[\frac{\partial^2}{\partial\theta^i \partial\theta^j} \log f_k(x_k; \theta_{m_k}) \right] & (25) \end{aligned}$$

If we opt for the Canonical Maximum Likelihood hypothesis as in El Maliani et al. (2011), then $\frac{\partial}{\partial\theta_m} c(u_1, \ldots, u_d; \theta_c) = 0$. It follows that $g_{\theta_c,\theta_m} = g_{\theta_m,\theta_c} = 0$. Thus, we obtain the Fisher-Rao metric

$$ds^2 = \sum_{i,j} g_{ij}(\theta) d\theta^i d\theta^j = g_{\theta_c,\theta_c} d\theta_c d\theta_c + \sum_{i=1}^{d} \sum_{k,l} g_{\theta_{m_k},\theta_{m_l}} d\theta_{m_l} d\theta_{m_k}. \tag{26}$$

It can be expressed by

$$ds^2 = ds^2_{copula} + \sum_{i=1}^{d} ds^2_{margins}, \tag{27}$$

and therefore the Fisher-Rao geodesic distance is a distance between the dependence structure of the two multivariate densities + a distance between the marginal distributions of these two multivariate densities.

However, since the Fisher-Rao distance is frequently intractable, one often considers related divergences such as Kullback–Leibler, symmetrized Jeffreys, Hellinger, or Bhattacharyya divergences which coincide with the quadratic form approximations of the Fisher-Rao distance between two close distributions, and which are computationally more tractable. It would be interesting to find the class of divergences that verifies such a decomposability. For instance, the Kullback–Leibler divergence does not: $KL(f, g) \neq KL(c_f, c_g) + \sum_{i=1}^{d} KL(f_i, g_i)$. However, if f and g have identical marginals, i.e. $\forall i \in \{1, \ldots, d\}$, $f_i = g_i$, then it can be shown Killiches et al. (2015) that $KL(f, g) = KL(c_f, c_g) = KL(c_f, c_g) + \sum_{i=1}^{d} KL(f_i, g_i)$.

How the choice of a particular distance will influence the clustering? A brief comparison of Fisher-Rao and its related divergences and the Wasserstein W_2 distance between bivariate Gaussian copulas is provided for illustration. Let $C_{R_A}^{\text{Gauss}}$, $C_{R_B}^{\text{Gauss}}$, $C_{R_C}^{\text{Gauss}}$ be three bivariate Gaussian copulas parameterized by the following correlation matrices

$$R_A = \begin{pmatrix} 1 & 0.5 \\ 0.5 & 1 \end{pmatrix}, R_B = \begin{pmatrix} 1 & 0.99 \\ 0.99 & 1 \end{pmatrix}, R_C = \begin{pmatrix} 1 & 0.9999 \\ 0.9999 & 1 \end{pmatrix}$$

respectively. Heatmaps of their densities are plotted in Fig. 10.

In Table 3, we report the distances $D(R_A, R_B)$ between $C_{R_A}^{\text{Gauss}}$ and $C_{R_B}^{\text{Gauss}}$, and the distances $D(R_B, R_C)$ between $C_{R_B}^{\text{Gauss}}$ and $C_{R_C}^{\text{Gauss}}$. We can observe that unlike Wasserstein W_2 distance, Fisher-Rao and related divergences consider that $C_{R_A}^{\text{Gauss}}$

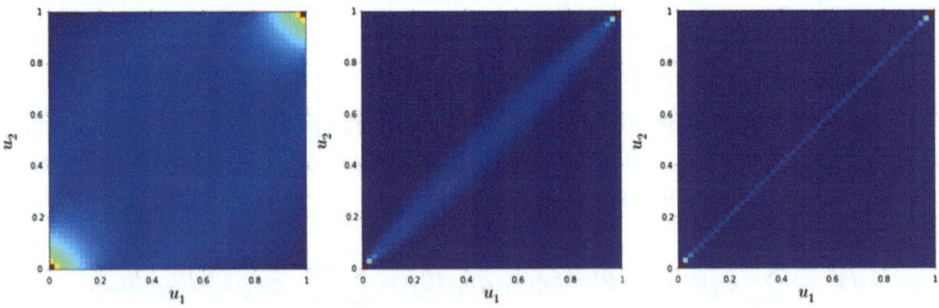

Fig. 10 Densities of $C_{R_A}^{\text{Gauss}}$, $C_{R_B}^{\text{Gauss}}$, $C_{R_C}^{\text{Gauss}}$ respectively; Notice that for strong correlations, the density tends to be distributed very close to the diagonal

Table 3 Distances in closed-form between Gaussians and their sensitivity to the correlation strength

	$D(\mathcal{N}(0,\Sigma_1),\mathcal{N}(0,\Sigma_2))$	$D(R_A,R_B)$		$D(R_B,R_C)$
Fisher-Rao Atkinson and Mitchell (1981)	$\sqrt{\frac{1}{2}\sum_{i=1}^{n}(\log\lambda_i)^2}$	2.77	<	3.26
$KL(\Sigma_1\|\|\Sigma_2)$	$\frac{1}{2}\left(\log\frac{\|\Sigma_2\|}{\|\Sigma_1\|}-n+tr(\Sigma_2^{-1}\Sigma_1)\right)$	22.6	<	47.2
Jeffreys	$KL(\Sigma_1\|\|\Sigma_2)+KL(\Sigma_2\|\|\Sigma_1)$	24	<	49
Hellinger	$\sqrt{1-\frac{\|\Sigma_1\|^{1/4}\|\Sigma_2\|^{1/4}}{\|\Sigma\|^{1/2}}}$	0.48	<	0.56
Bhattacharyya	$\frac{1}{2}\log\frac{\|\Sigma\|}{\sqrt{\|\Sigma_1\|\|\Sigma_2\|}}$	0.65	<	0.81
W_2 Takatsu et al. (2011)	$\sqrt{tr\left(\Sigma_1+\Sigma_2-2\sqrt{\Sigma_1^{1/2}\Sigma_2\Sigma_1^{1/2}}\right)}$	**0.63**	>	**0.09**

λ_i eigenvalues of $\Sigma_1^{-1}\Sigma_2$; $\Sigma=\frac{\Sigma_1+\Sigma_2}{2}$

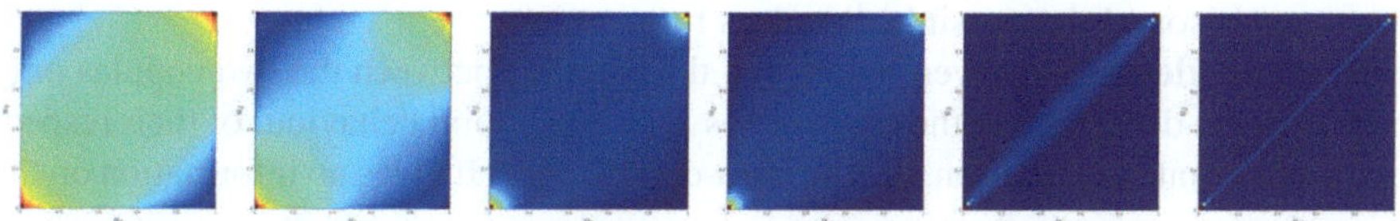

Fig. 11 Datasets of bivariate time series are generated from six Gaussian copulas with correlation .1, .2, .6, .7, .99, .9999

and $C_{R_B}^{\text{Gauss}}$ are nearer than $C_{R_B}^{\text{Gauss}}$ and $C_{R_C}^{\text{Gauss}}$. This may be an undesirable property for clustering since $C_{R_B}^{\text{Gauss}}$ and $C_{R_C}^{\text{Gauss}}$ both describe a strong positive dependence between the two variates whereas $C_{R_A}^{\text{Gauss}}$ describes only a mild positive dependence.

In financial applications, variates can be strongly correlated (for instance, the returns of different maturities in a term structure). In such cases, Fisher-Rao and related divergences yield a much different clustering than the one obtained from using a Wasserstein W_2 distance: Let's consider a dataset of N bivariate time series evenly generated from the six Gaussian copulas depicted in Fig. 11. When a clustering algorithm such as Ward is given a distance matrix computed from Fisher-Rao (displayed in Fig. 12), it will tend to gather in a cluster all copulas but the ones describing high dependence which are isolated. W_2 yields a more balanced and intuitive clustering where clusters contain copulas of similar dependence.

Thus, if the dependence is strong between the time series, the use of Fisher-Rao geodesic distance and related divergences may not be appropriate. They are relevant to find which samples were generated from the same set of parameters (clustering viewed as a generalization of the three-sample problem Ryabko 2010b) due to their local expression as a quadratic form of the Fisher Information Matrix determining the Cramér-Rao Lower Bound on the variance of estimators. To measure distance between copulas for clustering purpose, Wasserstein geometry may be more appropriate since it does not lead to these counter-intuitive clusters. We will investigate

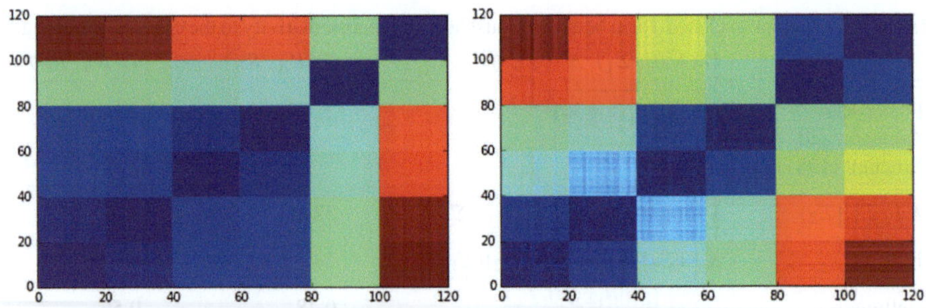

Fig. 12 Distance heatmaps for Fisher-Rao (*left*), W_2 (*right*); Using Ward clustering, Fisher-Rao yields clusters of copulas with correlations $\{.1, .2, .6, .7\}$, $\{.99\}$, $\{.9999\}$, W_2 yields $\{.1, .2\}$, $\{.6, .7\}$, $\{.99, .9999\}$

further this issue. We would also like to encompass the embedding of probability distributions into reproducing kernel Hilbert spaces (Sriperumbudur et al. 2009) in our comparison of the possible distances for copulas.

Inter-dependence. However, notice that the distance between the two copulas only measures the difference in the coordinates $x_1, \ldots, x_d$ joint behaviour of their respective multivariate distribution, i.e. the intra-dependence. It gives no information on the

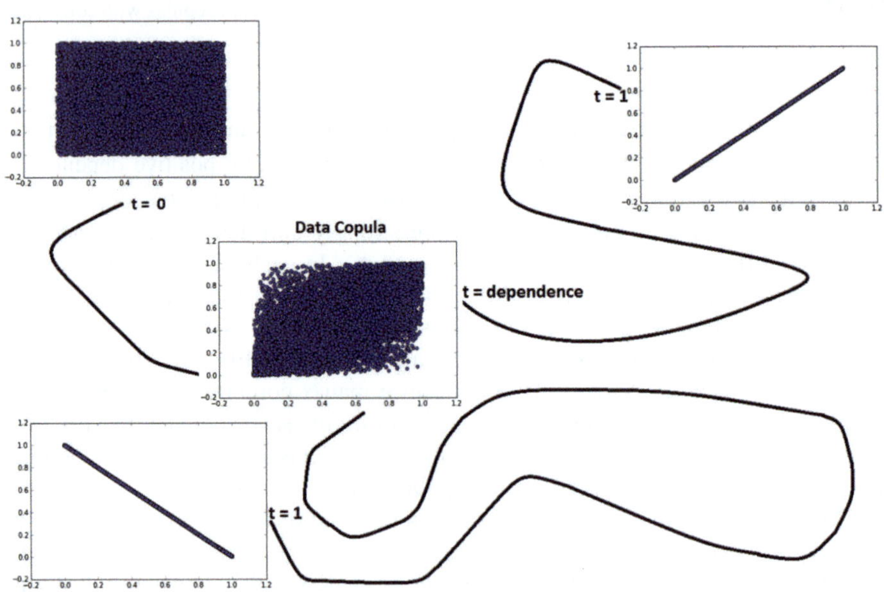

Fig. 13 Dependence can be seen as the relative distance between the independence copula and one or more target dependence copulas. In this picture, the target dependencies are "perfect dependence" and "perfect anti-dependence". The empirical copula (Data Copula) was built from positively correlated Gaussians, and thus is nearer to the "perfect dependence" copula (*top right corner*) than to the "perfect anti-dependence" copula (*bottom left corner*)

time series joint behaviour (how are they moving together?) To obtain such information, one could build the $2d$-variate copula of the two d-variate time series viewed as a single $2d$-variate time series and compare it to the $2d$-variate independence copula (this idea is depicted in Fig. 13). Such an approach, using optimal transport to compare copulas, is described in Marti et al. (2016). But, this construction captures a mixed information of intra-dependence (the coordinates joint behaviour) and inter-dependence (the multivariate time series joint behaviour), besides losing the notion of two different time series. It has been shown that copula is an inadequate tool to build distributions with multivariate marginals (Genest et al. 1995). In Li et al. (1996), authors propose an analogous tool called the linkage function to address these problems: the linkage function contains the information regarding the dependence structure among the underlying multivariate distributions (inter-dependence) but the dependence structure within the multivariate distributions (intra-dependence) is not included.

4 Discussion

In this work, we have presented a new modelling framework for studying financial time series. Clustering could allow to develop an alternative portfolio theory and more relevant risk measures. Several researchers have begun to explore this avenue of research. Until now they have used the Pearson correlation matrix as a similarity matrix for clustering the assets, and thus assuming the Gaussianity of the log-returns. We propose to replace the Pearson correlation matrix by a matrix whose coefficients measure more accurately the dependence and distributional similarities between the assets' returns which can follow any arbitrary joint distribution. For the Information Geometry theoretician, it boils down to design distances between dependent random variables. We think that an interesting approach could be achieved by developing a geometry based on the (copula, margins) representation for random variables, and maybe a (linkage, (copula, margins)) representation for random vectors. We have already started to experiment with (regularized) optimal transport and look forward to leverage information geometry distances to improve our clustering methodology of financial time series. We will be glad to obtain more feedback and hope that our problem was exposed clearly enough so other researchers can work on developing a proper geometry for these dependent (multivariate) distributions.

Acknowledgements Gautier Marti wants to thank Prof. Eguchi for helpful and encouraging remarks, Prof. Brigo for pointing us interesting research directions on dependence, copulas and optimal transport, and Frédéric Barbaresco for sending us relevant literature, historical references, and interesting discussions. We also want to thank our colleagues at Hellebore Capital, and the friendly feedbacks from Philippe Very. Finally, the authors thank the organizers of the workshop "Computational information geometry for image and signal processing" at the International Centre for Mathematical Sciences, Edinburgh, UK, for the invitation.

References

Allez, R., Bun, J., Bouchaud, J.-P. (2014). The eigenvectors of gaussian matrices with an external source. arXiv:1412.7108.

Ao, S. I., Yip, K., Ng, M., Cheung, D., Fong, P.-Y., Melhado, I., et al. (2005). Clustag: Hierarchical clustering and graph methods for selecting tag SNPs. *Bioinformatics*, *21*(8), 1735–1736.

Atkinson, C., Mitchell, A.F.S. (1981). Rao's distance measure. *Sankhyā: The Indian Journal of Statistics, Series A* (pp. 345–365).

Balakrishnan, S., Xu, M., Krishnamurthy, A., & Singh, A. (2011). *Noise thresholds for spectral clustering. NIPS, 2011*, 954–962.

Basalto, N., Bellotti, R., De Carlo, F., Facchi, P., Pantaleo, E., & Pascazio, S. (2007). Hausdorff clustering of financial time series. *Physica A: Statistical Mechanics and its Applications*, *379*(2), 635–644.

Bien, J., & Tibshirani, R. (2011). Hierarchical clustering with prototypes via minimax linkage. *Journal of the American Statistical Association*, *106*(495), 1075–1084.

Borysov, P., Hannig, J., & Marron, J. S. (2014). Asymptotics of hierarchical clustering for growing dimension. *Journal of Multivariate Analysis*, *124*, 465–479.

Bun, J., Allez, R., Bouchaud, J.-P., & Potters, M. (2015). Rotational invariant estimator for general noisy matrices. arXiv:1502.06736.

Chen, Z., & Van Ness, J. W. (1996). Space-conserving agglomerative algorithms. *Journal of Classification*, *13*(1), 157–168.

Cont, R. (2001). Empirical properties of asset returns: Stylized facts and statistical issues. *Quantitative Finance*, *1*(2), 223–236.

Costa, S. I. R., Santos, S. A., & Strapasson, J. E. (2014). Fisher information distance: A geometrical reading. *Discrete Applied Mathematics*, *197*, 59–69.

Dasu, T., Swayne, D. F., & Poole, D. (2005). Grouping multivariate time series: A case study. In *Proceedings of the IEEE Workshop on Temporal Data Mining: Algorithms, Theory and Applications, in conjunction with the Conference on Data Mining, Houston* (pp. 25–32).

Deheuvels, P. (1979) La fonction de dépendance empirique et ses propriétés. Un test non paramétrique d'indépendance. *Académie Royale de. Belgique. Bulletin de la Classe des Sciences (5)*, *65*(6), 274–292.

Deheuvels, P. (1981). An asymptotic decomposition for multivariate distribution-free tests of independence. *Journal of Multivariate Analysis*, *11*(1), 102–113.

Donnat, P., Marti, G., & Very, P. (2016). Toward a generic representation of random variables for machine learning. *Pattern Recognition Letters*, *70*, 24–31.

El Maliani, A. D., El Hassouni, M., Lasmar, N.-E., Berthoumieu, Y., & Aboutajdine, D. (2011). Color texture classification using rao distance between multivariate copula based models. *Computer analysis of images and patterns* (pp. 498–505). Berlin: Springer.

Fredricks, G. A., & Nelsen, R. B. (2007). On the relationship between Spearman's rho and Kendall's tau for pairs of continuous random variables. *Journal of Statistical Planning and Inference*, *137*(7), 2143–2150.

Genest, C., Quesada Molina, J. J., & Rodríguez Lallena, J. A. (1995). De l'impossibilité de construire des lois à marges multidimensionnelles données à partir de copules. *Comptes rendus de l'Académie des sciences. Série 1, Mathématique*, *320*(6), 723–726.

Hartigan, J. A. (1981). Consistency of single linkage for high-density clusters. *Journal of the American Statistical Association*, *76*(374), 388–394.

Khaleghi, A., Ryabko, D., Mary, J., & Preux, P. (2012). *Online clustering of processes*. (pp. 601–609).

Killiches, M., Kraus, D., & Czado, C. (2015). Model distances for vine copulas in high dimensions with application to testing the simplifying assumption. arXiv:1510.03671.

Krishnamurthy, A., Balakrishnan, S., Xu, M., & Singh, A. (2012). Efficient active algorithms for hierarchical clustering. In *International Conference on Machine Learning*.

Laloux, L., Cizeau, P., Bouchaud, J.-P., & Potters, M. (1999). Noise dressing of financial correlation matrices. *Physical Review Letters*, *83*(7), 1467.

Laloux, L., Cizeau, P., Potters, M., & Bouchaud, J.-P. (2000). Random matrix theory and financial correlations. *International Journal of Theoretical and Applied Finance*, *3*(03), 391–397.

Lange, T., Roth, V., Braun, M. L.,& Buhmann, J. M., (2004). Stability-based validation of clustering solutions. *Neural Computation*, *16*(6), 1299–1323.

Lemieux, V., Rahmdel, P. S., Walker, R., Wong, B. L. & Flood, M. (2014). Clustering techniques and their effect on portfolio formation and risk analysis (pp. 1–6).

Li, H., Scarsini, M., & Shaked, M. (1996). Linkages: A tool for the construction of multivariate distributions with given nonoverlapping multivariate marginals. *Journal of Multivariate Analysis*, *56*(1), 20–41.

Mantegna, R. N. (1999). Hierarchical structure in financial markets. *The European Physical Journal B-Condensed Matter and Complex Systems*, *11*(1), 193–197.

Mantegna, R. N., & Stanley, H. E. (1999). *Introduction to econophysics: Correlations and complexity in finance*. Cambridge: Cambridge University Press.

Marti, G., Nielsen, F., & Donnat, P. (2016). Optimal copula transport for clustering multivariate time series. *IEEE ICASSP*.

Marti, G., Very, P., Donnat, P., & Nielsen, F. (2015). A proposal of a methodological framework with experimental guidelines to investigate clustering stability on financial time series. *IEEE ICMLA*.

Meinshausen, N., & Bühlmann, P. (2010). Stability selection. *Journal of the Royal Statistical Society: Series B (Statistical Methodology)*, *72*(4), 417–473.

Murtagh, F., & Contreras, P. (2011). Methods of hierarchical clustering. arXiv:1105.0121.

Pantaleo, E., Tumminello, M., Lillo, F., & Mantegna, R. N. (2011). When do improved covariance matrix estimators enhance portfolio optimization? An empirical comparative study of nine estimators. *Quantitative Finance*, *11*(7), 1067–1080.

Plerou, V., Gopikrishnan, P., Rosenow, B., Nunes Amaral, L. A., Guhr, T., & Stanley, H. E. (2002). Random matrix approach to cross correlations in financial data. *Physical Review E*, *65*(6), 066126.

Pollard, D., et al. (1981). Strong consistency of *k*-means clustering. *The Annals of Statistics*, *9*(1), 135–140.

Potters, M., Bouchaud, J.-P., & Laloux, L. (2005). Financial applications of random matrix theory: Old laces and new pieces. arXiv:physics/0507111.

Ryabko, D. (2010a). Clustering processes (pp. 919–926).

Ryabko, D. (2010b). Clustering processes. In *Proceedings of the 27th International Conference on Machine Learning (ICML 2010)* (pp. 919–926). Haifa, Israel.

Shamir, O., & Tishby, N. (2007). Cluster stability for finite samples. In *NIPS*.

Shamir, O., & Tishby, N. (2008). Model selection and stability in k-means clustering. In *Learning theory*.

Singhal, A., & Seborg, D. E. (2002). Clustering of multivariate time-series data. In *American Control Conference, 2002. Proceedings of the 2002* (Vol 5, pp. 3931–3936). IEEE.

Sklar, A. (1959). Fonctions de répartition à n dimensions et leurs marges. Université Paris, 8.

Smola, A., Gretton, A., Song, L., & Schölkopf, B. (2007). A hilbert space embedding for distributions. *Algorithmic learning theory* (pp. 13–31). Berlin: Springer.

Sriperumbudur, B. K., Fukumizu, K., Gretton, A., Lanckriet, G. R. G., & Schölkopf, B. (2009). Kernel choice and classifiability for RKHS embeddings of probability distributions. In *NIPS* (pp. 1750–1758).

Takatsu, A., et al. (2011). Wasserstein geometry of gaussian measures. *Osaka Journal of Mathematics*, *48*(4), 1005–1026.

Terada, Y. (2013). Strong consistency of factorial k-means clustering. *Annals of the Institute of Statistical Mathematics*, *67*(2), 335–357.

Terada, Y. (2014). Strong consistency of reduced k-means clustering. *Scandinavian Journal of Statistics*, *41*(4), 913–931.

Tola, V., Lillo, F., Gallegati, M., & Mantegna, R. N. (2008). Cluster analysis for portfolio optimization. *Journal of Economic Dynamics and Control*, *32*(1), 235–258.

Tumminello, M., Lillo, F., & Mantegna, R. N. (2007). Shrinkage and spectral filtering of correlation matrices: A comparison via the kullback-leibler distance. arXiv:0710.0576.

Von Luxburg, U., Belkin, M., & Bousquet, O. (2008). Consistency of spectral clustering. *The Annals of Statistics*, *36*, 555–586.

Yang, K., & Shahabi, C. (2004). A PCA-based similarity measure for multivariate time series. In *Proceedings of the 2nd ACM International Workshop on Multimedia Databases* (pp. 65–74). ACM.

The Geometry of Orthogonal-Series, Square-Root Density Estimators: Applications in Computer Vision and Model Selection

Adrian M. Peter, Anand Rangarajan and Mark Moyou

1 Introduction

In its simplest form, information geometry (Murray and Rice 1993; Kass and Voss 1997; Amari and Nagaoka 2001; Arwini and Dodson 2008; Marriott and Salmon 2011) is identifiable with the use of differential geometry to characterize and analyze the space of probability distributions. In its relatively short, yet rich history, a number of probabilistic models have received the geometric treatment. However, the expositions have predominantly focused on exponential models (Efron 1975; Pistone and Rogantin 1999; Pistone and Cena 2007). This emphasis on exponential models can be easily justified given the multitude of theoretical results that exist and a slew of applications where popular models like Gaussians, beta, gamma, binomial, mixtures, etc. are the practitioner's go-to distributions. However, there are several applications in which exponential models—including their mixture forms—simply do not have the descriptive power to model the underlying data distribution. Here we discuss an alternative model where the square-root of the density function is expanded in an orthogonal series expansion (OSE). The coefficients of the basis expansion are readily interpreted as parameters indexing distributions on a statistical manifold with a well prescribed spherical geometry.

In most data-driven applications, we begin with a sample of data to which we apply various algorithmic procedures to estimate parameters, make inferences, and/or generate predictions. The applications we focus on here stem from computer vision. More specifically, we focus on the area of shape analysis where one of the primary objectives is to recognize 2D and 3D shape models. Our sample data in this context

A.M. Peter (✉) · M. Moyou
Department of Engineering Systems, Florida Institute of Technology, Melbourne, FL, USA
e-mail: apeter@fit.edu

A. Rangarajan
Department of Computer and Information Science and Engineering,
University of Florida, Gainesville, FL, USA

F. Nielsen et al. (eds.), *Computational Information Geometry*,
Signals and Communication Technology, DOI 10.1007/978-3-319-47058-0_9

come in the form of unlabeled point sets, one for each shape model. Though there exists a variety of shape representation frameworks ranging from working directly with unstructured point-sets (Chui and Rangarajan 2000; Guo et al. 2005) to weighted graphs (Siddiqi et al. 1998) and include curves, surfaces and other geometric models (Srivastava et al. 2005), the methods developed in the sequel use a density function representation. With advantages such as elimination of topology-based preprocessing (e.g. curve or surface extraction) or curtailing explicit correspondence discovery (Chen et al. 2010), the representation of geometric shapes as probabilistic distributions has yielded *state-of-the-art performance* in a myriad of shape analysis tasks; spanning the gamut from registration (Rangarajan et al. 1997; Chui and Rangarajan 2004; Peter et al. 2008; Jian and Vemuri 2011) to metric learning and shape classification (Thakoor et al. 2007; Moyou and Peter 2012; Moyou et al. 2014). Other benefits of a probabilistic representation include the inherent robustness to noise and localization error of the shape features and landmarks. The utility and accuracy of density function representations heavily rely on robust density estimation methods. We now introduce our approach to density estimation which expands the square-root of the density in a wavelet basis, which is a particular incarnation of the aforementioned orthogonal series expansion.

The use of wavelets as a density estimator was first explored in Doukhan (1988). Wavelet bases have the desirable property of being able to approximate a large class of functions ($\mathbb{L}^2$). Specifically for density estimation, wavelet analysis is often performed on normed spaces that have some notion of regularity like Besov, Hölder and Sobolev. From an empirical point of view, the utility of representing a density in a wavelet basis comes from the fact that they are able to achieve good global approximation properties due to their locally compact nature—a key property when it comes to modeling *shape densities* that contain bumps and/or abrupt variations. Until about 25 years ago, the basis expansions used in an OSE were essentially limited to Fourier bases (i.e. sines and cosines) Kronmal and Tarter (1968) or orthogonal polynomials (e.g. Schwartz 1967 and Izenman 1991). The main downfall of these bases is their infinite support, demanding a large number of terms in the series expansion to accurately approximate complex densities such as ones resulting from shape models (see Fig. 1).

The basic idea behind wavelet density estimation (for one-dimensional data) is to represent the density p as a linear combination of wavelet bases

$$p(x) = \sum_{j_0,k} \alpha_{j_0,k}\phi_{j_0,k}(x) + \sum_{j \geq j_0,k}^{\infty} \beta_{j,k}\psi_{j,k}(x) \tag{1}$$

where $x \in \mathbb{R}$, $\phi(x)$ and $\psi(x)$ are the *scaling* (a.k.a. father) and *wavelet* (a.k.a. mother) basis functions respectively, and $\alpha_{j_0,k}$ and $\beta_{j,k}$ are scaling and wavelet basis function coefficients; the j-index represents the current level and the k-index the integer translation value. (The translation range of k can be computed from the span of the

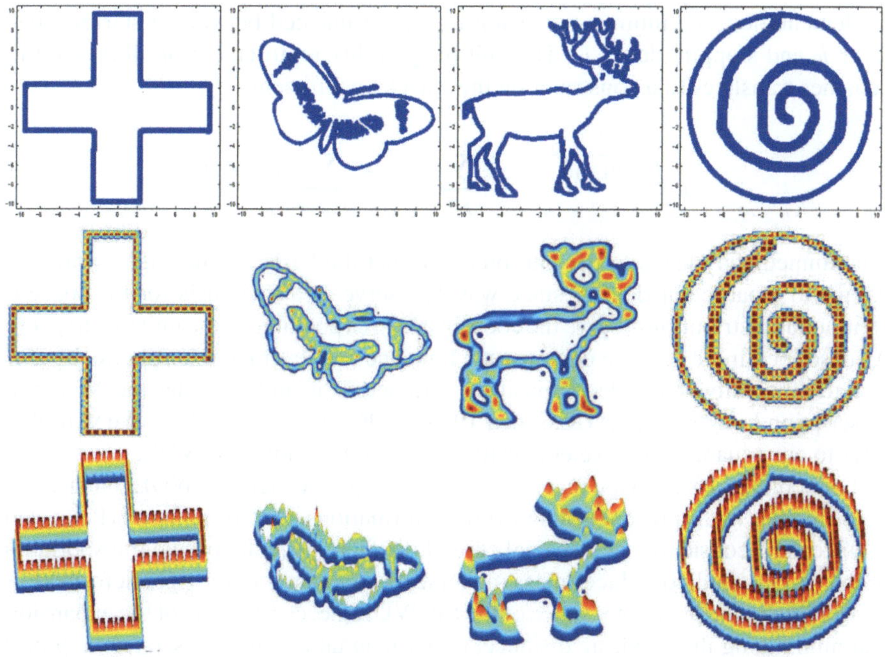

Fig. 1 Example wavelet densities estimated from points-sets of MPEG-7 shapes. *Top row* are point sets, cardinality from *left* to *right* 4,948; 5,578; 7,773; 11,984. *Second row* is a nadir view of the estimated densities using the following wavelet families (from *left* to *right*) Haar ($j_0 = 2$), Coiflet-4 ($j_0 = 1$), Symlet-10 ($j_0 = 0$) and Haar ($j_0 = 2$). *Third row* is the perspective view. Notice how the wavelet densities accurately represent the shapes

data and basis function support size Vannucci 1995). Our goal then is to estimate the coefficients of the wavelet expansion and obtain an estimator $\hat{p}$ of the density. This should be accomplished in a manner that retains the properties of the true density—notably the density should be non-negative and integrate to one.

To guarantee these properties, one typically resorts to estimating $\sqrt{p}$ as

$$\sqrt{p(x)} = \sum_{j_0,k} \alpha_{j_0,k}\phi_{j_0,k}(x) + \sum_{j \geq j_0,k}^{\infty} \beta_{j,k}\psi_{j,k}(x) \tag{2}$$

which directly gives $p = \left(\sqrt{p}\right)^2$. From previous work on wavelet density estimation of $\sqrt{p}$, one can estimate the coefficients as an inner product with the corresponding (orthogonal) basis (Pinheiro and Vidakovic 1997; Penev and Dechevsky 1997) or a maximum likelihood objective function which is minimized using a modified Newton's method (Peter and Rangarajan 2008) (our preferred method). For numerical implementation, the infinite expansion in (2) is truncated between a starting scale level j_0 and stopping level j_1. The unit integrability requirement of all probability densities translates to a constraint on the wavelet coefficients

$$\int \left(\sqrt{p(x)}\right)^2 dx = \sum_{j_0,k} \alpha^2_{j_0,k} + \sum_{j \geq j_0,k}^{j_1} \beta^2_{j,k} = 1. \tag{3}$$

This immediately leads to the interpretation that the basis coefficients—which are unique to a particular density since wavelets serve as a true basis for the space of continuous distributions—give the coordinates for a position on the unit hypersphere. As described in Sect. 4, we demonstrate how one can leverage this spherical geometry to efficiently select decomposition parameters j_0 and j_1 under the Minimum Description Length (MDL) (Rissanen 1978, 1996) framework. Note: We will often refer to our square-root wavelet density estimator as simply SR-WDE.

For applications in shape matching,[1] we leverage this rich SR-WDE to match 2D shape models in the presence of non-rigid deformations. Our formulation (Peter et al. 2008) can be considered as a special case of the more general optimal transportation problem. Our method induces mass movements to adjust for non-rigid deformations by working in the parameter space of the SR-WDE (i.e. coefficients of the expansion) and minimizing the geodesic distance, rather than taking the Wasserstein approach of working directly in the space of measures (Villani 2009). In fact, we illustrate how adopting current methods (e.g. Benamou et al. 2015) which employ deformations under Wasserstein shape interpolation are unable to morph shape densities in a geometrically meaningful way (Sect. 5.1) and lead to sub-optimal matching results.

To gain robustness to isometric transformations, we showcase an approach (in Sect. 5.2) for 2D and 3D modeling where we estimate the SR-WDE on the eigenspace of a shape's Laplace–Beltrami operator (LBO) (treating the shape itself as a manifold). Example eigenshapes resulting from 3D shape models are illustrated in Fig. 2. As before, under the SR-WDE representation, each shape density becomes a point on a unit hypersphere, whose geometry we leverage to calculate intrinsic statistics directly on the manifold of the shape representation. The similarity between shapes is computed using the closed-form distance between the densities on the hypersphere.

For shape matching, the indexing and retrieval accuracies are tested on a variety of 2D and 3D shape databases. Our methods are compared with other density-matching techniques for retrieval, e.g. D2 shape distributions (Osada et al. 2002), and feature-based methods such as Ohbuchi et al. (2008). Our MDL model selection approach is utilized to automatically select the wavelet basis expansion resolution levels and compared to other common criteria (AIC and BIC).

[1] Our use of the terminology shape matching refers to the notion of determining similarity between shapes. Shape matching can also refer to the act of finding non-rigid correspondence between shape models (for purposes such as registration), which we do not explicitly address.

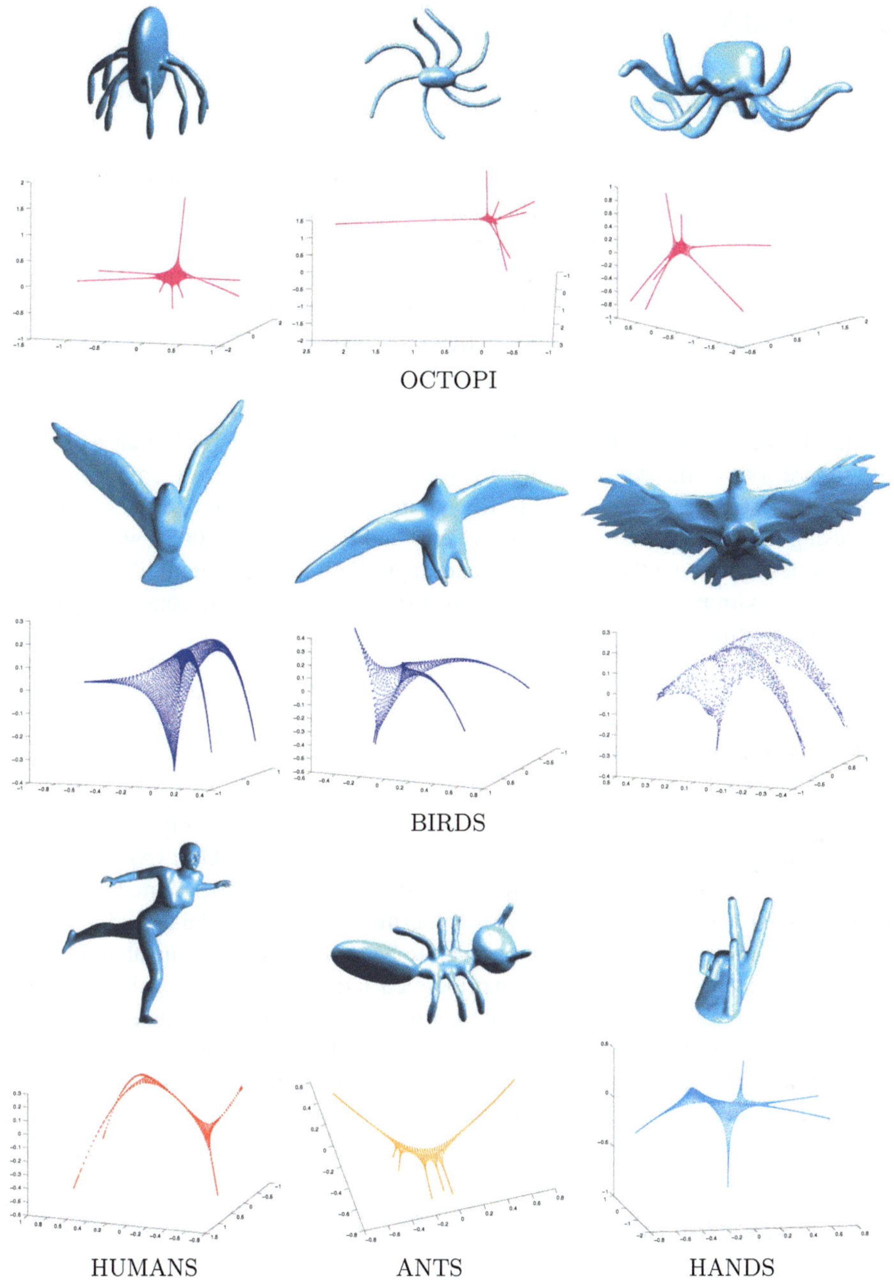

Fig. 2 Eigenshapes of various 3D models from the SHREC'12 dataset. These eigenshapes are formed from the eigenvector triplet (1, 2, 5). Notice, that shapes within a category have similar eigenshapes whereas across categories the eigenshapes are different. This shows the discriminating power of the low-order eigenvectors of the LBO

2 Related Work

2.1 Wavelet Density Estimation

Classical wavelet density estimation (Donoho et al. 1996; Hardle et al. 1998) does not try to explicitly ensure that the density is non-negative and usually suffers from negative values in the tails of the density. For example, in the work of Donoho et al. (1996), this artifact is introduced by the necessity to threshold the coefficients. The method we adopt, estimates the square root of the density $\sqrt{p}$ rather than p. Estimating $\sqrt{p}$ has several advantages: (i) non-negativity is guaranteed by the fact $p = \left(\sqrt{p}\right)^2$ (ii) integrability to one is easy to maintain even in the presence of thresholding, and (iii) the square root is a variance stabilizing transform (Montgomery 2004). The initial works (Pinheiro and Vidakovic 1997; Penev and Dechevsky 1997) that estimated the square root of the density using a wavelet basis expansion rely on a projection method to estimate the coefficients $\{\alpha_{j_0,k}\}$ and $\{\beta_{j,k}\}$. We estimate these parameters using the method first presented in Peter and Rangarajan (2008), which avoids these issues by casting the density estimation problem in a maximum likelihood setting. The maximum likelihood model also ensures the asymptotic consistency of the estimated coefficients.

2.2 Model Selection

Several model selection criteria have been proposed, but arguably the following are the most commonly used: Akaike information criterion (AIC) (Akaike 1973), Bayesian information criterion (BIC) (Schwarz 1978) and Minimum Description Length (MDL) (Rissanen 1978, 1996). A fourth—Bayesian model selection (BMS) (Kass and Raftery 1995)—has been proven to be asymptotically equivalent to MDL (Balasubramanian 1997). We review a principled and geometric approach to selecting the model order of all orthogonal series estimators using the Minimum Description Length (MDL) criterion (Peter and Rangarajan 2011). Specifically, we focus on wavelet density estimators and derive several insightful model selection properties motivated by the information geometry of such spaces.

2.3 Shape Matching

The literature in shape modeling and matching spans a broad spectrum of representations and their corresponding metrics. There are several surveys, e.g. Shilane et al. (2004), that succinctly describe shape representations such as unstructured point-sets or curves. They also detail the myriad of similarity measures that provide a means by which to compare shapes under a common representation. Because we incorporate a linear assignment solver to handle non-rigid deformations, our method is situated in close proximity to techniques that use transportation and assignment problem formulations (Luenberger 1984) to obtain their distance measures. One popular measure is the Earth Mover's Distance (EMD) (Rubner et al. 2000), which is a metric between general mass distributions of objects and is also known as the Wasserstein metric (first order) or Mallow's distance (Levina and Bickel 2001). Given two distributions $\mathbf{x}$ and $\mathbf{y}$, the goal becomes to find a matrix $f_{i,j}$ that establishes a flow between all features x_i and y_j in $\mathbf{x}$ and $\mathbf{y}$. Feasible flows must satisfy row sum, column sum and total sum constraints. Obtaining the flow and subsequently the EMD is generally based on the solution to the transportation problem (Hitchcock 1941). Hence, one of the main differences between our approach and EMD is that we solve a matching problem in contrast to the transportation problem. The EMD also requires one to decide on the features as well as the appropriate weighting of each feature per object. For some applications these choices may already be readily apparent, but for most this requires an added level of effort and investigation. Our method simply works on the point sets that naturally arise either from sampling or preprocessing. More recent uses of the (second order) Wasserstein metric (Cuturi and Doucet 2015; Solomon et al. 2015; Benamou et al. 2015) have also demonstrated applications in shape analysis, however, as discussed in Sect. 5.1, these formulations are not suited well for matching when geometric transformations exists between the comparison models.

One of the earliest LBO spectral techniques was the Shape-DNA approach by Reuter et al. (2006). In this work, the authors proposed to use the eigenvalues of the LBO as an isometry-invariant shape descriptor. Sun et al. (2009) developed their retrieval signature based on observing the heat diffusion over time along the surface. Their now well-known descriptor, the Heat Kernel Signature (HKS), captures information about the intrinsic localized shape geometries via the heat kernel, characterizing the shape up to an isometry.

Rustamov (2007) forms a descriptor vector from the eigenvalues and eigenvectors of the Laplace–Beltrami operator. The deformation invariant representation of the surface is referred to as the GPS embedding. The GPS embedding scales each eigenvector with the square root of their corresponding eigenvalue (analogous to the scaling employed for eigenvectors of operators on finite dimensional vector spaces). Whereas Rustamov uses up to 25 eigenvectors to construct a histogram descriptor using pairwise distances between the GPS points, we use only three low order eigenvectors and directly estimate a density function on the eigenspace.

More recent developments by Bronstein et al. have detailed a scale invariant heat kernel signature (SI-HKS) (Bronstein and Kokkinos 2010) and Shape Google (Bronstein et al. 2011). In the Shape Google framework, they algorithmized the practical use of the HKS for retrieval applications by incorporating the common bag-of-words framework from text retrieval. While these efforts also utilize the LBO, our work discussed here is the first to apply the SR-WDE directly on the isometry invariant eigenspace.

Hou et al. (2012) showcase another technique based on the bag-of-features framework called bag-of-feature graphs. A shape is represented by constructing graphs derived from the HKS. Given a vocabulary of geometric words, corresponding to each word they build a graph that records spatial information between features, weighted by their similarity to this word. Khoury et al. (2012) present a 3D model retrieval method based on creating an index of closed curves in $\mathbb{R}^3$ generated from the center of a 3D model. They use the commute-time mapping function which is derived from the eigenvectors of the LBO. Each curve describes a small region of the 3D model and is robust to several transformations. In order to describe the whole mesh, the method uses a set of indexed curves and shapes are matched based on the distance between the sets of curves. Several alternative 3D model indexing approaches and shape descriptors are discussed in surveys such as Tangelder and Veltkamp (2008).

Our techniques presented here draw their inspiration from our shape analysis framework introduced in Peter and Rangarajan (2009), which uses geodesic distances on the manifold of Gaussian mixture models (GMMs) to establish a shape similarity metric. In this previous work, we represented shapes as mixture models and used the Fisher–Rao metric derived directly from the representation to obtain intrinsic distances on the manifold of parametric mixtures. Like this method, the present techniques also leverage the geometry that results directly from the shape representation. However, when using GMMs it is not feasible to use the resulting metric for retrieval because the geodesics are not in closed-form. (GMMs present a large computational burden of solving for geodesic distances on arbitrary, high-dimensional manifolds.) With the present method, we have a well understood geometry with an easy to compute metric—simply the angular distance on an unit hypersphere. There is also an interesting line of investigation that adopts the Bregman divergence (Liu et al. 2010; Nielsen and Nock 2014) as preferred measure of similarity between densities and demonstrate favorable computational efficiencies. However, these techniques are beyond the present scope of discussion.

3 Square-Root Wavelet Density Estimator

The idea of representing shapes as densities is usually brought to fruition in two ways. Either the density is directly estimated from the shape's discrete samples (Wang et al. 2008) or some other feature is first extracted from the shape and then the density is fit to these features (Osada et al. 2002; Rubner et al. 2000); our method falls in line with the former. To our knowledge, this is the first time a *wavelet density estimator* has been used to directly represent shapes. Previous uses of wavelets in shape analysis (Chuang and Kuo 1996) have been mainly restricted to extracting descriptors of contour shapes.

Many of the issues of estimating a bona fide density can be overcome by first estimating $\sqrt{p(x)}$ and then obtaining the desired density as $\left(\sqrt{p}\right)^2$ (Penev and Dechevsky 1997; Pinheiro and Vidakovic 1997; Peter and Rangarajan 2008). For two dimensional densities the wavelet expansion of the square root of the density is given by

$$\sqrt{p(\mathbf{x})} = \sum_{j_0,\mathbf{k}} \alpha_{j_0,\mathbf{k}} \phi_{j_0,\mathbf{k}}(\mathbf{x}) + \sum_{j \geq j_0,\mathbf{k}}^{j_1} \sum_{w=1}^{3} \beta^w_{j,\mathbf{k}} \psi^w_{j,\mathbf{k}}(\mathbf{x}) \tag{4}$$

where $\mathbf{x} \in \mathbb{R}^2$, j_1 is some stopping scale level for the multiscale decomposition and $(k_1, k_2) = \mathbf{k} \in \mathbb{Z}^2$ is a multi-index that represents the spatial location of the basis. (The translation range of $\mathbf{k}$ can be computed from the span of the data and basis function support size.) The father and mother basis are tensor product combinations of their one dimensional counterparts, i.e.

$$\begin{aligned} \phi_{j_0,\mathbf{k}}(\mathbf{x}) &= 2^{j_0} \phi(2^{j_0} x_1 - k_1) \phi(2^{j_0} x_2 - k_2) \\ \psi^1_{j,\mathbf{k}}(\mathbf{x}) &= 2^j \phi(2^j x_1 - k_1) \psi(2^j x_2 - k_2) \\ \psi^2_{j,\mathbf{k}}(\mathbf{x}) &= 2^j \psi(2^j x_1 - k_1) \phi(2^j x_2 - k_2) \\ \psi^3_{j,\mathbf{k}}(\mathbf{x}) &= 2^j \psi(2^j x_1 - k_1) \psi(2^j x_2 - k_2). \end{aligned} \tag{5}$$

The goal is to estimate the set of coefficients $\left\{\alpha_{j_0,\mathbf{k}}, \beta^w_{j,\mathbf{k}}\right\}$ and reconstruct the density using (4). For 3D density estimation, the tensor product constructions of the father and mother's one dimensional bases results in eight combinations, i.e.

$$\begin{aligned} \phi_{j_0,\mathbf{k}}(\mathbf{x}) &= 2^{\frac{3j_0}{2}} \phi(2^{j_0} x_1 - k_1) \phi(2^{j_0} x_2 - k_2) \phi(2^{j_0} x_3 - k_3) \\ \psi^1_{j,\mathbf{k}}(\mathbf{x}) &= 2^{\frac{3j}{2}} \phi(2^j x_1 - k_1) \psi(2^j x_2 - k_2) \psi(2^j x_3 - k_3) \\ &\vdots \\ \psi^6_{j,\mathbf{k}}(\mathbf{x}) &= 2^{\frac{3j}{2}} \psi(2^j x_1 - k_1) \psi(2^j x_2 - k_2) \phi(2^j x_3 - k_3) \\ \psi^7_{j,\mathbf{k}}(\mathbf{x}) &= 2^{\frac{3j}{2}} \psi(2^j x_1 - k_1) \psi(2^j x_2 - k_2) \psi(2^j x_3 - k_3). \end{aligned} \tag{6}$$

The 3D SR-WDE follows directly via appropriate modification of Eq. (4). An efficient maximum likelihood method to estimate the coefficients $\left\{\alpha_{j_0,\mathbf{k}}, \beta^w_{j,\mathbf{k}}\right\}$, with fast convergence, is discussed in Peter and Rangarajan (2008), as well a more substantial review of the necessary wavelet theory. Due to the increased indexing notation for two and three dimensional wavelet expansions, we will typically resort to one dimensional arguments, as in Sect. 1, with it being understood that all results directly translate to these higher dimensions. Since we produce bona fide two and three dimensional density estimators, our SR-WDE can be independently applied in any application that can benefit from these robust probabilistic models. As such, the authors have made an open source implementation of them available for general purpose use.[2]

[2]ICE Lab Software: http://research.fit.edu/ice/?q=node/26.

3.1 Geometry of Wavelet Densities

Equation (3) showed that a natural by-product of working with the square root of the density and then expanding it with an orthonormal wavelet expansion was that it imposed a constraint on the basis coefficients; namely the sum of squared coefficient values must equal one. Hence, the basis coefficients serve a dual role as the coordinates for a position on the unit hypersphere uniquely indexing different distributions. The ordering of the coefficients in the coordinate vector can be taken in any arrangement but it must be consistent across all densities. The dimensionality of the hypersphere is determined by the cardinality of the set containing all the coefficients. The hypersphere geometry of the densities can be more rigorously justified when we analyze the $\sqrt{p(x)}$ representation under the theoretical basis of information geometry (Amari and Nagaoka 2001; Srivastava et al. 2007). In this context, the Fisher information matrix (FIM) serves as the metric tensor on the manifold of a parametric family of distributions. One of the algebraic forms of the FIM is given by

$$g_{u,v} = 4 \int \frac{\partial \sqrt{p(x|\mathbf{\Theta})}}{\partial \theta^u} \frac{\partial \sqrt{p(x|\mathbf{\Theta})}}{\partial \theta^v} dx \tag{7}$$

where $\mathbf{\Theta} = \{\theta^1, \ldots, \theta^m\}$ denotes the parameters of the distribution and u and v indicate the row and column index, i.e. for a family with m parameters the FIM is $m \times m$. Under an orthonormal expansion of $\sqrt{p(x|\mathbf{\Theta})}$, Eq. (7) reduces to the canonical metric tensor of a unit hypersphere embedded in an $m + 1$ Euclidean space. Rather than use the metric tensor to intrinsically compute geodesics on the hypersphere (an undertaking which would require us to parametrize the manifold), we can accomplish the same computation by realizing that the constraint $\sum_{i=1}^{m+1} (\theta^i)^2 = 1$ also implies the unit hypersphere geometry. Hence, closed-form geodesics distances can be simply computed using the usual angle measure between two unit vectors. Such is the case in our framework where $\sqrt{p(x|\mathbf{\Theta})}$ has been expanded in a orthonormal wavelet basis with the coefficients of the expansion serving as the parameters of the density, i.e. $\mathbf{\Theta} = \{\alpha_{j_0,k}, \beta_{j,k}\}$. Two shapes represented as wavelet densities end up as two points on the hypersphere, see Fig. 3. Since this is a unit hypersphere with the wavelet coefficients for each shape playing the role of two unit vectors, the angle between these unit vectors immediately gives the geodesic distance between the shapes. More concretely, the coefficients of the probability density $\{\alpha_{j_0,k}, \beta_{j,k}\}$ serve as the coordinates $c = [\alpha_{j_0,1}, \ldots, \alpha_{j_0,m}, \beta_{j,1,}, \ldots, \beta_{n,m}]$ indexing the location of a density on a unit hypersphere; then the distance between two distributions p_1 and p_2 indexed by their coordinates c_1 and c_2, respectively, is given by

$$d(p_1, p_2) = \cos^{-1}(c_1^T c_2). \tag{8}$$

It is also interesting to note that we can obtain this *same* inner product interpretation required in (8) by taking the approach of working with a similarity measure directly between the densities, instead of analyzing the geometry implied by the coef-

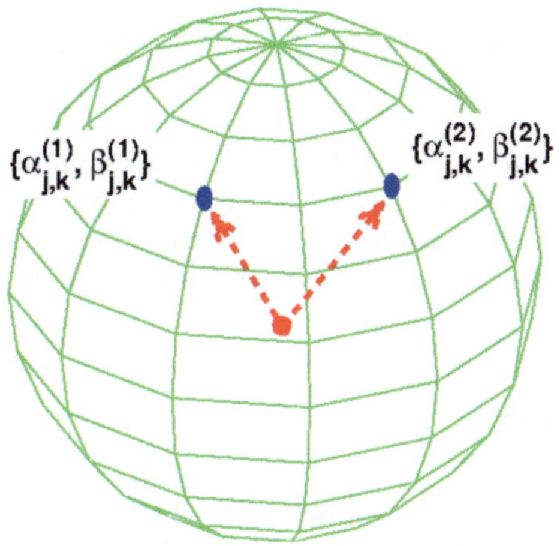

Fig. 3 Hypersphere of densities. Unit integrability for densities requires $\sum_{j_0,k} \alpha_{j_0,k}^2 + \sum_{j \geq j_0,k}^{j_1} \beta_{j,k}^2 = 1$, also the FIM is reduced to the canonical metric of the unit hypersphere when $\sqrt{p}$ is expanded in an orthonormal basis. This places the shapes represented by the densities on unit hypersphere with coordinates given by the wavelet coefficients. The above figure shows two densities, see coefficient superscript, on the hypersphere—their geodesic distance is the angle between the unit vectors

ficient constraints and the metric tensor. In particular, using the Hellinger divergence (Beran 1977) to calculate the distance between two densities p_1 and p_2 gives

$$\begin{aligned} D_H(p_1, p_2) &= \int_{\mathbb{R}^2} (\sqrt{p_1} - \sqrt{p_2})^2 dx \\ &= 2 - 2\left[\sum_{j_0,k} \alpha_{j_0,k}^{(1)} \alpha_{j_0,k}^{(2)} + \sum_{j \geq j_0,k}^{j_1} \beta_{j,k}^{(1)} \beta_{j,k}^{(2)}\right] \end{aligned} \tag{9}$$

where $\{\alpha^{(1)}, \beta^{(1)}\}$ and $\{\alpha^{(2)}, \beta^{(2)}\}$ are the wavelet parameters of p_1 and p_2 respectively. Notice that we can factor out a -2 and drop the constant without effecting the qualities of the measure. This reduces (9) to an inner product between the coefficients of the densities, hence essentially giving the same measure as the one we derived above by analyzing the geometry of the space of distributions ($\cos^{-1}(\cdot)$ is not present). We refer the reader to Rubner et al. (2000) for a summary of other distance measures between densities.

The spherical geometry detailed here differs from the one that results when working in the space of measures. One can develop a square-root geometry on the space of measures for $\sqrt{p}$ (Bhattacharyya 1943; Srivastava et al. 2007); however, this Hilbert-space geometry is restricted to the positive hyperoctant of the unit hypersphere. Whereas, the parametric square-root model of our SR-WDE utilizes the full unit hypersphere. Under the SR-WDE model one technical issue arises: the identifiability of the unique MLE solution. A sign ambiguity is inherently present due to the fact that both the strictly positive and strictly negative versions of a coefficient set satisfy sum-of-squares constraint, i.e.

$$\sum_{j_0,k} \alpha_{j_0,k}^2 + \sum_{j\geq j_0,k}^{j_1} \beta_{j,k}^2 = \sum_{j_0,k} \left(-\alpha_{j_0,k}\right)^2 + \sum_{j\geq j_0,k}^{j_1} \left(-\beta_{j,k}\right)^2 = 1. \tag{10}$$

Practically, this is does not pose a challenge as one can simply evaluate both the plus and minus version of the coefficient sets when calculating the geodesic distance in (8) to determine which one provides the minimal distance. For simply reconstructing the density, this is a non-issue as the sign ambiguity is nullified when we set $p = (\sqrt{p})^2$.

4 Model Selection

The basic premise behind the resulting functional form of model criteria like AIC, BIC, and MDL is to assign a goodness-of-fit measure (via the likelihood of the observed data sample) and a complexity penalty that can depend on the number of parameters in the model as well as the sample size. The AIC criterion is given by

$$AIC = -2 \ln p(E|\hat{\mathbf{\Theta}}) + 2k \tag{11}$$

and BIC

$$BIC = -2 \ln p(E|\hat{\mathbf{\Theta}}) + k \ln(N), \tag{12}$$

where E is the evidence (current observed data samples), $\hat{\mathbf{\Theta}}$ the maximum likelihood estimate (MLE) of the parameters, N the number of samples, and k the cardinality of the model parameters. For example, $k = 2$ for a linear model where the parameters correspond to (m, b), i.e. the slope and intercept of the line. In the context of SR-WDE addressed in this chapter, k will represent the number of coefficients per the multiresolution decomposition structure. For each criterion, the best model is the minimizer of these measures. Both AIC and BIC reward paucity of parameters as a penalty is paid for large values of k. Since BIC's second term also incorporates the sample size, it tends to prefer smaller complexity models (versus AIC) for sample sizes greater than eight. After eight samples, the second term of BIC, $k \ln(N)$, always has a lower value than AIC's second term, $2k$.

The complexity of a model under AIC and BIC is only measured by the cardinality of the parameters. This is the basic departure point of these (and others) versus MDL: they fail to take into account the functional form (how the parameters interact in the model) of the models. The MDL criterion given by

$$MDL = -\ln p(E|\hat{\mathbf{\Theta}}) + \frac{k}{2} \ln \left(\frac{N}{2\pi}\right) + \ln \int \sqrt{\det g_{u,v}(\mathbf{\Theta})} d\mathbf{\Theta} \tag{13}$$

has an extra term (the third term) which penalizes based on the volume occupied by the model's manifold in the space of probability distributions (more on this follows shortly). Chronologically, Eq. (13) is the more recent version of MDL

(Rissanen 1996). The original MDL (Rissanen 1978) was similar to AIC and BIC in that it only contained the first two terms in (13), thus lacking a penalty based on the functional form. Our experiments in Sect. 6 will demonstrate the usefulness of incorporating the additional volume term.

In practically all useful models, the Riemannian volume term in (13) must be computed by truncating the parameter space and using numerical techniques such as Monte Carlo integration. When one uses an orthogonal series density estimator, this term is *known in closed-form*. MDL was originally developed using coding theory arguments that are based on the notion of finding the shortest code to describe the observed data (Grünwald 2005), the more regularity in the data the shorter the code. Shorter code lengths can be shown to be inversely proportional to the likelihood of observing the data, i.e. higher probabilities are associated with shorter code lengths and smaller probabilities with large code lengths. Hence the use of the terminology 'minimum description length' to find the best model. The criterion as given in (13) is an approximation to the code length for the maximum-likelihood code (Rissanen 1996). We now illustrate how MDL can be re-derived using differential geometry. It will allow us to transition from describing the second and third terms of (13) as penalties for the number of parameters and functional form, respectively. Instead, we will see that together they determine a volume ratio designed to measure the ellipsoidal volume around the maximum likelihood estimate relative to the total volume occupied by the model in the space of probability densities.

4.1 MDL from Differential Geometry

In this section we briefly recap the geometric development of MDL as first presented by Balasubramanian (1997). The author refers to the model selection criterion as the *razor*. It is asymptotically equivalent to MDL. The derivations begin from a Bayesian approach by considering the posterior of a parametric model class $\mathcal{M}$

$$p(\mathcal{M}|E) = \frac{p(\mathcal{M}) \int p(\boldsymbol{\Theta})p(E|\boldsymbol{\Theta})d\Theta}{p(E)} \tag{14}$$

where $\boldsymbol{\Theta} \in \mathbb{R}^d$ are the parameters of the model class. Hence, $p(\boldsymbol{\Theta})$ is the prior distribution on the parameters and $p(E|\boldsymbol{\Theta})$ is the likelihood. When comparing two candidate model classes $\mathcal{M}_1$ and $\mathcal{M}_2$, we can drop $p(E)$ since it is common factor and we can also omit the prior on the models, $p(\mathcal{M}_i)$, by assuming they are equally likely. (To avoid aberrant cases, we assume throughout that the parameter spaces of candidate models are compact.) These assumptions reduce (14) to $p(\mathcal{M}|E) \propto \int p(\boldsymbol{\Theta})p(E|\boldsymbol{\Theta})d\Theta$. It was show in Balasubramanian (1997), Myung et al. (2000) that the Jeffrey's prior (Jeffreys 1961)

$$p(\boldsymbol{\Theta}) = \frac{\sqrt{\det g_{u,v}(\boldsymbol{\Theta})}}{\int \sqrt{\det g_{u,v}(\boldsymbol{\Theta})} d\boldsymbol{\Theta}} \tag{15}$$

is the appropriate prior to choose when the desire is to: treat all parameters equally (uniform), be invariant to reparametrizations of the parameter space, and geometrically count only *distinguishable* volumes on the parameter domain. (The notion of distinguishability was rigorously derived in Balasubramanian 1997.) Finally we assume the observed data $E = \{x_i\}_{i=1}^{N}$ are i.i.d., hence $p(E|\boldsymbol{\Theta}) = \prod_{i=1}^{N} p(x_i|\boldsymbol{\Theta})$. With the aforementioned substitutions, the razor is given as

$$R(\mathcal{M}) = \frac{\int \sqrt{\det g_{u,v}(\boldsymbol{\Theta})} \exp\left\{-N\left(\frac{-\ln p(E|\boldsymbol{\Theta})}{N}\right)\right\} d\boldsymbol{\Theta}}{\int \sqrt{\det g_{u,v}(\boldsymbol{\Theta})} d\boldsymbol{\Theta}}. \tag{16}$$

In order to use the razor for practical evaluation of candidate models, the integral in the numerator of (16) must be approximated around the maximum likelihood estimate of the parameters, $\hat{\boldsymbol{\Theta}}$. (The integral approximation technique is commonly referred to as the Laplace approximation.) To a second order approximation, this yields the final version of the razor

$$\begin{aligned} \rho(\mathcal{M}) &= -\ln R(\mathcal{M}) = -\ln p(E|\hat{\boldsymbol{\Theta}}) + \frac{k}{2}\ln\left(\frac{N}{2\pi}\right) \\ &+ \ln \int \sqrt{\det g_{u,v}(\boldsymbol{\Theta})} d\boldsymbol{\Theta} + \frac{1}{2}\ln\left(\frac{\det \tilde{g}_{u,v}(\boldsymbol{\Theta})}{\det g_{u,v}(\boldsymbol{\Theta})}\right) \end{aligned} \tag{17}$$

where $\tilde{g}_{u,v}$ is the *empirical Fisher information* computed from our observed sample values. Notice that the first three terms of (17) correspond to the MDL criterion in (13). The last term considers the ratio of the expected Fisher to the empirical Fisher, which has the property that as $N \to \infty$, $\tilde{g}_{u,v} \to g_{u,v}$ (empirical Fisher approaches expected Fisher), so this term vanishes, giving us back the MDL Eq. (13).

To better understand the connection of MDL to the Riemannian volumes associated with a model class, we can rewrite (17) as

$$\rho(\mathcal{M}) = -\ln p(E|\hat{\boldsymbol{\Theta}}) + \ln\left(\frac{\mathcal{V}_{\mathcal{M}}}{\mathcal{V}_{\hat{\boldsymbol{\Theta}}}}\right). \tag{18}$$

The numerator of the second term is the total Riemannian volume, $\mathcal{V}_{\mathcal{M}} = \int \sqrt{\det g_{u,v}(\boldsymbol{\Theta})} d\boldsymbol{\Theta}$, of the probabilistic manifold (i.e. total volume of the model class). The denominator $\mathcal{V}_{\hat{\boldsymbol{\Theta}}} = \left(\frac{2\pi}{N}\right)^{\frac{k}{2}} G(\hat{\boldsymbol{\Theta}})$, where $G(\hat{\boldsymbol{\Theta}}) = \left(\frac{\det g_{u,v}(\hat{\boldsymbol{\Theta}})}{\det \tilde{g}_{u,v}(\hat{\boldsymbol{\Theta}})}\right)^{\frac{1}{2}}$, is a term that measures appreciable volume of distinguishable distributions around the maximum likelihood estimate that comes close to the truth (close in the sense that the model is able to predict the evidence E with high probability). As observed above, this data dependent term has the property that $G(\hat{\boldsymbol{\Theta}}) \to 1$ as $N \to \infty$. Hence the ellipsoidal volume around the MLE can be approximated by $\tilde{\mathcal{V}}_{\hat{\Theta}} \approx \left(\frac{2\pi}{N}\right)^{\frac{k}{2}}$. Given this approximation, we have

$$\rho(\mathcal{M}) = MDL = -\ln p(E|\hat{\Theta}) + \ln\left(\frac{\mathcal{V}_{\mathcal{M}}}{\tilde{\mathcal{V}}_{\hat{\Theta}}}\right). \tag{19}$$

Hence it can be seen that MDL penalizes models that have excessively small distinguishable volumes close to the truth (small $\mathcal{V}_{\hat{\Theta}}$) or those that occupy a large volume in the space of distributions (large $\mathcal{V}_{\mathcal{M}}$). The volumes in the second term of (19) are an intrinsic property of the model and together are often referred to as the *geometric complexity* of the model. MDL selects those models that have a low geometry complexity by picking those models with "*highest maximum likelihood per the relative ratio of the distinguishable distributions*" (Myung et al. 2000).

4.2 MDL and the Geometry of Square-Root Wavelet Densities

Up to now we have discussed the derivation and interpretations of the MDL criterion for an arbitrary parameter manifold of a probabilistic model class. We now turn our attention to the application of the MDL criterion to select the decomposition levels for our wavelet density estimation framework described in Sect. 3. It is worth reiterating that the fundamental idea of the closed-form MDL criterion holds true for all valid orthogonal series expansions, and not just the present focus on wavelets. Hence, *we would like to be able to use Eq.* (13) *to decide how to pick the best* j_0 *and* j_1. The number of parameters, k in (13), for a particular choice of j_0 and j_1 is given by the cardinality of the coefficient set over all levels of the decomposition, i.e. $k = \#\{\Theta\} = \#\left\{\alpha_{j_0,l}, \beta_{j,l}\right\}$. As discussed in Sect. 3.1, the coefficients are coordinates for the location of the density on the unit hypersphere embedded in a k-dimensional space. Thus each candidate model, given by choice of j_0 and j_1, is a unit hypersphere and computing the Riemannian volume $\mathcal{V}_{\mathcal{M}}$ in (19) amounts to calculating the *surface area* of a unit hypersphere. With this understanding, we now have a systematic procedure to select the best j_0 and j_1:

1. For each value of j_0 and j_1 estimate the wavelet density coefficients of the expansion (Peter and Rangarajan 2008). This will give you the likelihood term needed for (19).
2. The cardinality of the coefficient set resulting for the selection of j_0 and j_1 will provide the value of k needed to compute volumes $\mathcal{V}_{\mathcal{M}}$ and $\mathcal{V}_{\hat{\Theta}}$ (the remaining terms of the MDL).
3. The optimal $\left\{j_0^*, j_1^*\right\}$ is the one that minimizes (19).

Though systematic, the above process fails to take full advantage of the theoretical consequences associated with the use of wavelets. For example, there are significant computational savings by leveraging the nested subspace structure of wavelet bases.

Another issue is that we must address an anomaly that arises when computing the volume of a unit hypersphere as the dimensions increase: $\mathcal{V}_{\mathcal{M}} \to 0$ as $k \to \infty$. The following subsections expand on these topics.

4.3 MDL is Invariant to Multiresolution Analysis

The first observation we make is that the *MDL criterion is invariant to multiresolution decompositions (consisting of scaling and wavelet functions) in comparison to their corresponding single level scaling counterparts.* This is a significant result that enables us to perform our model search over j_0 instead of j_0 and j_1. This result directly follows from the nested subspace property of wavelet bases and the dyadic relationship of the basis functions at different levels.

In order to establish the invariance of MDL to multiresolution analysis (MRA) versus an appropriate single level scaling-function expansion, we have to establish that the goodness-of-fit and geometric complexity terms are identical for both. First let us establish equivalence of the goodness-of-fit as measured by the log likelihood. Consider a wavelet density estimate using only scaling functions from an arbitrary level j. These form a basis for V_j (see Strang and Nguyen 1997). However, functions expanded using scaling functions from level j can be equivalently represented using both scaling and wavelet bases that span level $j-1$, V_{j-1} and W_{j-1} respectively. Then V_{j-1} can be recursively broken down again and again. The recursive decomposition relationship is given by

$$\begin{aligned} V_j &= V_{j-1} \bigoplus W_{j-1} \\ &= V_{j-2} \bigoplus W_{j-2} \bigoplus W_{j-1} \, . \\ &= V_{j_0} \bigoplus \bigoplus_{l=j_0}^{j-1} W_l \end{aligned} \tag{20}$$

Hence, densities estimated using only scaling functions have an equivalent representation in a multiresolution hierarchy. Since the estimated densities (either from only level j or MRA from j_0 to $j-1$) are equivalent, their corresponding log likelihoods would be the same. So two models, one with only scaling functions and one with an equivalent MRA representation, give the same goodness-of-fit measure for the MDL criterion.

To show that geometric complexities are identical, we have to establish that an expansion using only scaling functions has the same number of coefficients as its corresponding MRA. This is clearly true by the very nature of the dyadic relationships between levels in a MRA: *basis functions at a coarser level $j-1$ have twice the support of those at level j, hence half the number of coefficients.* The number of coefficients at a particular level is associated with the number of translations needed to span a defined spatial support. Theoretically, an infinite number of translations are used, but for any finite sample set the span of translations needed to cover the data will also be finite. The cardinality of the coefficient set from a level j with only scaling functions would equal the cardinality of coefficients from the coarser level $j-1$ that has both scaling and wavelet bases, i.e.

$$\begin{array}{rl} k = & \#\left\{V_j\right\} \\ = & \#\left\{V_{j-1}\right\}+\#\left\{W_{j-1}\right\}=\frac{k}{2}+\frac{k}{2} \\ = & \#\left\{V_{j-2}\right\}+\#\left\{W_{j-2}\right\}+\#\left\{W_{j-1}\right\}=\frac{k}{4}+\frac{k}{4}+\frac{k}{2}, \\ = & \#\left\{V_{j_0}\right\}+\sum_{l=j_0}^{j-1}\#\left\{W_l\right\} \end{array} \tag{21}$$

where we have slightly abused the notation $\#\{\cdot\}$ to count the number of coefficients for a chosen basis level's function space. Since the value of k essentially determines the geometric complexity, it will be identical for single level decomposition at level j or a MRA from j_0 up to $j-1$. (The number of samples N is also a factor in the $\mathcal{V}_{\hat{\boldsymbol{\Theta}}}$ term of geometric complexity, but it will be the same for all models so can be ignored in this analysis.)

With both the goodness-of-fit and geometric complexity shown to be the same for MRA versus single-level scaling function bases, *it is sufficient for density estimation to use only scaling functions* and to search for the best model by iterating over various starting levels j_0. So is MRA for wavelet density estimation not needed? It depends. If your goal is to simply obtain a reconstruction of the density, then it can be argued that scaling functions alone are enough. But if one's goal is sparsity among the coefficients (which is what MRA is designed for), then a different mechanism that measures this property must be incorporated into the model selection framework. Such a measure would include wavelet thresholding as part of the criterion for selecting the model. This is an avenue of future research.

4.4 Closed-Form Computation of $\mathcal{V}_{\mathcal{M}}$

In practice, the application of the MDL Eq. (19) almost always requires numerical integration to compute $\mathcal{V}_{\mathcal{M}}$, the Riemannian volume of the statistical manifold. This involves derivation of the Fisher information metric (FIM), appropriate truncation of the parameter space to perform the integration and other numerical adjustments to ensure that the FIM does not become singular. For very high-dimensional parameter spaces, one has to employ Monte Carlo integration methods. Only for very simple models is $\mathcal{V}_{\mathcal{M}}$ in an analytic form; sometimes even the FIM is not in closed-form and may require an additional numerical integration step. One significant advantage of our SR-WDE framework is that all of our models have a unit hypersphere geometry (again this is true for all orthogonal series expansions). Hence, $\mathcal{V}_{\mathcal{M}}$ is known in closed-form. It is merely the surface area ($\mathcal{V}_{\mathcal{S}}$) of a unit hypersphere of dimension $k-1$ where $k=\#\{\boldsymbol{\Theta}\}=\#\left\{\alpha_{j_0,l}\right\}$. (Choosing the j_0 decomposition level determines the coefficient set, the cardinality of which is k.) One would intuitively expect the volume of a manifold to increase as the number of dimensions increase. However, the unit hypersphere exhibits an odd property in that it decreases in volume (and surface area) as the dimensions increase (Scott 2001).

The surface area of a unit hypersphere $\mathcal{S}$ is given by

$$\mathcal{V}_{\mathcal{S}} = \begin{cases} \frac{k\pi^{\frac{k}{2}}}{(\frac{k}{2})!} & , k \text{ even} \\ 2^k \pi^{\frac{k-1}{2}} \frac{(\frac{k-1}{2})!}{(k-1)!} & , k \text{ odd} \end{cases}. \tag{22}$$

As shown in Fig. 4, the maximum surface area is reached at dimension seven, and then the surface area rapidly decreases to zero. Recall that the geometric complexity assesses a cost based on the ratio of the manifold volume to the ellipsoidal volume around the MLE, i.e. the penalty is $\ln\left(\frac{\mathcal{V}_{\mathcal{S}}}{\mathcal{V}_{\hat{\Theta}}}\right)$. If the $\mathcal{V}_{\mathcal{S}}$ shrinks to zero so fast that it is smaller than $\mathcal{V}_{\hat{\Theta}}$, then our penalty term is not valid since it would become negative. Having $\mathcal{V}_{\hat{\Theta}} > \mathcal{V}_{\mathcal{S}}$ tells us that the model is misspecified (Navarro 2004). Geometrically we can visualize this as the ellipsoidal volume around the MLE protruding out of the smaller model manifold. In practice, one has to be careful to consider the trade-off between the number of samples and the number of parameters. A valid region of well-specified models is easily achieved when we consider $\mathcal{V}_{\hat{\Theta}} = \left(\frac{2\pi}{N}\right)^{\frac{k}{2}}$. Once we reach above seven samples, i.e. $N \geq 7$, the ellipsoidal volume starts to decline exponentially as the number of parameters k increases. Since we need the number of samples to be generally greater than the number of parameters to avoid an ill-posed density estimation problem, we can easily satisfy our requirement of needing $\mathcal{V}_{\hat{\Theta}} < \mathcal{V}_{\mathcal{S}}$. In Fig. 5, we illustrate $\ln(\mathcal{V}_{\mathcal{S}})$ versus $\ln\left(\mathcal{V}_{\hat{\Theta}}\right)$ over a range of sample cardinalities and number of parameters. Notice that for a sufficiently high number of samples relative to the number of parameters (i.e. dimensions of the unit

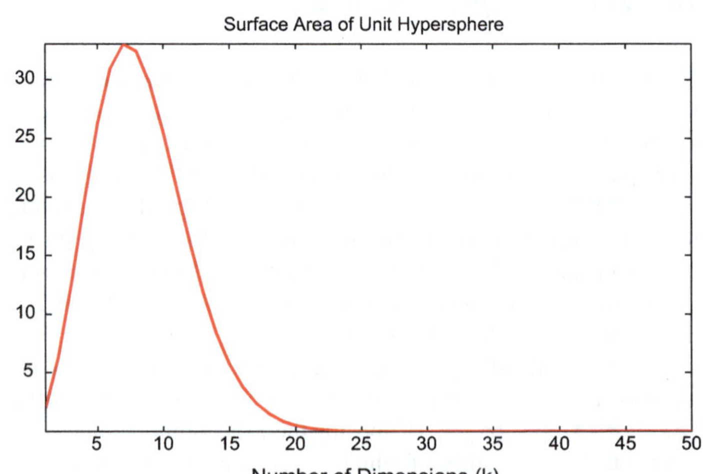

Fig. 4 Surface area of unit hypersphere. Maximum surface area is at dimension seven

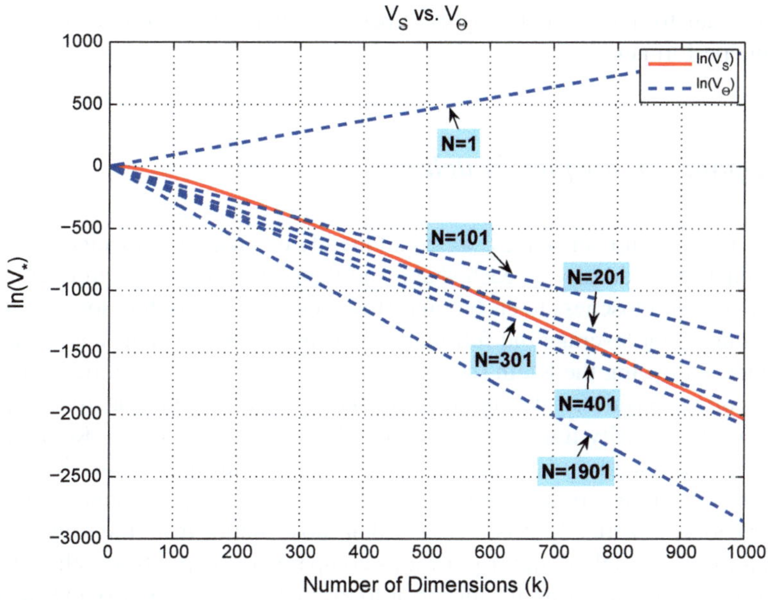

Fig. 5 Riemannian volume comparisons, $\ln(\mathcal{V}_S)$ (*solid line*) versus $\ln\left(\mathcal{V}_{\hat{\Theta}}\right)$ (*dashed line*). Misspecified models occur when $\mathcal{V}_{\hat{\Theta}} > \mathcal{V}_S$. For sufficiently high number of samples we see that $\mathcal{V}_{\hat{\Theta}} < \mathcal{V}_S$ as desired

hypersphere), there is a sharp decrease of $\mathcal{V}_{\hat{\Theta}}$ as desired. It is worth noting that to guarantee uniqueness of the estimated density, the coefficients of the expansion should be restricted to the positive orthant of the unit hypersphere. This requires the volume term be divided by a 2^k factor. We can easily account for misspecified models under this restriction by simply increasing the number of samples.

5 2D and 3D Shape Matching

In this section, we detail the use of our SR-WDE for applications in 2D and 3D shape matching. The overarching theme is the same: given point-set representation of shape models, we fit SR-WDE to each and then establish similarities between shapes by simply matching the densities using the closed-form distance on the unit hypersphere. In the first scenario, we focus on 2D shape matching, in the presence of non-rigid deformations, using optimal transport techniques (Peter et al. 2008). We then detail our LBO-shape matching approach (Moyou and Peter 2012; Moyou et al. 2014) which presents a unifying framework for both 2D and 3D shape matching by first estimating the LBO of a shape model and then estimating the SR-WDE on the low-order eigenvectors of LBO. By first obtaining the eigenvectors of the LBO and then estimating the SR-WDE on the eigenvector coordinates, we gain invariance to isometric transformations that may exist between shapes and considerably improve matching performance.

5.1 Optimal Transport Matching

Given a pair of point-set shapes, we could merely estimate the wavelet coefficients of the square-root density of each shape and then take their inner product to get a measure of their closeness to each other. However, this approach is somewhat naïve in that it does not leverage the full mathematical formalisms that relate one shape to another. Following the Klein school of thought (Klein 1872), similarity between shapes is often considered after quotienting out some transformation group, typically the group of similarity transformations (Dryden and Mardia 1998). Removing the transformations enables us to analyze effects that are intrinsic to the shapes. Non-rigid transformations are the most general, basically encompassing any continuous transformation. Practically it is expected that most shapes from the same category should differ by "smaller" non-rigid warps compared to shapes from other arbitrary categories; hence correcting for this prior to evaluating the similarity metric should enhance its discriminability. In our framework, we could incorporate non-rigid alignment in one of two ways: perform non-rigid alignment of the point sets prior to fitting the wavelet density or fit the density to the data and then adjust for non-rigid deformations by warping the densities. The former method usually involves adopting a spline based model to represent the non-rigid transformation (Bookstein 1989) and can involve iterative optimization to solve for the spline parameters. Though these methods are able to model a large class of non-rigid deformations, they do not possess the computational efficiency needed for querying systems. Our method takes the second option of warping the densities which we accomplish by locally translating wavelet coefficients.

We now give a simple example to illustrate how warping the densities by local translations can increase recognition. Suppose two shapes have been affine aligned and there only remains a non-rigid warp between the two. We model the non-rigid deformation, in the infinitesimal, as local translations. Figure 6 shows the estimated densities of two hypothetical shapes, see (a) and (b). The coefficients for the basis functions of each shape are indicated by a red bar. The density function shown in (a) only differs by a translation to density (b). Notice that if we were to stack the coefficients in a vector (from bottom left to top right) for each density and perform an inner product between them, the resulting value would be zero. This leads to high geodesic distance, $\cos^{-1}(0) = \frac{\pi}{2}$. However, if we simply slide the wavelet bases of one shape to align to locations on the other, our inner product would then yield a very high correlation indicating the true similarity between the shapes. Also we must be careful that whatever mechanism we use to translate the bases does not alter the values of their coefficients and compromise the properties of a bona fide density, i.e. (3) must hold to maintain unit integrability. The most straightforward way to accommodate these objectives is to reformulate our similarity metric under the action of a

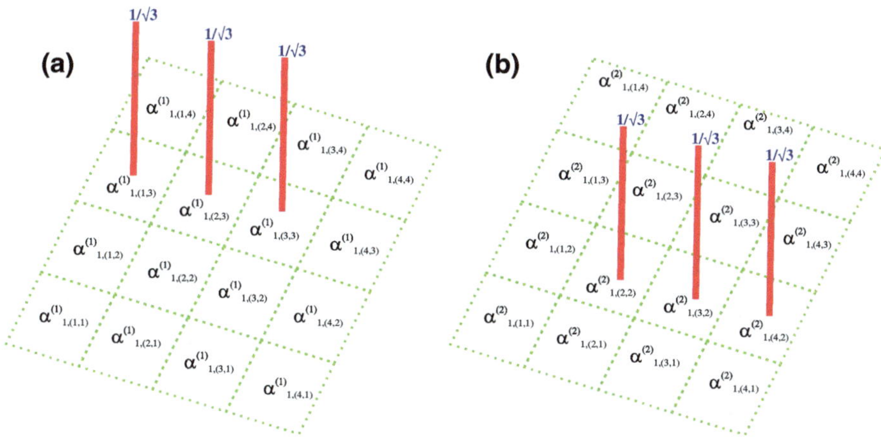

$$c_1 = \left[0\,0\,\tfrac{1}{\sqrt{3}}\,0\,0\,0\,\tfrac{1}{\sqrt{3}}\,0\,0\,0\,\tfrac{1}{\sqrt{3}}\,0\,0\,0\,0\,0\right]^T$$
$$c_2 = \left[0\,0\,0\,0\,0\,\tfrac{1}{\sqrt{3}}\,0\,0\,0\,\tfrac{1}{\sqrt{3}}\,0\,0\,0\,\tfrac{1}{\sqrt{3}}\,0\,0\right]^T$$

Fig. 6 Local non-rigid effects and the need for linear assignment. **a** is density p_1 of the first shape, with only scaling coefficients, $c_1 = \left[\alpha^{(1)}_{j,\mathbf{k}}\right]^T$, shown. **b** is the second shape with density p_2 with coefficients $c_2 = \left[\alpha^{(2)}_{j,\mathbf{k}}\right]^T$. Locally the point sets only differed by a translation which resulted in the densities differing by a translation. Without linear assignment the coefficient vectors of these would give a inner product of 0 and consequently large geodesic distance on the hypersphere. Linear assignment can correctly recover the local translation and then the geodesic distance will be small, reflecting the true similarity between the shapes (color figure online)

permutation group on the ordering of the coefficients. These specific requirements can be addressed within a linear assignment construct (Luenberger 1984); thus our deformation model can be interpreted as a "sliding grammar" wherein we only allow wavelets at each level j to independently slide to get a good match. The independent sliding assumption at each level implies that the "probability density mass" corresponding to each wavelet is independent of the rest. Consequently, this allows us to independently slide each wavelet to get a best match while maintaining the unit integrability constraint. While this justifies the independence assumption, "deformation grammars" more complex than sliding could be considered, e.g. splitting coefficients. However, we restricted ourselves to only sliding the wavelets leaving more exotic rules for future research. Even though each wavelet is allowed to slide, we cannot allow the sliding wavelets to collide and end up at the same spatial location. This imposes a permutation constraint on the sliding wavelets, see Fig. 7. Thus our new objective to minimize becomes

$$\begin{aligned} D(p_1, p_2; \pi) = -2 + 2\Big[&\textstyle\sum_{j_0\mathbf{k}} \alpha^{(1)}_{j_0,\mathbf{k}}\alpha^{(2)}_{j_0,\pi(\mathbf{k})} \\ &+ \textstyle\sum_{j\geq j_0,\mathbf{k}}^{j_1} \beta^{(1)}_{j,\mathbf{k}}\beta^{(2)}_{j,\pi(\mathbf{k})}\Big] \end{aligned} \tag{23}$$

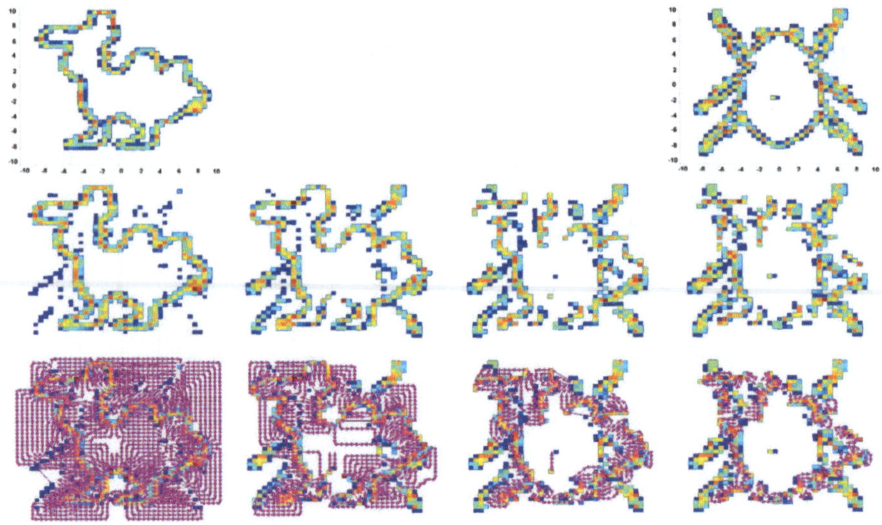

Fig. 7 Effects of λ on linear assignment. *Top row far left* is target shape and *far right* is the source. *Second row* shows for small λ the source shape is almost perfectly transformed to the target while for large λ the source shape retains original shape; λ values from *left* to *right* 10, 250, 500, and 1000. *Third row* illustrates the wavelet coefficients movement in row two (best viewed in color). The densities were estimated using the Haar family with $j_0 = 1$ (color figure online)

where $\pi(\mathbf{k})$ is a permutation operator that takes as input the wavelet spatial index $\mathbf{k}$ and returns a new index $\mathbf{k}'$ at the same level. (Since the wavelet coefficients can all be reversed to get the same density, there's an overall sign symmetry which is accounted for in the linear assignment algorithm by running it twice—once with the set of coefficients $\{\alpha_{j_0,\mathbf{k}}, \beta_{j,\mathbf{k}}\}$ and a second time with $\{-\alpha_{j_0,\mathbf{k}}, -\beta_{j,\mathbf{k}}\}$.) The space of possible permutations is large and hence this objective needs to be regularized to yield useful results. Otherwise, every *source* shape's coefficients could be re-ordered to be in the shape of the *target*; this is a detriment to recognition since any shape can essentially match another. To overcome this effect, we penalize large spatial movements by incorporating a cost based on the Euclidean distance between the centers of basis functions. This restricts large movements of the coefficients forcing them to be only locally translated. Incorporating this penalty gives our final objective function

$$E(\pi) = D(p_1, p_2; \pi) + \lambda \Big[\sum\nolimits_{j_0,\mathbf{k}} \|\mathbf{r}(j_0,\mathbf{k}) - \mathbf{r}(\pi(j_0,\mathbf{k}))\|^2 + \sum\nolimits_{j,\mathbf{k}} \|\mathbf{r}(j,\mathbf{k}) - \mathbf{r}(\pi(j,\mathbf{k}))\|^2 \Big] \tag{24}$$

where $\mathbf{r}(j, \mathbf{k})$ is a location operator—essentially giving us the center of the wavelet basis at $(j, \mathbf{k})$—which has two inputs, the level j (and this includes j_0), the wavelet spatial index $\mathbf{k}$ and returns a spatial location $\mathbf{r} \in \mathbb{R}^2$. The basic idea here is that as the regularization parameter λ is increased, the objective increasingly favors shorter wavelet sliding movements and hence smaller deformations. The optimal permutation π^* can be obtained by setting up the cost matrix

$$\mathcal{C} = c_1 c_2^T + \lambda d \tag{25}$$

where c_i is a vectorized representation of all the density wavelet coefficients for shape i and the matrix d contains pairwise distances between the wavelet basis locations. Figure 7 illustrates the effect of λ on the linear assignment and hence the similarity metric.

Recently, Wasserstein metrics have had a resurgence due to the more efficient, entropy-regularized optimization of the quadratic metric (Cuturi 2013). Several techniques (Flamary et al. 2014; Solomon et al. 2015; Benamou et al. 2015) employing this entropic regularization approximation have demonstrated interesting anecdotal results for shape analysis. Yet, none have evaluated the performance of the quadratic Wasserstein distance metric as meaningful shape retrieval metric. Since the Wasserstein metric works on the space of measures, we conjuncture that working strictly in the space of measures is not appropriate when we want to consider the underlying geometric distortions that may exist in the ground space—the ground space is random variable domain on which we have fit the densities. We illustrate our point in Fig. 8, where the current state-of-the-art (Benamou et al. 2015) shape interpolation

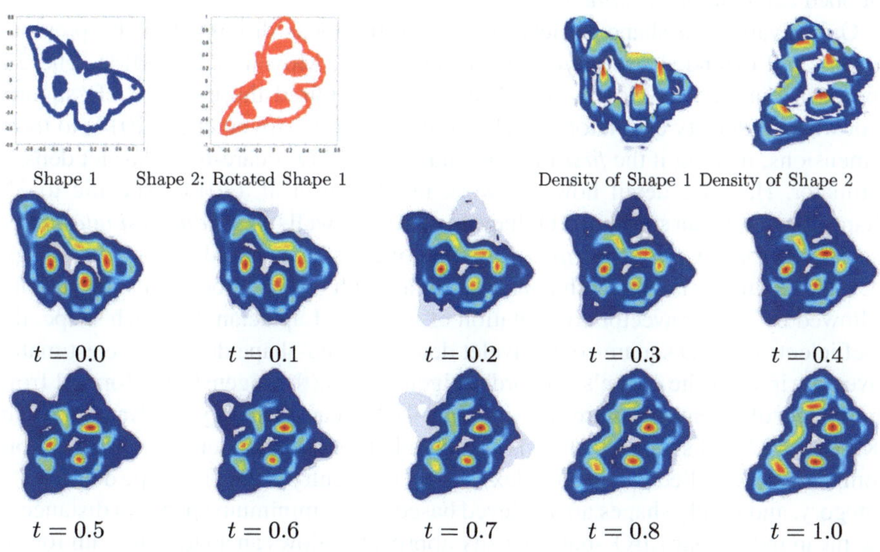

Fig. 8 Wasserstein barycentric geodesics. The standard Wasserstein metric is unable to correctly capture geometric transformations that may exist between shapes. Using the current state-of-the-art Wasserstein shape interpolation from Benamou et al. (2015), we see that intermediate densities do not geometrically change. Instead, the measure weights are arbitrarily allowed to morph from one shape to another without respecting geometric structures of the shape

method, Wasserstein barycentric geodesics, is used morph two shape densities. These straight-line Wasserstein geodesics do not produce meaningful intermediate shapes. We anticipate this will lead to suboptimal performance as a shape metric. In the future, we plan to directly compare retrieval performance of our optimal transport method on the parameter space of the densities to the Wasserstein measure space approach. The answer may lie in the closely-related early works such as Rangarajan et al. (1996), Gold and Rangarajan (1996), where the authors optimize over the space of joint densities—almost identical to the Wasserstein formulation—to establish shape correspondence, while simultaneously solving for the deformation model.

5.2 *Isometric Invariant Matching*

For large scale shape matching applications, we desire that any candidate technique consider requirements such as: compressible shape representations, rich discrimination capabilities with categorical flexibility, transformation invariance, computational scalability, etc. Our proposed shape retrieval framework satisfies many of these sought-after characteristics, like supporting multiresolution sparse representations, inherent robustness to isometric transformations, and having a simple closed-form similarity measure. We achieve these notable advantages as a result of using our SR-WDE model on the eigenspace the shape's Laplace–Beltrami operator (LBO) (dubbed *LBO-shape densities*).

Our advances in shape retrieval builds on the ideas discussed in the previous section and extended by Moyou and Peter (2012), Moyou et al. (2014). The 2D approach outlined in Moyou and Peter (2012) required the use of a 2D square-root wavelet density estimator, which was extended in Moyou et al. (2014) to three dimensions, making it the *first* implementation of a 3D square-root wavelet density estimator. Here we detail how the work in Moyou et al. (2014) (specific to 3D shape matching) can subsume 2D shape matching as well—*providing a single unified framework for 2D and 3D shapes.* Retrieval of 2D shapes based on the LBO-Shape Density technique begins with constructing a graph on the unordered shape points, followed by an eigenvector computation of its graph Laplacian. For each shape, the coefficients of a 3D square-root wavelet density (LBO-shape density) are estimated given a triplet of the shape's low order eigenvectors (the eigenshapes formed from the triplets of eigenvectors are shown in Fig. 9). For each category, the Karcher Mean (Karcher 1977) of shape densities is computed on the hypersphere—creating a more compact index (called the mean index), where each entry is the prototype density of a category, and finally shapes are retrieved based on the minimum spherical distance to the mean index. Our LBO-shape density approach follows an analogous path for 3D shapes, we first compute the low order eigenvectors of the shape's LBO (see Fig. 2)—the only thing changed from 2D is that we now use the cotangent approximation of the LBO. Again, a triplet of eigenvectors is used to estimate a 3D square-root wavelet density for each shape. After which, the mean index is generated by executing the Karcher mean algorithm on the estimated densities per shape category. The minimum

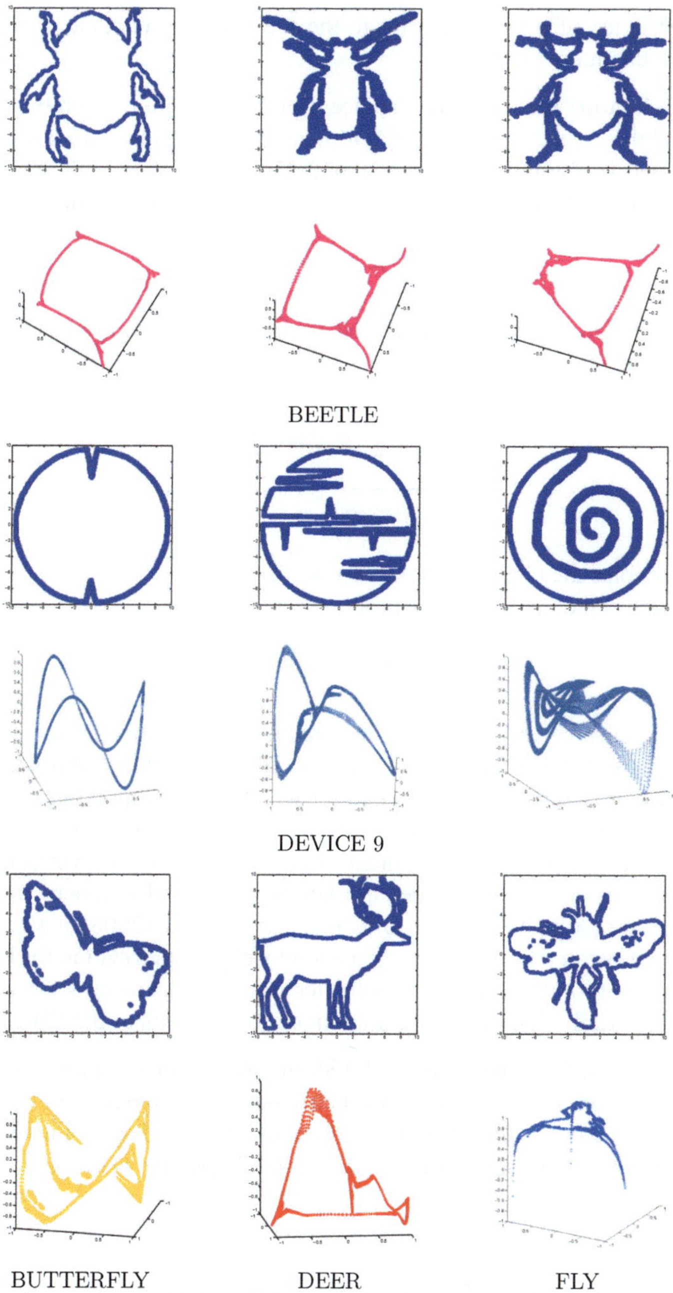

Fig. 9 Eigenshapes of various 2D models from the MPEG 7 dataset formed from the eigenvectors the graph Laplacian. The original 2D shapes are shown in rows 1, 3 and 5 in *blue*. The image below each original shape is its corresponding eigenshape, which are formed using the eigenvector triplet (1, 2, 5). To match 2D shapes we estimate wavelet densities directly on these eigenshapes *(LBO-shape densities)*. Notice, that shapes within a category have similar eigenshapes whereas across categories the eigenshapes are distinctly different (color figure online)

spherical distance of a shape density to the mean index serves as a classification criterion for the query shapes.

Laplace–Beltrami Operator The Laplace–Beltrami operator generalizes the Laplacian of Euclidean spaces to Riemannian manifolds. Consider a function f to be a C^2 real-valued function defined on a smooth Riemannian manifold M with metric tensor g. The coordinate-free Laplace–Beltrami operator Δ is defined as

$$\Delta f = -\text{div}\,(\text{grad}\, f)\,, \tag{26}$$

where div and grad are the divergence and gradient on the manifold M (Isaac 1984). In local coordinates, this reduces to

$$\Delta f = \underbrace{\frac{1}{\sqrt{|g|}} \sum_i \partial_i \sqrt{|g|}}_{divergence} \underbrace{\sum_j g^{ij} \partial_j f}_{gradient}\,. \tag{27}$$

where g^{ij} denotes the elements of the inverse of g.

Solving the standard *eigenvalue* problem

$$\Delta v = -\lambda v, \tag{28}$$

for λ (the eigenvalue of Δ) and v (the eigenvector corresponding to λ), provides the necessary tools for a variety of geometric analysis on the manifold. (Note: For the LBO, $\lambda = 0$ is always an eigenvalue for which its corresponding eigenvector is constant and hence discarded in most applications, including ours). The Laplace–Beltrami operator over a compact manifold S is bounded and symmetric positive semi-definite; its set of eigenvalues are non-negative real numbers and its set of eigenvectors are countable (Zeng et al. 2012). In Jones et al. (2008), it was proven that the LBO eigenvectors are theoretically a good local parametrization for Riemannian manifolds, affirming their use as coordinates in our subsequent density estimation step. We now briefly describe two common approximations of the LBO.

Graph Laplacian Construction for 2D Shape Retrieval The *Laplacian matrix* of a graph G or *graph Laplacian* is a symmetric positive semidefinite matrix given as $L = D - A$, where A is the adjacency matrix and D s the diagonal matrix of vertex degrees. The spectral decomposition of the graph Laplacian is given as

$$L\phi = \lambda\phi, \tag{29}$$

where λ is an eigenvalue of L with a corresponding eigenvector ϕ. The eigenvalues of the graph Laplacian are non-negative and constitute a discrete set. The spectral properties of L are used to embed the points into a lower dimensional space, and gain insight into the geometry of the shape (Isaac and Roberts 2011). The LBO-shape density technique possesses no-equal point set cardinality requirements and topological constraints because we use the unordered shape points to construct undirected graphs. The number of points in each shape varies, therefore the eigenvectors of their corresponding graph Laplacians will be of different lengths. By estimating wavelet densities on these eigenvectors we eliminate the point-set cardinality requirements.

Cotangent Construction for 3D Shape Retrieval The cotangent Laplacian approximation uses a linear finite element method (FEM) to discretize the LBO and was first detailed in Dziuk (1988), Pinkall et al. (1993). To compute the solution to (28), we verify it using its weak form,

$$\langle \Delta v, \varphi_i \rangle = -\lambda \langle v, \varphi_i \rangle \;\; \forall i \tag{30}$$

for some test function φ_i under the L_2 inner product. Finding the eigenvectors of the LBO with linear FEM amounts to solving the the *generalized eigenvalue problem*

$$A_{\cot} v_i = -\lambda B v_i, \tag{31}$$

with

$$\begin{aligned} A_{\cot}(i,j) &:= \begin{cases} \frac{\cot \kappa_{i,j} + \cot \xi_{i,j}}{2} & \text{edge } (i,j) \\ -\sum_{k \in N(i)} A_{\cot}(i,k) & i = j \end{cases} \\ B(i,j) &:= \begin{cases} \frac{|t_1| + |t_2|}{12} & \text{edge } (i,j) \\ \frac{\sum_{k \in N(i)} |t_k|}{6} & i = j, \end{cases} \end{aligned} \tag{32}$$

where $|t_i|$ is the area of triangle t_i. The variables t_1 and t_2 are the triangles that share edge (i, j); B is the mass matrix; $A_{\cot}$ is the stiffness matrix with cotangent weights, $\kappa_{i,j}$ and $\xi_{i,j}$, are the angles opposite to the edge (i, j) (see Reuter et al. 2009 for a more extensive description of the discretization). Through this formulation, the weighted inner product induced by B, i.e. $\langle f, g \rangle_B = f^T B g$ for $f, g \in \mathbb{L}^2$, generalizes the $\mathbb{L}^2$ inner product (i.e. $B := I$); it is intrinsic to the surface on which it is defined; and is adapted to the local sampling of M through the variation of triangle areas (Patané 2013). The mass matrix B also accurately encodes the geometry of the input surface through the area of its triangles and leads to FEM distances having higher robustness to topological and scale changes, irregular sampling and noise. Note that this formulation of the LBO only works on closed manifolds which was the focus of this initial research effort; extensions to other LBO discretizations will be done in future research. It is well known that the eigenvectors of the decomposition of a matrix are recovered up to a sign factor of ± 1, and any of these methods (Umeyama 1988; Park et al. 2000; Caelli and Kosinov 2004; White and Wilson 2007) can be used to sufficiently address this problem.

Eigenvectors corresponding to smaller eigenvalues correlate to lower frequency mode properties of the shape, and larger eigenvalue-eigenvector pairs capture high frequency modes of the surface—see Fig. 10 for a visualization of these characteristics. In our extensive empirical evaluations, we have found that using triplet combi-

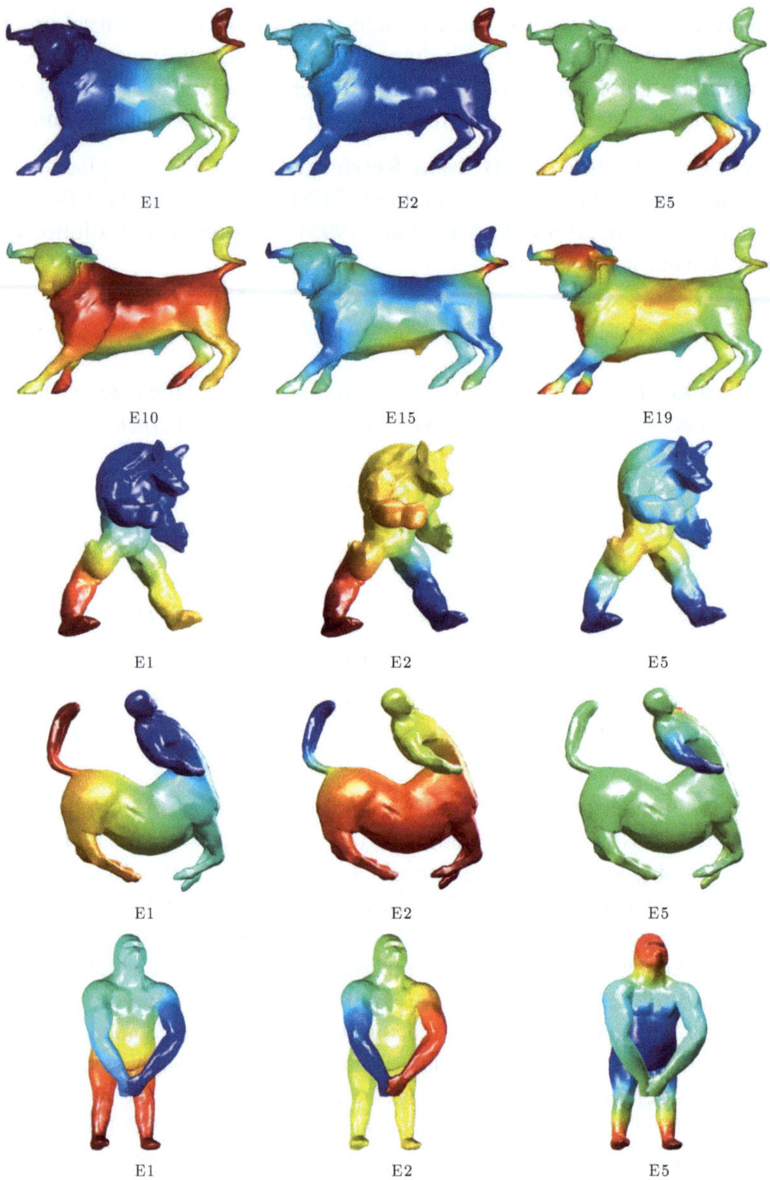

Fig. 10 Example eigenvectors of a bull (Shape COSEG, Four Leg category), armadillo, centaur, and gorilla (SHREC'11). The subtitles indicate which eigenvector is plotted on the shape, for instance E10 is the 10th eigenvector. The low order eigenvectors represent the lower frequency modes of the shape, and are more robust to noise and deformations

Algorithm 1 Numerical computation of Karcher mean on manifold M. For the present context $M = S^m$, m-dimensional unit hypersphere, the Exp and Log maps are defined in (33) and (34), respectively. κ is a small step size parameter.

Input: $\rho_1, \rho_2, \ldots, \rho_m \in M$
Output: $\mu \in M$
Let $\mu^0 = \rho_1$
While $\|\gamma^\tau - \gamma^{\tau-1}\| > \epsilon$

$$\gamma^\tau = \frac{\kappa}{m} \sum_{i=1}^{m} \mathrm{Log}_{\mu^{\tau-1}}(\rho_i)$$
$$\mu^\tau = \mathrm{Exp}_{\mu^{\tau-1}}(\gamma^\tau)$$

nations from the 10 lowest order eigenvectors are sufficient to discriminate among the various shape categories. Once the desired triplet of eigenvectors are selected, our objective now becomes to estimate the trivariate distribution $p(v_1, v_2, v_3)$ on the eigenvectors in a wavelet basis. This can be readily accomplished using the SR-WDE detailed previously in Sect. 3.

Retrieval Using Intrinsic Means on the Wavelet Hypersphere Given exemplars from a particular shape category, we can obtain a prototype representation of that category by computing a mean model. In the proposed framework, this notion translates to computing a mean density function from the LBO-shape densities of a particular category. Keep in mind, with our framework, both 2D and 3D shape matching require estimation of 3D densities, resulting in similar hypersphere representations of the densities. Hence, they share the same retrieval mechanics—a unique and unifying property of LBO-shape densities.

Since the densities are points on the manifold (hypersphere), obtaining a mean density function requires us to compute the generalized Karcher mean (Karcher 1977). To execute this intrinsically on the manifold, we employ the Exponential (Exp) and Logarithm (log) maps on the manifold (available as analytic formulas for the hypersphere), and implement the simple optimization procedure detailed in Algorithm 1 (more details in Pennec 2006). In the algorithm and equations below, we let $\rho_i = \left\{\alpha^{(i)}_{j_0,l}, \beta^{(i)}_{j,l}\right\}$ represent the vectorized set of estimated wavelet coefficients associated with the ith shape.

The Exp map takes a vector γ on the tangent space at $\rho_1, \gamma \in T_{\rho_1}\left(S^{n-1}\right)$, and returns a point ρ_2 on the hypersphere

$$\rho_2 = \mathrm{Exp}_{\rho_1}(\gamma) = \cos(|\gamma|)\,\rho_1 + \sin(|\gamma|)\,\frac{\gamma}{|\gamma|}. \tag{33}$$

Conversely, the Log map takes a point ρ_2 on the hypersphere and returns a vector on the tangent space at ρ_1, by letting

$$\tilde{\rho} = \rho_2 - \langle \rho_2, \rho_1 \rangle \rho_1 \tag{34}$$

$$\gamma = \mathrm{Log}_{\rho_1}(\rho_2) = \tilde{\rho}\frac{\cos^{-1}(\langle \rho_1, \rho_2 \rangle)}{\sqrt{\langle \tilde{\rho}, \tilde{\rho} \rangle}}.$$

For the shape retrieval problem with multiple categories, our classification approach is to compute a mean density for each of the classes (mean index) using the associated densities in each class. Given a query shape we estimate its wavelet density from the corresponding eigenspace and compute the distance of the query density to each entry in the mean index using the closed-form distance on the hypersphere

$$d(\rho, \mu_i) = \cos^{-1}(\rho^T \mu_i), \tag{35}$$

where ρ is coefficient set for the query shape density and μ_i is the set of coefficients associated with the mean shape density of the ith class. The category label of the closest mean density is assigned to the query shape. It is worth noting, that all of our analysis is taking place intrinsically on the manifold of our shape representation, an added advantage over other methodologies that decouple the representation and matching.

6 Experimental Results

In the previous sections, we detailed how the square-root wavelet density estimator and its associated spherical geometry can be leveraged for model selection (choosing the starting j_0 resolution level), optimal transport 2D shape matching, and the unifying 2D and 3D LBO-shape density framework. We now present experimental validation of each these three techniques.

6.1 Model Selection

The experiments evaluated the capability of the model selection methods to adequately select the best probability density for a given set of sample data, while judiciously balancing the desire for accuracy and model complexity. We first validate our approach on a complex set of 1D densities as in Marron and Wand (1992) and utilize a variety of compactly supported wavelet families in the density estimation. From each 1D density, 2000 samples were drawn and used in the parameter estimation process. For shape analysis, we illustrate the utility of MDL criterion to select the optimal density function representation and matching of shape point sets.

The MDL criterion of Eq. (13) (denoted MDL-3 in results) was applied to the selection of the best j_0 level for the wavelet density estimator. For comparative analysis, we computed several other information-theoretic model selection criteria:

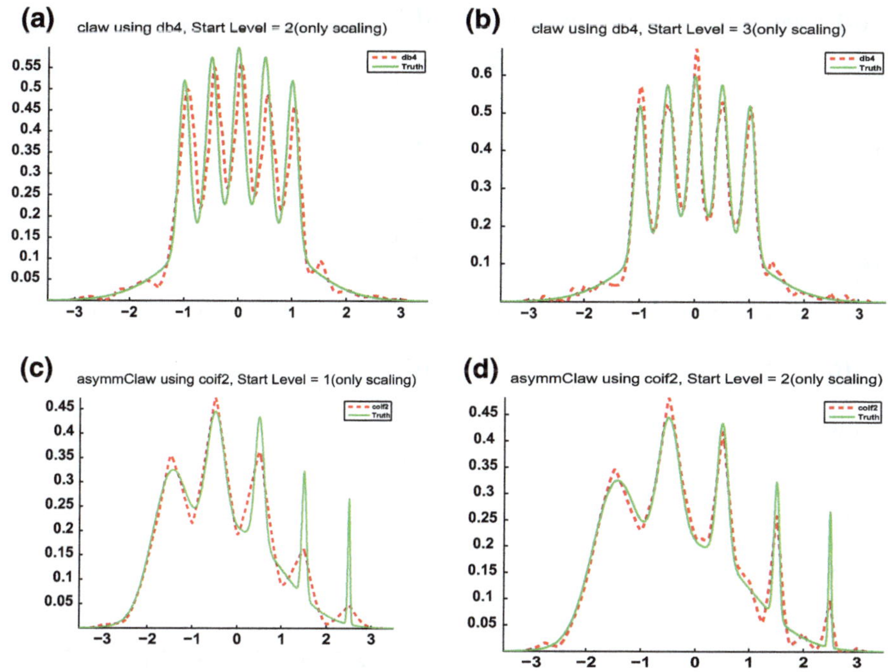

Fig. 11 Model selection using MDL-3 versus MDL-2. MDL-3 is able to select more complex models than MDL-2. **a** $j_0^* = 2$ by MDL-2. **b** $j_0^* = 3$ by MDL-3. **c** $j_0^* = 1$ by MDL-2. **d** $j_0^* = 2$ by MDL-3. Wavelet family DB4 used for **a** and **b**, COIF2 used for **c** and **d**

the original two-term MDL (MDL-2) which lacks the model class Riemannian volume, AIC and BIC. In addition, since the true densities are known, we calculated three standard discrepancy measures: mean-squared error (MSE), Hellinger divergence (HELL) and L_1 loss. The best starting level j_0^* was selected as the minimum of these measures for $j_0 \in [-1, 6]$. A larger value of j_0 indicates a more complex model since it corresponds to a finer resolution level in the wavelet decomposition. Extensive comparisons were conducted using basis functions from the Daubechies, Symlet and Coiflet families for each of criteria. MDL-3 and MDL-2 generally agreed on best levels across densities and families. There are a few cases in which MDL-3 (with the additional volume term) selected more complex models than MDL-2. In each of these cases, the selection of the higher complexity model was justified by the need to accurately capture the abrupt variations of the true data-generating densities. A high-level summarization of the general trends in the experiments is visually illustrated in Figs. 11, 12 and 13. In Fig. 11, we see examples of two such cases where the MDL-3 selected value of j_0 provides a better suited model. Thus the inclusion of the full geometric complexity $\ln\left(\frac{\mathcal{V}_S}{\mathcal{V}_{\hat{\mathbf{\Theta}}}}\right)$ can aid in the selection of more accurate models.

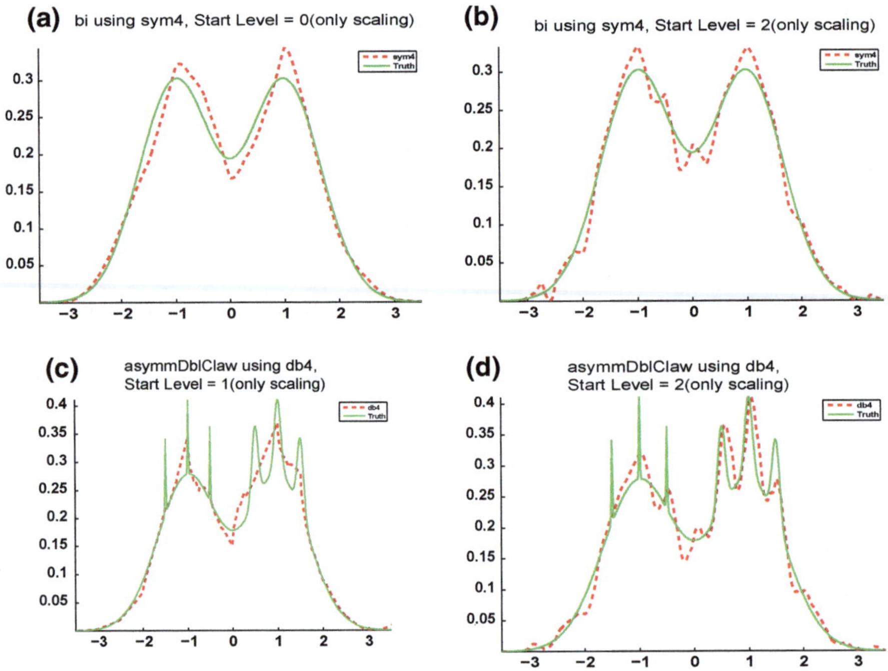

Fig. 12 Model selection using MDL-3 versus AIC. AIC generally selects more complex models than MDL-3. This can be helpful for complex densities like in **c** and **d**, but can also over estimate smooth ones like in **a** and **b**. **a** $j_0^* = 0$ by MDL-3. **b** $j_0^* = 2$ by AIC. **c** $j_0^* = 1$ by MDL-3. **d** $j_0^* = 2$ by AIC. Wavelet family SYM4 used for **a** and **b**, DB4 used for **c** and **d**

In general, our MDL criterion also agrees with the AIC and BIC. As expected, AIC tends to pick slightly more complex models than MDL and BIC. This is because AIC does not incur a penalty dependent on the sample size. This slight over estimation can be a benefit when considering complex densities but it can also over compensate, see Fig. 12. AIC selects a more complex model than necessary for the bimodal density [see (a) and (b)]. It starts to favor trends in the data, degrading its generalization capability. However, for a complex density like the asymmetric double claw, the AIC selection is a better model. BIC tends to somewhat underestimate the models, selecting less complex models than necessary to accurately represent the densities (see Fig. 13a and b).

In real-world applications, the MSE, HELL, and L_1 are not useful model selection criteria since the true underlying densities are not accessible or unknown. They also lack the trade-off between goodness-of-fit and complexity, only using the former as the performance measure. Hence, biased error measures like the MSE, tend to pick more complex models, Fig. 13c and d. Since we know the true densities, the global agreement between these error measures and the information-theoretic model selection criteria showcases the power of these methods—without knowledge of true

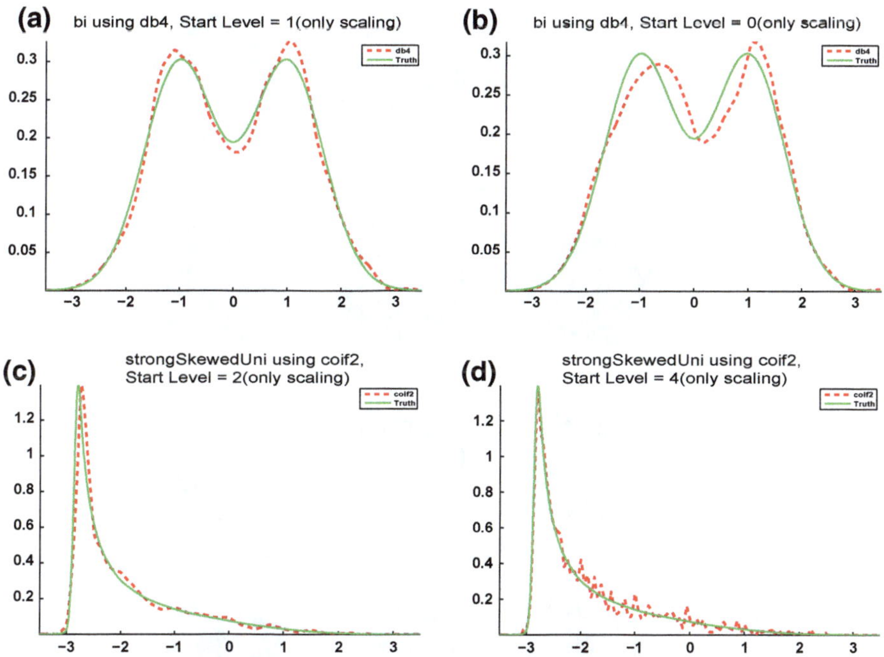

Fig. 13 Model selection using MDL-3 versus BIC and MSE. BIC tends to favor less complex models than MDL-3, see **a** and **b**. The MSE generally overfits since it is only a goodness-of-fit measure. **a** $j_0^* = 1$ by MDL-3. **b** $j_0^* = 0$ by BIC. **c** $j_0^* = 2$ by MDL-3. **d** $j_0^* = 4$ by MSE. Wavelet family DB4 used for **a** and **b**, COIF2 used for **c** and **d**

densities, they are able to select models that best describe the data while balancing the complexity of the model.

For applications to 2D shape analysis, Fig. 14 illustrates the j_0 levels chosen by MDL-3, MDL-2, AIC and BIC for three MPEG-7 shapes (only a subset of the entire experiment). Here the density functions were estimated from point set representation of the shapes. The shape matching techniques discussed throughout this chapter can easily utilizing our model selection approach to select the appropriate j_0 for the SR-WDE.

6.2 2D Optimal Transport Matching

The presented technique was evaluated on the MPEG-7 database (Latecki et al. 2000). The original data set consists of 70 different categories with 20 observations per category for a total of 1400 binary images. Each image consists of a single shape. One of the main strengths of our method is its accessibility and ease of use. The first part involves simply taking the data samples for each object and using

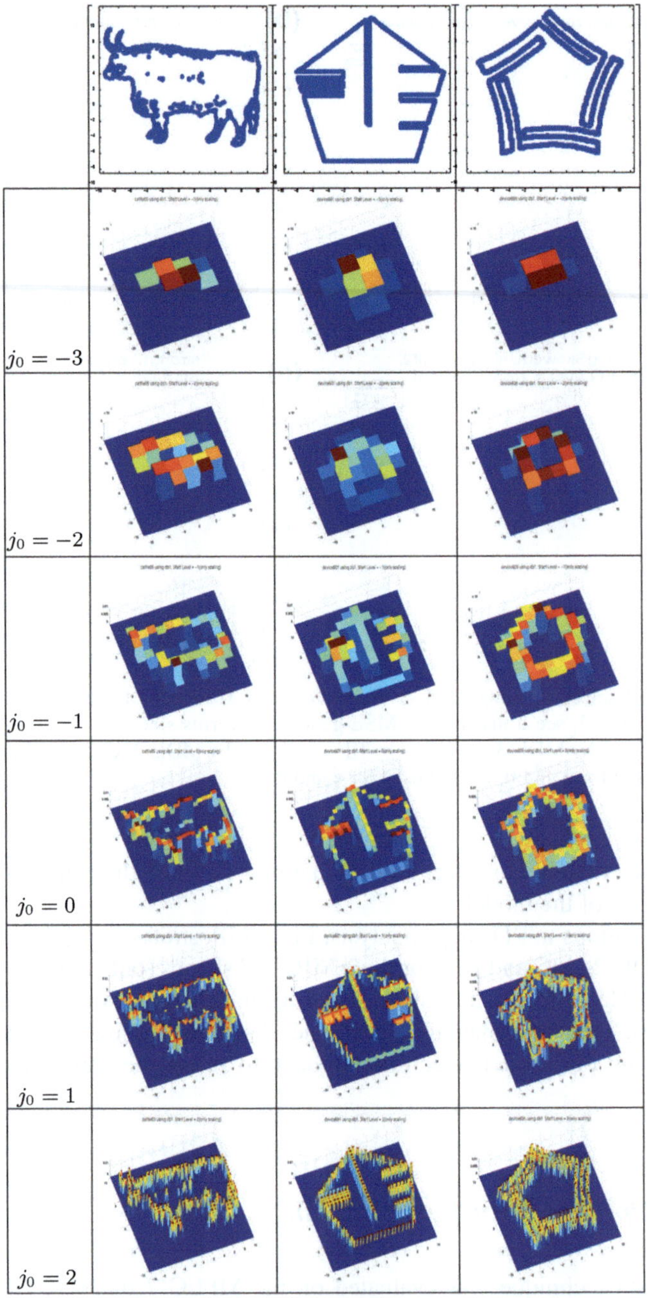

Fig. 14 Model selection for 2D densities using MDL-3, MDL-2, AIC and BIC. Row 1 represents MPEG-7 shapes Cattle-05 (8,671 points), Device6-01 (8,947 points), and Device6-08 (11,301 points), respectively. Remaining rows show densities estimated from these point sets at different j_0 levels. For all three shapes, MDL-3 and AIC selected $j_0 = 1$, while MDL-2 and BIC selected $j_0 = 0$

them to estimate $\left\{\alpha_{j_0,\mathbf{k}}, \beta^w_{j,\mathbf{k}}\right\}$ for the wavelet expansion of $\sqrt{p}$. In the context of shape indexing this phase is completely off-line, i.e. wavelet densities for the entire database can be estimated once and before the actual similarity computation takes place. Next, to compare two shapes, we first use the regularized linear assignment (24) to handle non-rigid effects and then use closed-form distance on unit hypersphere to obtain the similarity measure between them. We compare the performance of our optimal transport method to D2 shape distributions (Osada et al. 2002). In Osada et al. (2002), the authors then use pairwise distances between shape points to construct a 1D histogram for each shape; this serves as a unique shape signature. Instead of using histograms, we estimate a 1D wavelet density for each shape. Distance metrics between shapes can be obtained by using a variety of 1D density dissimilarity measures. In addition to the Hellinger divergence, Eq. (9), we computed three other measures:

- Bhattacharyya: $D(p_1, p_2) = 1 - \int \sqrt{p_1 p_2} dx$
- χ^2: $D(p_1, p_2) = \int \frac{(p_1 - p_2)^2}{p_1 + p_2}$
- L_2: $D(p_1, p_2) = \left(\int (p_1 - p_2)^2 dx\right)^{\frac{1}{2}}$

Performance on the MPEG-7 is most commonly evaluated using the bulls-eye criterion (Latecki et al. 2000; McNeill and Vijayakumar 2006). Each shape is used as a query shape and the top 40 matches are retrieved from all 1400 shapes (the test shape is not removed). For a single query, the maximum possible correct retrievals are 20 coinciding with the number of shapes in each category. Table 1 lists the recognition rates using several density similarity measures for both optimal transport matching and D2 shape distributions. Our optimal transport matching significantly outperforms D2 shape distributions. This gives credence to the idea of working with feature representations that mimic the true visual properties of shapes, i.e. D2 shape distributions represent objects using a 1D signature derived from the 2D points whereas our method represents shapes using 2D densities which are visually similar to the shapes. The three different metrics computed for the optimal transport technique illustrate how λ impacts recognition performance. A judicious choice for λ can be made by optimizing over a training set.

Table 1 MPEG-7 recognition rate. Our optimal transport method out performs D2 Shape Distributions (Osada et al. 2002). In our method, the choice of λ affects the recognition rate. See text for explanation of metrics. (LA $\equiv$ linear assignment, EDP $\equiv$ Euclidean distance penalty)

Optimal Transport Matching		
Metrics	$\lambda = 500$	$\lambda = 2250$
Geodesic w/ LA	81.7%	**85.25%**
Geodesic + EDP	32.6%	12.1%
EDP	32.5%	11.8%

D2 Shape Distributions	
Metrics	
χ^2	59.3%
Hellinger	58.6%
Bhattacharyya	58.6%
L_2	56.6%

6.3 LBO-Shape Density Matching

We demonstrate performance on SHREC'12 (Li et al. 2012) of a wavelet density-based indexing framework. In our processing pipeline, we first computed LBO eigenvectors followed by density estimation and indexing. Our specific approach for 3D shape retrieval was to obtain the low order eigenvectors (1, 2, 5) of each shape's cotangent approximation to the LBO after removal of the constant eigenfunction. First, the coefficients of a wavelet density were calculated on the triplet of eigenvectors using different wavelets (see Table 2). Then, these coefficients were used to compute the Karcher mean (atlas in LBO space) for each category, and form the mean index. Finally, shapes were classified using the minimum geodesic distance to the mean index and retrieval lists were ordered. The wavelets used in our experiments are members of the Daubechies family: Haar (Daubechies 1), Daubechies 4 and Daubechies 2 where the number indicates the order of vanishing moments. For the SHREC'12 database, Table 2 details our performance against the indexed heat curve approach in Khoury et al. (2012). Our approach outperforms the competing method across all the evaluation metrics except for the E-Measure. The effect of not conducting pose normalization as a pre-processing step is studied (see Haar-UN column in Table 2). The high performance results, even in the absence of pose normalization are directly due to the use of the LBO and its invariance under isometric transformations followed by a wavelet density distance in LBO space.

Table 2 Performance our LBO-shape density technique on SHREC'12

SHREC'12					
LBO shape density					IHC (Khoury et al. 2012)
Wavelet	Haar	DB4	Haar-UN	DB2	
Res. Lev.	3	2	3	2	
NN	0.969	0.931	0.939	0.931	0.810
FT	0.969	0.931	0.939	0.931	0.570
ST	0.969	0.944	0.973	0.951	0.710
E (10)	0.918	0.882	0.889	0.882	–
E (32)	0.429	0.423	0.430	0.423	0.590
DCG	0.980	0.958	0.967	0.959	–
AP	0.973	0.942	0.954	0.943	–

7 Conclusion

This chapter has detailed theoretical and practical benefits of using orthogonal series expansions of the square root of the density function. Theoretically we have demonstrated the spherical information geometry associated with these models and how one can utilize intrinsic analysis on this space to address important issues such as model order selection of the series expansion and statistical calculations on the space of densities such as computing mean densities. Throughout we have employed wavelets as the preferred basis functions, yielding a number of practical benefits for our shape matching applications: faithful density reconstructions, sparse multiscale representations, and efficient optimization. Our square-root wavelet density estimator (SR-WDE) representation of shapes was applied in two different manners: (1) matching shapes with non-rigid deformations in an optimal transport formulation, and (2) approximating the Laplace–Beltrami operator on the shape manifold and the estimating densities on the low order eigenspace for isometric invariant matching. We demonstrated how when employing the SR-WDE, the decomposition levels of the expansion can be principally chosen via the use of MDL criterion, which is in closed-form for the spherical manifold. Several experiments validated our techniques with state-of-the-art performances demonstrated for many cases. In the future, we will continue to explore other applications beyond shape analysis for the SR-WDE. As a general purpose density estimator, it is a valuable tool for any application requiring estimation of robust probabilistic models.

References

Akaike, H. (1973). Information theory and an extension of the maximum likelihood principle. In B. N. Petrov & F. Csaki (Eds.), *Proceedings of the 2nd International Symposium on Information Theory* (pp. 267–281).

Amari, S. I., & Nagaoka, H. (2001). *Methods of information geometry*. Providence: American Mathematical Society.

Arwini, K., & Dodson, C. (2008). *Information geometry: Near randomness and near independence*. New York: Springer.

Balasubramanian, V. (1997). Statistical inference, Occam's razor, and statistical mechanics on the space of probability distributions. *Neural Computation*, *9*(2), 349–368.

Benamou, J. B., Carlier, G., Cuturi, M., Nenna, L., & Peyré, G. (2015). Iterative Bregman projections for regularized transportation problems. *SIAM Journal on Scientific Computing*, *37*(2), A1111–A1138.

Beran, R. (1977). Minimum Hellinger distance estimates for parametric models. *The Annals of Statistics*, *5*(3), 445–463.

Bhattacharyya, A. (1943). On a measure of divergence between two statistical populations defined by their probability distributions. *Bulletin of the Calcutta Mathematical Society*, *35*, 99–109.

Bookstein, F. L. (1989). Principal warps: Thin-plate splines and the decomposition of deformations. *IEEE Transactions on Pattern Analysis Machine Intelligence*, *11*(6), 567–585.

Bronstein, A. M., Bronstein, M. M., Guibas, L. J., & Ovsjanikov, M. (2011). Shape Google: Geometric words and expressions for invariant shape retrieval. *ACM Transactions on Graphics*, *30*, 1–20.

Bronstein, M. M., & Kokkinos, I. (2010). Scale-invariant heat kernel signatures for non-rigid shape recognition. In *IEEE Conference on Computer Vision and Pattern Recognition (CVPR)*. (pp. 1704–1711)
Caelli, T., & Kosinov, S. (2004). An eigenspace projection clustering method for inexact graph matching. *IEEE Transactions on Pattern Analysis and Machine Intelligence*, *26*, 515–519.
Chen, T., Vemuri, B. C., Rangarajan, A., & Eisenschenk, S. J. (2010). Group-wise point-set registration using a novel CDF-based Havrda-Charvát divergence. *International Journal of Computer Vision*, *86*(1), 111–124.
Chuang, G. C. H., & Kuo, C. C. J. (1996). Wavelet descriptor of planar curves: Theory and applications. *IEEE Transactions on Image Processing*, *5*(1), 56–70.
Chui, H., & Rangarajan, A. (2000). A new algorithm for non-rigid point matching. In *IEEE Conference on Computer Vision and Pattern Recognition (CVPR)* (Vol. 2, pp. 44–51). IEEE Press.
Chui, H., & Rangarajan, A. (2004). Unsupervised learning of an atlas from unlabeled point-sets. *IEEE Transactions on Pattern Analysis and Machine Intelligence*, *26*, 160–172.
Cuturi, M. (2013). Sinkhorn distances: Lightspeed computation of optimal transport. In C. Burges, L. Bottou, M. Welling, Z. Ghahramani, & K. Weinberger (Eds.), *Neural information processing systems* (pp. 2292–2300).
Cuturi, M., & Doucet, A. (2015). Fast computation of Wasserstein barycenters. In *International Conference on Machine Learning* (pp. 685–693).
Donoho, D. L., Johnstone, I. M., Kerkyacharian, G., & Picard, D. (1996). Density estimation by wavelet thresholding. *Annals of Statistics*, *24*(2), 508–539.
Doukhan, P. (1988). Formes de Töeplitz associées à une analyse multiechélle. *Comptes Rendus de l' Académie des Sciences*, *306*, 663–666.
Dryden, I. L., & Mardia, K. V. (1998). *Statistical shape analysis*. New York: Wiley.
Dziuk, G. (1988). Finite elements for the Beltrami operator on arbitrary surfaces. *Partial differential equations and calculus of variations, lecture notes in mathematics* (Vol. 1357, pp. 142–155). New York: Springer.
Efron, B. (1975). Defining the curvature of a statistical problem (with applications to second order efficiency). *The Annals of Statistics*, *3*(6), 1189–1242.
Flamary, R., Courty, N., Rakotomamonjy, A., & Tuia, D. (2014). *2014*. Workshop on Optimal Transport and Machine Learning (December: Optimal transport with Laplacian regularization. In Neural Information Processing Systems.
Gold, S., & Rangarajan, A. (1996). Softassign versus softmax: Benchmarks in combinatorial optimization. In D. S. Touretzky, M. C. Mozer, & M. E. Hasselmo (Eds.), *Advances in neural information processing systems* (Vol. 8, pp. 626–632). Cambridge: MIT Press.
Grünwald, P. (2005). A tutorial introduction to the minimum description length principle. In P. Grünwald, I. Myung, & M. Pitt (Eds.), *Advances in minimum description length: Theory and applications*. Cambridge: MIT Press.
Guo, H., Rangarajan, A., & Joshi, S. (2005). 3-D diffeomorphic shape registration on hippocampal data sets. In *Medical Image Computing and Computer Assisted Intervention (MICCAI)* (pp. 984–991).
Hardle, W., Kerkyacharian, G., Pickard, D., & Tsybakov, A. (1998). *Wavelets, approximation, and statistical applications* (Vol. 129). Lecture notes in statistics. New York: Springer.
Hitchcock, F. L. (1941). The distribution of a product from several sources to numerous localities. *Journal of Mathematical Physics*, *20*, 224–230.
Hou, T., Hou, X., Zhong, M., & Qin, H. (2012). Bag-of-feature-graphs: A new paradigm for non-rigid shape retrieval. In *International Conference on Pattern Recognition (ICPR)* (pp. 1513–1516).
Isaac, C. (1984). *Eigenvalues in Riemannian geometry* (2nd ed., Vol. 115). San Diego: Academic Press Professional, Inc.
Isaacs, J., & Roberts, R. (2011). Metrics of the Laplace-Beltrami eigenfunctions for 2D shape matching. In *IEEE International Conference on Systems, Man and Cybernetics* (pp. 3347–3352).
Izenman, A. (1991). Recent developments in nonparametric density estimation. *Journal of the American Statistical Association*, *86*(413), 205–224.

Jeffreys, H. (1961). *Theory of probability* (3rd ed.). New York: Oxford University Press.
Jian, B., & Vemuri, B. (2011). Robust point set registration using Gaussian mixture models. *IEEE Transactions on Pattern Analysis and Machine Intelligence*, *33*(8), 1633–1645.
Jones, P. W., Maggioni, M., & Schul, R. (2008). Manifold parametrizations by eigenfunctions of the Laplacian and heat kernels. *Proceedings of the National Academy of Sciences*, *105*, 1803–1808.
Karcher, H. (1977). Riemannian center of mass and mollifier smoothing. *Communications on Pure and Applied Mathematics*, *30*(5), 509–541.
Kass, R. E., & Raftery, A. E. (1995). Bayes factors. *Journal of the American Statistical Association*, *90*, 773–795.
Kass, R., & Voss, P. (1997). *Geometrical foundations of asymptotic inference*. New York: Wiley-Interscience.
Khoury, R., Vandeborre, J. P., & Daoudi, M. (2012). Indexed heat curves for 3D-model retrieval. In *ICPR* (pp. 1964–1967).
Klein, F. (1872). Vergleichende Betrachtungen über neuere geometrische Forsuchungen. Erlangen.
Kronmal, R., & Tarter, M. (1968). The estimation of probability densities and cumulatives by fourier series methods. *Journal of the American Statistical Association*, *63*, 925–952.
Latecki, L. J., Lakämper, R., & Eckhardt, U. (2000). Shape descriptors for non-rigid shapes with a single closed contour. In *CVPR* (pp. 424–429).
Levina, E., & Bickel, P. (2001). The earth mover's distance is the Mallows distance: Some insights from statistics. *International Conference on Computer Vision*, *2*, 251–256.
Li, B., Schreck, T., Godil, A., Alexa, M., Boubekeur, T., Bustos, B., et al. (2012). SHREC'12 track: Sketch-based 3D shape retrieval. In *Eurographics Workshop on 3D Object Retrieval* (pp. 109–118).
Liu, M., Vemuri, B., Amari, S. I., & Nielsen, F. (2010). Total Bregman divergence and its applications to shape retrieval. In *IEEE Conference on Computer Vision and Pattern Recognition (CVPR)* (pp. 3463–3468).
Luenberger, D. (1984). *Linear and nonlinear programming*. Reading: Addison-Wesley.
Marriott, P., & Salmon, M. (2011). *Applications of differential geometry to econometrics*. Cambridge: Cambridge University Press.
Marron, S. J., & Wand, M. P. (1992). Exact mean integrated squared error. *The Annals of Statistics*, *20*(2), 712–736.
McNeill, G., & Vijayakumar, S. (2006). Hierarchical Procrustes matching for shape retrieval. In *CVPR* (pp. 885–894).
Montgomery, D. C. (2004). *Design and analysis of experiments*. New York: Wiley.
Moyou, M., & Peter, A. M. (2012). Shape analysis on the hypersphere of wavelet densities. In *21st International Conference on Pattern Recognition* (pp. 2091–2094).
Moyou, M., Ihou, K. E., & Peter, A. M. (2014). LBO-shape densities: Efficient 3D shape retrieval using wavelet density estimation. In *22nd International Conference on Pattern Recognition (ICPR)* (pp. 52–57).
Murray, M., & Rice, J. (1993). *Differential geometry and statistics*. London: Chapman and Hall/CRC.
Myung, I. J., Balasubramanian, V., & Pitt, M. A. (2000). Counting probability distributions: Differential geometry and model selection. *Proceedings of the National Academy of Sciences*, *97*, 11170–11175.
Navarro, D. J. (2004). A note on the applied use of MDL approximations. *Neural Computation*, *16*, 1763–1768.
Nielsen, F., & Nock, R. (2014). Optimal interval clustering: Application to Bregman clustering and statistical mixture learning. *IEEE Signal Processing Letters*, *21*(10), 1289–1292.
Ohbuchi, R., Osada, K., Furuya, T., & Banno, T. (2008). Salient local visual features for shape-based 3D model retrieval. In *Shape Modeling International (SMI)* (pp. 93–102).
Osada, R., Funkhouser, T., Chazelle, B., & Dobkin, D. (2002). Shape distributions. *ACM Transactions on Graphics*, *21*(4), 807–832.

Park, S., Lee, K., & Lee, S. (2000). A line feature matching technique based on an eigenvector approach. *Computer Vision and Image Understanding (CVIU)*, *77*(3), 263–283.

Patané, G. (2013). wFEM heat kernel: Discretization and applications to shape analysis and retrieval. *Computer Aided Geometric Design*, *30*(3), 276–295.

Penev, S., & Dechevsky, L. (1997). On non-negative wavelet-based density estimators. *Journal of Nonparametric Statistics*, *7*, 365–394.

Pennec, X. (2006). Intrinsic statistics on Riemannian manifolds: Basic tools for geometric measurements. *Journal of Mathematical Imaging and Vision*, *25*(1), 127–154.

Peter, A. M., & Rangarajan, A. (2008). Maximum likelihood wavelet density estimation for image and shape matching. *IEEE Transactions on Image Processing*, *17*(4), 458–468.

Peter, A. M., & Rangarajan, A. (2009). Information geometry for landmark shape analysis: Unifying shape representation and deformation. *Transactions on Pattern Analysis and Machine Intelligence*, *31*(2), 337–350.

Peter, A. M., & Rangarajan, A. (2011). An information geometry approach to shape density minimum description length model selection. In *Information Theory in Computer Vision and Pattern Recognition - Workshop held at ICCV 2011* (pp. 1432–1439).

Peter, A. M., Rangarajan, A., & Ho, J. (2008). Shape L'Âne rouge: Sliding wavelets for indexing and retrieval. In *IEEE Conference on Computer Vision and Pattern Recognition (CVPR)* (pp. 1–8).

Pinheiro, A., & Vidakovic, B. (1997). Estimating the square root of a density via compactly supported wavelets. *Computational Statistics & Data Analysis*, *25*(4), 399–415.

Pinkall, U., Juni, S. D., & Polthier, K. (1993). Computing discrete minimal surfaces and their conjugates. *Experimental Mathematics*, *2*, 15–36.

Pistone, G., & Cena, A. (2007). Exponential statistical manifold. *Annals of the Institute of Statistical Mathematics*, *59*(1), 27–56.

Pistone, G., & Rogantin, P. (1999). The exponential statistical manifold: Mean parameters, orthogonality and space transformations. *Bernoulli*, *5*(4), 721–760.

Rangarajan, A., Gold, S., & Mjolsness, E. (1996). A novel optimizing network architecture with applications. *Neural Computation*, *8*(5), 1041–1060.

Rangarajan, A., Chui, H., & Bookstein, F. (1997). The softassign Procrustes matching algorithm. In Information Processing (Ed.), *in Medical Imaging (IPMI'97)* (pp. 29–42). New York: Springer.

Reuter, M., Wolter, F. E., & Peinecke, N. (2006). Laplace-Beltrami spectra as 'Shape-DNA' of surfaces and solids. *Computer-Aided Design*, *38*, 342–366.

Reuter, M., Biasotti, S., Giorgi, D., Patanè, G., & Spagnuolo, M. (2009). Discrete Laplace-Beltrami operators for shape analysis and segmentation. *Computers & Graphics*, *33*(3), 381–390.

Rissanen, J. (1978). Modeling by shortest data description. *Automatica*, *14*, 465–471.

Rissanen, J. (1996). Fisher information and stochastic complexity. *IEEE Transactions on Information Theory*, *42*, 40–47.

Rubner, Y., Tomasi, C., & Guibas, L. J. (2000). The earth mover's distance as a metric for image retrieval. *International Journal of Computer Vision*, *40*(2), 99–121.

Rustamov, R. M. (2007). Laplace-Beltrami eigenfunctions for deformation invariant shape representation. In *Proceedings of the Fifth Eurographics Symposium on Geometry Processing, Barcelona, Spain, July 4–6, 2007* (pp. 225–233).

Schwartz, S. (1967). Estimation of probability density by an orthogonal series. *The Annals of Mathematical Statistics*, *38*(4), 1261–1265.

Schwarz, G. (1978). Estimating the dimension of a model. *The Annals of Statistics*, *6*(2), 461–464.

Scott, D. W. (2001). *Multivariate density estimation: Theory, practice, and visualization*. New York: Wiley-Interscience.

Shilane, P., Min, P., Kazhdan, M., & Funkhouser, T. (2004). The Princeton shape benchmark. In *Shape Modeling International (SMI)*.

Siddiqi, K., Shokoufandeh, A., Dickinson, S. J., & Zucker, S. W. (1998). Shock graphs and shape matching. In *ICCV* (pp. 222–229).

Solomon, J., de Goes, F., Peyré, G., Cuturi, M., Butscher, A., Nguyen, A., et al. (2015). Convolutional Wasserstein distances: Efficient optimal transportation on geometric domains. In *SIGGRAPH*.

Srivastava, A., Joshi, S. H., Mio, W., & Liu, X. (2005). Statistical shape analysis: Clustering, learning, and testing. *IEEE Transactions on Pattern Analysis and Machine Intelligence*, *27*(4), 590–602.

Srivastava, A., Jermyn, I., & Joshi, S. (2007). Riemannian analysis of probability density functions with applications in vision. In *IEEE Computer Vision and Pattern Recognition (CVPR)* (pp. 1–8).

Strang, G., & Nguyen, T. (1997). *Wavelets and filter banks*. Wellesley: Wellesley-Cambridge Press.

Sun, J., Ovsjanikov, M., & Guibas, L. (2009). A concise and provably informative multi-scale signature based on heat diffusion. In *SGP* (pp. 1383–1392)

Tangelder, J. W., & Veltkamp, R. C. (2008). A survey of content based 3D shape retrieval methods. *Multimedia Tools and Applications*, *39*(3), 441–471.

Thakoor, N., Gao, J., & Jung, S. (2007). Hidden Markov model-based weighted likelihood discriminant for 2D shape classification. *IEEE Transactions on Image Processing*, *16*(11), 2707–2719.

Umeyama, S. (1988). An eigendecomposition approach to weighted graph matching problems. *IEEE Transactions on Pattern Analysis Machine Intelligence*, *10*, 695–703.

Vannucci, M. (1995). Nonparametric density estimation using wavelets. Technical report DP 95-26, ISDS, Duke University. http://www.isds.duke.edu.

Villani, C. (2009). *Optimal transport: Old and new*. New York: Springer.

Wang, F., Vemuri, B. C., Rangarajan, A., & Eisenschenk, S. J. (2008). Simultaneous nonrigid registration of multiple point sets and atlas construction. *IEEE Transactions on Pattern Analysis and Machine Intelligence*, *30*(11), 2011–2022.

White, D., & Wilson, R. (2007). Spectral generative models for graphs. In *14th International Conference on Image Analysis and Processing. ICIAP 2007* (pp. 35–42).

Zeng, W., Guo, R., Luo, F., & Gu, X. (2012). Discrete heat kernel determines discrete Riemannian metric. *Graphical Models*, *74*(4), 121–129.

Dimensionality Reduction for Measure Valued Evolution Equations in Statistical Manifolds

Damiano Brigo and Giovanni Pistone

1 Introduction

In this paper we propose a dimensionality reduction method for infinite dimensional measure valued evolution equations such as the Fokker–Planck (or forward Kolmogorov) partial differential equation or the Kushner–Stratonovich resp. Duncan–Mortensen–Zakai stochastic partial differential equations of nonlinear filtering, with potential applications to signal processing, quantitative finance, physics and quantum theory evolution equations, among many other areas.

This problem naturally shows up when one has to compute the probability distribution of the solution of a stochastic differential equation, or the conditional probability distribution of the solutions of a stochastic differential equation given a related observation process (filtering). Areas where such problems originate naturally are given in signal processing and stochastic filtering in particular, in quantitative finance, in heat flows, in quantum theory and potentially many others, as we discuss in Sect. 2.

Our method is based on the projection coming from a duality argument built in the non-parametric infinite-dimensional exponential statistical manifold structure developed by G. Pistone and co-authors, whose rich history is summarized in Sect. 3.

Dimensionality reduction and finite dimensional approximations will be based on projection on subspaces, so that the study of subspaces is fundamental. We first consider general subspaces in Sect. 4, trying also to clarify non-parametric exponential and mixture subspaces, and then move to finite dimensional subspaces in Sect. 5.

Clearly the choice of the finite dimensional manifold on which one should project the infinite dimensional equation is crucial, and we propose finite dimensional

D. Brigo (✉)
Department of Mathematics, Imperial College London,
180 Queen's Gate, London SW7 2AZ, UK
e-mail: damiano.brigo@gmail.com; damiano.brigo@imperial.ac.uk

G. Pistone
de Castro Statistics, Collegio Carlo Alberto, Via Real
Collegio 30, 10024 Moncalieri, Italy

F. Nielsen et al. (eds.), *Computational Information Geometry*,
Signals and Communication Technology, DOI 10.1007/978-3-319-47058-0_10

exponential and mixture families. This same problem had been studied, especially in the context of nonlinear filtering, by D. Brigo and co-authors. In those works the L^2 structure on the space of square roots of densities (based on the map $p \mapsto \sqrt{p}$, leading to the Hellinger distance) or of densities themselves (based on the map $p \mapsto p$, leading to the L^2 direct metric) was used, and no infinite dimensional manifold environment space for the equation to be projected was introduced. In fact, the main difficulty here is the fact the cone L^2_+ has empty relative interior unless the sample space is finite. Here we re-examine such works when adopting the exponential statistical manifold as an infinite dimensional environment, which allows for a deeper understanding of the geometric structures at play. We will see earlier in Sect. 3 that the statistical manifold approach and the Hellinger approach lead to the same metric in the finite dimensional manifold, whereas the L^2 direct approach leads to a different metric. This different "direct metric" works well with a specific type of finite dimensional mixture families, but since the direct metric structure is not compatible with the finite dimensional metric induced by the statistical manifold we will not pursue it further here but leave it for further work.

Going back to Sect. 5, in that section we further clarify how the finite dimensional and infinite dimensional terminology for exponential and especially mixture spaces are related. In the case of mixtures, one has to be careful in distinguishing mixtures generated by convex combinations of given distributions and sets of distributions that are closed under convex mixing.

Section 6 considers the finite dimensional projected differential equation for the approximated evolution in a number of cases, in particular the heat equation and the Fokker–Planck equation, and shows how this is derived in detail under the statistical manifold structure introduced earlier. For the particular case of the Fokker–Planck equation we discuss the interpretation of the projected, finite dimensional law as law of a different process, thus providing a tool for designing stochastic differential equations whose solutions densities evolve in a given finite dimensional family. We also discuss how one can measure the goodness of the approximation, show that projection in the statistical manifold structure is equivalent with the assumed density approximation for exponential families, and finally prove that if the sufficient statistics of the exponential family are chosen among the backward diffusion operator eigenfunctions then the projected equation provides the maximum likelihood estimator of the Fokker Planck equation solution.

Section 7 concludes the paper, hinting at further research problems.

This paper is a substantial update of our 1996 preprint Brigo and Pistone (1996).

2 Infinite Dimensional Measure Valued Evolution Equations

Stochastic Differential Equations (SDEs) are used in many areas of mathematics, physics, engineering and social sciences. SDEs represent extensions of ordinary differential equations to systems that are perturbed by random noise. In many problems, and we will see two important examples below, it is important to characterize the evolution in time of the probability law of the solution X_t of the SDE. This probability law, whose density is denoted usually by p_t, satisfies typically a partial differential equation (PDE) called Fokker–Planck (or forward Kolmogorov) equation or a stochastic partial differential equation (SPDE) called Kushner–Stratonovich (or Duncan–Mortensen–Zakai in an unnormalized version) equation, depending on the problem. Such measure-valued evolution equations are typically infinite dimensional, in that their solution curves in time $t \mapsto p_t$ do not stay in an a-priori given finite-dimensional parametric family, or in a finite dimensional manifold, unless very special conditions are satisfied. This implies that PDEs and SPDEs cannot be reduced exactly to ODEs or SDEs respectively, but that finite dimensional approximations of these equations need to be considered. One way to obtain finite dimensional approximations is choosing a finite dimensional subspace of the space where the equations for p_t are written, and project the original PDE or SPDE for p_t onto the subspace, using suitable geometric structures, thus obtaining a finite dimensional approximation that is driven by the best local approximation of the relevant vector fields. In this paper our aim is to clarify what kind of geometric structures can make the above approach fully rigorous. Most past works on dimensionality reduction of measure valued equations, see for example Hanzon (1987), Brigo et al. (1998, 1999), Armstrong and Brigo (2016) to name a few, use the L^2 space as a framework to implement the above projection. Here we will use the statistical manifold developed by G. Pistone and co-authors instead.

2.1 *The Fokker–Planck or Forward Kolmogorov Equation*

Let us start our formal analysis by introducing the complete probability space $(\Omega, \mathcal{F}, \mathbb{P})$, with a filtration $\{\mathcal{F}_t,\ t \geq 0\}$, on which we consider a stochastic process $\{X_t, t \geq 0\}$ of diffusion type, solution of a SDE in $\mathbb{R}^N$. Let the SDE describing X be of the following form

$$dX_t = f_t(X_t)dt + \sigma_t(X_t)dW_t, \tag{1}$$

where $\{W_t, t \geq 0\}$ is an M-dimensional standard Brownian motion independent of the initial condition X_0, and the drift f_t and diffusion coefficient σ_t are respectively an N-dimensional vector function and an $N \times M$ matrix function. We define $a(x) := \sigma_t(x)\sigma_t(x)'$ the $N \times N$ diffusion matrix, where the prime symbol denotes transposition. In the following to contain notation we will often neglect the time argument in f_t and a_t. The equation above is an Itô stochastic differential equation. The following set of assumptions will be in force throughout the paper.

(A) Initial condition: We assume that the initial state X_0 is independent of the process W and has a density p_0 w.r.t. the Lebesgue measure on $\mathbb{R}^n$, with finite moments of any order, and with p_0 almost surely positive.

(B) Local strong existence: $f \in C^{1,0}$, $a \in C^{2,0}$, which means that f is once continuously differentiable wrt x and continuous wrt t and a is twice continuously differentiable wrt x and continuous wrt t. These assumptions imply in particular local Lipschitz continuity.

(C) Growth/Non-explosion: there exists $K > 0$ such that

$$2x' f_t(x) + \|a_t(x)\| \leq K\,(1 + |x|^2),$$

for all $t \geq 0$, and for all $x \in \mathbb{R}^N$.

Under assumptions (A), (B) and (C) $\exists!$ solution $\{X_t\,,\ t \geq 0\}$ to the state equation, see Stroock and Varadhan (1979, Theorem 10.2.1).

(D) We assume that the law of X_t is absolutely continuous and its density $p_t(x)$ at x has regularity $C^{2,1}$ in (x, t) and satisfies the Fokker–Planck equation (FPE):

$$\frac{\partial p_t}{\partial t} = \mathcal{L}_t^* p_t, \tag{2}$$

where the backward diffusion operator $\mathcal{L}_t$ is defined by

$$\mathcal{L}_t = \sum_{i=1}^{N} f_i \frac{\partial}{\partial x_i} + \frac{1}{2} \sum_{i,j=1}^{N} a_{i,j} \frac{\partial^2}{\partial x_i \partial x_j},$$

and its dual (forward) operator is given by

$$\mathcal{L}_t^* p = -\sum_{i=1}^{N} \frac{\partial}{\partial x_i}(f_i p) + \frac{1}{2} \sum_{i,j=1}^{N} \frac{\partial^2}{\partial x_i \partial x_j}(a_{i,j} p).$$

We assume also $p_t(x)$ to be positive for all $t \geq 0$ and almost all $x \in \mathbb{R}^N$.

Assumption (D) holds for example under conditions given by boundedness of the coefficients f and a plus uniform ellipticity of a_t, see Stroock and Varadhan (1979, Theorem 9.1.9). Different conditions are also given in Friedman (1975, Theorem 6.4.7).

Situations where knowledge of the Fokker–Planck solution is important occur for example in signal processing and quantitative finance, among many other fields. Consider the following two examples.

2.2 *Stochastic Filtering with Discrete Time Observations*

In a filtering problem with discrete time observations, the SDE above (1) for X is an unobserved signal, of which we observe in discrete time a function h perturbed by noise, namely a process

$$Y_{t_k} = h(X_{t_k}) + V_{t_k}$$

where $t_0 = 0, t_1, \ldots, t_k, \ldots$ are discrete times at which observations Y arrive. The process V is a second Brownian motion, independent of the process W driving the signal X, and models the noise that perturbs our observation h. The filtering problem consists of estimating X_{t_k} given observations $Y_{t_0}, Y_{t_1}, \ldots, Y_{t_k}$ for all $k = 1, 2, \ldots$. It was shown in Brigo et al. (1999, Sect. 6.2), that one can find a suitable finite dimensional exponential family (including the observation function h among the exponent functions) such that the correction step (Bayes formula) at each arrival of new information is exact. What really brings about the infinite dimensional nature of the problem is the prediction step: between observation, the density of the signal evolves according to the FPE for X, and it is this FPE, and the operator $\mathcal{L}^*$ in particular, that leads to infinite dimensionality. Therefore, to study infinite dimensionality in filtering problems with discrete time observations, it suffices to study the Fokker–Planck equation, see again Brigo et al. (1999, Sect. 6.2) for the details.

2.3 *Filtering with Continuous Time Observations and Quantum Physics*

Consider again the filtering problem, but assume now that observations arrive in continuous time and are given by a stochastic process

$$dY_t = h(X_t)dt + dV_t.$$

In this case the solution of the filtering problem is no longer a PDE but a SPDE driven by the observation process dY. The SPDE features the same operator $\mathcal{L}^*$ as the FPE and is infinite dimensional. The SPDE exists in a normalized or unnormalized form, and has been studied extensively. It has been shown that even for toy systems like the cubic sensor ($N = 1, M = 1, f_t = 0, \sigma_t = 1, h(x) = x^3$) the SPDE solution is infinite dimensional (Hazewinkel et al. 1983). Finite dimensional approximations based on finite dimensional exponential and mixture families, building on the L^2 structure on the space of densities or their square roots to build a projection, have been considered in Hanzon (1987), Brigo et al. (1998, 1999), Armstrong and Brigo (2016). Nonlinear filtering equations are not of interest merely in signal processing. Several authors have noticed analogies between the filtering SPDEs hinted at above and the evolution equations in quantum physics, see for example Mitter (1979). Moreover, the related projection filter developed by D. Brigo and co-authors has been applied to quantum electrodynamics, see for example van Handel and Mabuchi (2005).

The SPDE case driven by rough paths such as dY is of particular interest because it combines the geometry in the state space for X and Y and the geometry in the space of probability measures associated with X conditional on Y's history. In this paper we are focusing on the latter but in presence of SPDEs one may have to work with the former as well. One of the problems in this case is choosing the right type of projection also from the state space geometry point of view and see how the optimality of the SPDE projected solution compares with the local optimality in the projection of the separate drift and diffusion coefficient vector fields of the SPDE. This is related to the different projections suggested in Armstrong and Brigo (2015) for evolution equations driven by rough paths. For such equations there is more than one possible projection, depending on the notion of optimality one chooses, which is related to the rough paths properties.

2.4 Valuation of Securities with Volatility Smile in Mathematical Finance

In Mathematical Finance, often one models stochastic local volatility for a given asset price S via a two-dimensional SDE under the pricing measure

$$\begin{aligned} dS_t &= rS_t dt + \sqrt{\xi_t}\, v(S_t) dW_t \\ d\xi_t &= k(\theta - \xi_t)dt + \eta\sqrt{\xi_t} dV_t \\ d\langle W, V\rangle_t &= \rho\, dt \end{aligned} \tag{3}$$

where r, k, θ, η are positive constants, $\rho \in [-1, 1]$, and v is a regular function. In case $v(S) = S$ one has the Heston model, whereas for more general v's one has a stochastic-local volatility model. One may also extend the model with a third stochastic process for the short rate r, introducing a stochastic process r_t of diffusion type replacing the constant risk free rate r, obtaining a three dimensional diffusion. We assume below r is constant.

To calibrate the model one has to fit a number of vanilla options. To do this, it is important to know the distribution of S_T at different maturities $T > 0$. In general, this can be deduced by the solution p_t of the FPE for the two-dimensional diffusion process $X_t = [S_t, \xi_t]'$ by integrating with respect to the second component. However, the solution of the FPE for this X is not know in general and is infinite dimensional. It may therefore be important to be able to find a good finite dimensional approximation for this density in order to value vanilla options in a way that leads to an easier calibration process.

2.5 The Anisotropic Heat Equation in Physics

We have mentioned earlier that the L^2 structure has been used in the past to project infinite-dimensional measure valued evolution equations for densities $t \mapsto q_t$. This structure has been invoked with the maps $q \mapsto \sqrt{q}$ (Brigo et al. 1998, 1999) or even $q \mapsto q$ (Armstrong and Brigo 2016; Brigo 2011), as we will explain more in detail below. It should be noted that the approach $q \mapsto q$ corresponds to the classical variational approach to parabolic equations, see e.g. the textbook by Brezis (2011, Chap. 8–10). A typical example of such approach is the equation whose weak form is

$$\frac{d}{dt}\int p_t(x)f(x)\,dx + \int \sum_{ij} a_{ij}(x)\left(\frac{\partial}{\partial x_i}p_t(x)\right)\left(\frac{\partial}{\partial x_j}f(x)\right)\,dx = 0, \tag{4}$$

where both the density p_t and the test function f belong to a Sobolev's space. This corresponds to the operator's form $\frac{\partial}{\partial t}p_t = \mathcal{L}^* p_t$, with

$$\mathcal{L}^* p(x) = \sum_{ij}\frac{\partial}{\partial x_j}\left(a_{ij}(x)\frac{\partial}{\partial x_i}p(x)\right).$$

This special case is the heat equation in the anisotropic case when the specific heat is constant, and is an important example of infinite dimensional evolution equation we aim at approximating with a finite dimensional evolution. We will keep this equation as an ongoing working example, and we will refer to it as our *running example* throughout the paper.

Going back to (4), in the following we will discuss an extension of the exponential statistical bundle to the case where the densities are (weakly) differentiable and belong to a weighted Sobolev's space, see Lods and Pistone (2015, Sect. 6).

All the above examples from signal processing in engineering, from social sciences, from physics and quantum physics should be enough to motivate the study of finite dimensional approximations of the FPE or of the filtering SPDE. We will tackle the FPE in the following sections, but many other applications are possible.

We now move to introduce the environment space where the above equations will be examined, the nonparametric infinite dimensional exponential statistical manifold of Giovanni Pistone and co-authors.

3 Information Geometric Background

In this section we review the construction of Information Geometry (IG) via the exponential statistical manifold, as originally developed by Pistone and Sempi (1995). More precisely, we will refer to an updated version of the theory we call (exponential) statistical bundle. Among other applications, we will include a qualification intended to deal with the special case of differentiable densities on a real space where we take a Gaussian probability density as background measure μ. This is referred to shortly as Gaussian space.

3.1 The Exponential Statistical Manifold and the L^2 Approach

We start with an introduction and we shall move to formal definitions in Sect. 3.2. This approach to IG considers the space of all positive densities of a measured sample space $(X, \mathcal{X}, \mu)$ which are (in an information-theoretic sense) near a given positive density p. The idea is representing each element q of this space with the chart

$$s_p : q \mapsto \log \frac{q}{p} - \mathbb{E}_p \left[\log \frac{q}{p} \right] = \log \frac{q}{p} + \mathrm{D}\,(p \,\|q)\,. \tag{5}$$

We define Banach spaces denoted B_p and domains $\mathcal{E}$ and $\mathcal{S}_p$, such that the mappings $s_p : \mathcal{E} \to \mathcal{S}_p \subset B_q$, $p \in \mathcal{E}$, defined in Eq. (5), form the *affine* atlas of a manifold modeled on the Banach spaces B_p, $p \in \mathcal{E}$. An atlas is affine if all change-of-chart transformation are affine functions. The Banach space B_p, the domain $\mathcal{E}$, and the domain $\mathcal{S}_p$ are formally defined in Sect. 3.2. We shall show a crucial property of the model Banach spaces B_p, $p \in \mathcal{E}$, namely they are all isomorphic to each other.

Each B_p is a vector space of p-centered random variables, so that the patches are easily shown to be of an exponential form, precisely each $s_p^{-1} = \mathrm{e}_p : \mathcal{S}_p \to \mathcal{E}$ is given by

$$\mathrm{e}_p(u) = \exp\left(u - K_p(u)\right) \cdot p, \quad u \in \mathcal{S}_p \subset B_p,$$

where $K_q(u) = \log \mathbb{E}_q\left[\mathrm{e}^u\right]$ will be defined more precisely later on in Definition 1.

The affine manifold so constructed is not a Riemannian manifold as the Banach spaces B_p are not Hilbert spaces. Instead, the theory specifies a second set of Banach spaces ${}^*B_p$, $p \in \mathcal{E}$, in natural duality with the B_p's, and a second affine atlas of the form

$$\eta_p(q) = \frac{q}{p} - 1 \in {}^*B_p, \quad q \in \mathcal{E}, \tag{6}$$

discussed by Cena and Pistone (2007).

The result is a non parametric version of S.-i. Amari's IG, see Amari (1987), Amari and Nagaoka (2000). Natural vector bundles based on this (dually) affine Banach manifold can be defined together with the proper parallel transports, leading to a first and second order calculus based on connections derived from such transports. We do not develop this aspect here, see the overview by Pistone (2013).

In the application we consider below, the base space is the Lebesgue space on $\mathbb{R}^d$ and the reference measure is given by the standard Gaussian density. Recent results allow to qualify the theory by considering densities which are differentiable in the sense of distribution and belong to a particular Sobolev space. This is interesting here because it gives the base to discuss partial differential equations in the variational form, see a few results in Lods and Pistone (2015).

Many expressions of the density other than Eq. (5) have place in the literature, for example the use of a deformed logarithm, see e.g. Naudts (2011). The most classical is the L^2-embeddings based on the map $q \mapsto \sqrt{q} \in L^2(\mu)$ that was used by

Brigo et al. (1998, 1999) in discussing the approximation of nonlinear filters. This mapping is actually a mapping from the set of densities to the Hilbert manifold of the unit sphere, so that a natural set of charts is given by the charts of the manifold of the unit sphere of $L^2(\mu)$. Viewed as such, this mapping is not a chart, but it can be still used to pull-back the L^2 structure in order to project on finite dimensional submanifolds. The relation between the exponential manifold and the L^2 unit ball manifold is discussed by Gibilisco and Pistone (1998), whereas Brigo et al. (1998) view the infinite dimensional evolution equation environment as the whole L^2 and so avoid the thorny question of defining an inifinite dimensional manifold structure related to the Hilbert structure. A more refined approach would be either considering an infinite dimensional manifold structure different from the L^2 structure, as we do here, or using a moving enveloping manifold for the finite dimensional exponential case (Brigo et al. 1999) from which one can project to the chosen finite dimensional exponential submanifold of densities.

In a context quite similar to our own, a new type of chart has been introduced by Newton (2012, 2013, 2015), namely $q \mapsto q - 1 + \log q - \mathbb{E}_\mu\left[\log q\right]$. This map is restricted to densities which are in $L^2(\mu)$ and such that $\log p \in L^1(\mu)$. As this domain does not fit well with our exponential manifold, we postpone its study to further research.

Recently, the larger framework of signed measures has been discussed with applications to Statistics, see Ay et al. (2016) and their forthcoming book on *Information Geometry* announced in Schwachhöfer et al. (2015).

As a further option, the identity representation $q \mapsto q \in L^2(\mu)$ has been shown to be of interest in our problem by Brigo (2011), Armstrong and Brigo (2016). This amounts to assuming that densities are square integrable and to using the L^2 norm directly for densities, rather than their square roots. This metric is called the "direct L^2 metric" in Armstrong and Brigo (2016). The image of this mapping is no longer a subset of the unit sphere in L^2, and this has consequences when projecting evolution equations for unnormalized probability densities onto finite dimensional manifolds, in that the projection will not take care of normalization. The identity representation above could possibly be interpreted using the charts $q \mapsto \frac{q}{p} - 1$ of Eq. (6) which belongs to ${}^*B_p \subset L_0^2(p)$, but we do not consider this angle here. We just point out that the direct metric approach leads to a different metric and projection than the exponential statistical manifold, whereas the statistical manifold structure agrees with the L^2 Hellinger structure. We will see this explicitly later on in Sect. 6.8.

We now proceed to present formal definitions of our approach.

3.2 Model Spaces

In a Banach manifold each chart of the atlas takes values in a Banach space. The model Banach spaces need not be equal, but they do need to be isomorphic on each connected component. It is the approach used for example by S. Lang in his textbook

(Lang 1995). We begin by recalling our definition of model spaces as introduced first in Pistone and Sempi (1995) with the purpose of defining a Banach manifold on the set $\mathcal{P}_>$ of strictly positive densities on a given measure space.

For each $p \in \mathcal{P}_>$ the Young function $\Phi(x) = \cosh x - 1$ defines the Orlicz spaces $L^\Phi(p)$ of random variables U such that $\mathbb{E}_p[\Phi(\alpha U)] < +\infty$ for an $\alpha > 0$. On Orlicz spaces see for example the monograph by Musielak (1983). The vector space $L^\Phi(p)$ is the same as the set of random variables such that, for some $\epsilon > 0$, $\mathbb{E}_p\left[\mathrm{e}^{tU}\right] < \infty$ if $t \in]-\epsilon, +\epsilon[$. In other words, the space is characterized by the existence of the moment generating function in a neighborhood of 0. This functional setting is implicit in the classical statistical theory. In fact, parametric exponential families are statistical models of the form

$$p(x;\theta) = \exp\left(\sum_{j=1}^{d} \theta_j U_j - \kappa(\boldsymbol{\theta})\right) \cdot p,$$

where the so-called sufficient statistics U_j, $j = 1, \ldots, d$, necessarily belong to the Orlicz space $L^\Phi(p)$, see e.g. L.D. Brown monograph (Brown 1986). We will later adopt the notation c for the sufficient statistics, in line with previous works by Brigo and co-authors on finite dimensional approximations. More generally, given a closed subspace $\mathcal{V}_p \subset L^\Phi(p)$, a $\mathcal{V}_p$-exponential family is the set of positive densities of the form $\mathrm{e}^{U-\kappa(U)} \cdot p$.

We define the subspaces of centered random variables

$$B_p = L_0^\Phi(p) = \left\{U \in L^\Phi(p) \middle| \mathbb{E}_p[U] = 0\right\}$$

to be used as model space at the density p. The norm of these spaces is the induced Orlicz norm from $L^\Phi(p)$.

A critical issue of this choice of model spaces is the fact the Banach spaces B_p are not reflexive and bounded functions are not dense if the sample space does not consist of a finite number of atoms. Technically, the Φ-function lacks a property called Δ_2 in the literature on Orlicz spaces. Precisely, if Ψ the convex conjugate of Φ, $\Psi(y) = \int_0^y (\Phi')^{-1}(v)\,dv$, $y > 0$, then the Orlicz space $L^\Psi(p)$ is Δ_2, so that it is separable and moreover its dual is identified with $L^\Phi(p)$ in the pairing

$$L^\Phi(p) \times L^\Psi(p) \ni (U, V) \mapsto \langle U, V\rangle_p = \mathbb{E}_p[UV].$$

Moreover, a random variable U belongs to $L^\Phi(p)$ if $Ł^\Psi(p) \ni V \mapsto \mathbb{E}_p[UV]$ is a bounded linear map. We write ${}^*B_p = L_0^\Psi(p)$ so that there is separating duality $B_p \times {}^*B_p \ni (U, V) \mapsto \mathbb{E}_p[UV]$. In this duality, the space ${}^*B_p$ is identified with the elements of the pre-dual of B_p which are centered random variables.

If the sample space is not finite, not all B_p are isomorphic, but we have the following crucial result, see Pistone and Rogantin (1999), Cena and Pistone (2007), Santacroce et al. (2015). Before the theorem we need a definition.

Definition 1 1. For each $p \in \mathcal{P}_>$, the *moment generating functional* is the positive lower-semi-continuous convex function $G_p \colon B_p \ni U \mapsto \mathbb{E}_p\left[\mathrm{e}^U\right]$ and the *cumulant generating functional* is the non-negative lower semicontinuous convex function $K_p = \log G_p$. The interior of the common proper domain $\{U|G\}_p(U) < +\infty^\circ = \left\{U\middle|K_p(U) < \infty\right\}^\circ$ is an open convex set $\mathcal{S}_p$ containing the open unit ball (for the Orlicz norm).

2. For each $p \in \mathcal{P}_>$, the *maximal exponential family* at p is

$$\mathcal{E}(p) = \left\{\mathrm{e}^{u-K_p(u)} \cdot p\middle|u \in \mathcal{S}_p\right\}. \tag{7}$$

3. Two densities $p, q \in \mathcal{P}_>$ are *connected by an open exponential arc*, $p \smile q$, if there exists a one-dimensional exponential family containing both in the interior of the parameters interval. Equivalently, for a neighborhood I of [0, 1]

$$\int_\Omega p^{1-t} q^t \, d\mu = \mathbb{E}_p\left[\left(\frac{q}{p}\right)^t\right] = \mathbb{E}_q\left[\left(\frac{p}{q}\right)^{1-t}\right] < +\infty, \quad t \in I\,.$$

Theorem 1 (Portmanteau Theorem) *Let $p, q \in \mathcal{P}_>$. The following statements are equivalent:*

1. *$p \smile q$ (i.e. p and q are connected by an open exponential arc);*
2. *$q \in \mathcal{E}(p)$;*
3. *$\mathcal{E}(p) = \mathcal{E}(q)$;*
4. *$\log \frac{q}{p} \in L^\Phi(p) \cap L^\Phi(q)$;*
5. *$L^\Phi(p) = L^\Phi(q)$ (i.e. they both coincide as vector spaces and their norms are equivalent);*
6. *There exists $\varepsilon > 0$ such that $\frac{q}{p} \in L^{1+\varepsilon}(p)$ and $\frac{p}{q} \in L^{1+\varepsilon}(q)$.*

It follows from this structural result that the manifold we are going to define has connected components which are maximal exponential families. Hence we restrict our study to a given maximal exponential family $\mathcal{E}$, where the mention of a reference density is not required any more.

3.3 Exponential Statistical Manifold, Statistical Bundles

Let $\mathcal{E}$ be a maximal exponential family. The spaces B_p, $p \in \mathcal{E}$, are isomorphic under the affine mappings ${}^e\mathbb{U}_p^q B_p \ni U \mapsto U - \mathbb{E}_q[U] \in B_q$, $p, q \in \mathcal{E}$ and the pre-dual spaces ${}^*B_p$, $p \in \mathcal{E}$, are isomorphic under the affine mappings ${}^m\mathbb{U}_p^q {}^*B_p \ni U \mapsto \frac{q}{p}U \in {}^*B_q$, $p, q \in \mathcal{E}$. Such families of isomorphism are the relevant parallel transports in our construction. Precisely, ${}^e\mathbb{U}_p^q$ is the *exponential* transport and ${}^m\mathbb{U}_p^q$ is the *mixture* transport and they are dual semigroups,

$$\left\langle {}^e\mathbb{U}_p^q U, V\right\rangle_q = \left\langle U, {}^m\mathbb{U}_q^p V\right\rangle_p \quad \text{and} \quad \langle W, V\rangle_q = \left\langle {}^e\mathbb{U}_p^q W, {}^m\mathbb{U}_p^q V\right\rangle_p,$$

for $U \in B_p$, $V, W \in B_q$.

We review below some basic topics from Pistone (2013) and Lods and Pistone (2015).

Definition 2 1. The *exponential manifold* is the maximal exponential family $\mathcal{E}$ with the affine atlas of global charts $(s_p : p \in \mathcal{E})$,

$$s_p(q) = \log\frac{q}{p} - \mathbb{E}_p\left[\log\frac{q}{p}\right].$$

2. The *statistical exponential bundle* $S\mathcal{E}$ is the manifold defined on the set

$$\left\{(p, V)\middle| p \in \mathcal{E}, V \in B_p\right\}$$

by the affine atlas of global charts

$$\sigma_p : (q, V) \mapsto \left(s_p(q), {}^e\mathbb{U}_q^p V\right) \in B_p \times B_p, \quad p \in \mathcal{E}$$

3. The *statistical predual bundle* $^*S\mathcal{E}$ is the manifold defined on the set

$$\left\{(p, W)\middle| p \in \mathcal{E}, W \in {}^*B_p\right\}$$

by the affine atlas of global charts

$${}^*\sigma_p : (q, W) \mapsto \left(s_p(q), {}^m\mathbb{U}_q^p W\right) \in B_p \times {}^*B_p, \quad p \in \mathcal{E}$$

It should be noted that the full statistical manifold on positive densities actually splits into connected components which are exponential manifolds $\mathcal{E}$ and that all the charts of the affine atlases have global domains.

The statistical bundle $S\mathcal{E}$ is a specific version of the tangent bundle of the exponential manifold. In fact, if we define $\mathrm{e}_p = s_p^{-1}$, we have $\mathrm{e}_p(U) = \mathrm{e}^{U - K_p(U)} \cdot p$ and for each regular curve $p(t) = \mathrm{e}^{U(t) - K_p(U(t))} \cdot p$, $U(\cdot) \in C^1(I; B_p)$ the velocity of the expression in the s_p chart is $\dot{U}(t) \in B_p$; viceversa, for each $U \in B_p$ we have the regular curve $t \mapsto \mathrm{e}^{tU - K_p(tU)} \cdot p$.

The general notions of velocity and gradient take a specific form in the statistical bundle. Let $t \mapsto p(t)$ be a regular curve in the exponential manifold and let $f : \mathcal{E} \to \mathbb{R}$ be a regular function.

Definition 3 1. The *score* of the curve $t \mapsto p(t)$ is the curve $t \mapsto (p(t), Dp(t)) \in S\mathcal{E}$ such that

$$\frac{d}{dt}\mathbb{E}_{p(t)}[V] = \left\langle V - \mathbb{E}_{p(t)}[V], Dp(t)\right\rangle_{p(t)}$$

for all $V \in L^{\Psi}(p)$, $p \in \mathcal{E}$.

2. The *statistical gradient* of f is the section $\mathcal{E} \ni p \mapsto (p, \operatorname{grad} f(p)) \in {}^*S\mathcal{E}$ such that for each regular curve

$$\frac{d}{dt} f(p(t)) = \langle \operatorname{grad}\ f(p(t)), Dp(t) \rangle_{p(t)} .$$

In most cases we are able to identify the score as $Dp(t) = \frac{\dot{p}(t)}{p(t)} = \frac{d}{dt} \log p(t)$. We turn now to the regularity properties of the cumulant generating funtion.

Proposition 1 (Properties of the CGF) *Let K_p be the cumulant generating functional at $p \in \mathcal{E}$ and let $\mathcal{S}_p$ be the interior of the proper domain.*

1. *$K_p \colon \mathcal{S}_p \to \mathbb{R}$ is 0 at 0, otherwise is strictly positive; it is convex and infinitely Fréchet differentiable. The value at 0 of the differential of order n in the direction $U_1, \ldots, U_n \in B_p$ is the value of the n-th joint cumulant under p of the random variable $U_1, \ldots, U_n$.*
2. *The value at $U \in \mathcal{S}_p$ of the differential of order n in the direction $U_1, \ldots, U_n \in B_p$ is the value of the n-th joint cumulant under $q = e_p(U) = e^{U - K_p(U)} \cdot p$ of the random variable $U_1, \ldots, U_n$, namely*

$$\mathrm{D}^n\, K_p\,(U)\ [U_1, \ldots, U_n] = \frac{\partial^n}{\partial t_1 \cdots \partial t_n} \log \mathbb{E}_q \left[e^{t_1 U_1 + \cdots + t_n U_n} \right] \bigg|_{t=0} .$$

3. *In particular, $\frac{q}{p} - 1 \in {}^*B_p$ and*

$$\mathrm{D}\, K_p(U)\ [V] = \mathbb{E}_q\ [V] = \left\langle \frac{q}{p} - 1, V \right\rangle_p \tag{8}$$

$$\mathrm{D}^2\, K_p(U)\ [U_1, U_2] = \operatorname{Cov}_q\ (U_1, U_2) = \left\langle {}^e\mathbb{U}_p^q U_1, {}^e\,\mathbb{U}_p^q U_2 \right\rangle_q . \tag{9}$$

Equations (8) and (9) above show that the geometry of the exponential manifold is fully encoded in the cumulant generating function K_p. The relevant abstract structure is called Hessian manifold, cf Hirohiko Shima's monograph (Shima 2007).

3.4 Maximal Exponential Families of Gaussian Type

In this section we study the specific case of the statistical manifold whose components allow for including the Gaussian density (the Gaussian space case), or a generalised Gaussian density. The aim is to develop a framework where partial differential equations are naturally defined.

Let M be the standard Gaussian density (Maxwell density) on the d-dimensional real space. The maximal exponential family $\mathcal{E}\,(M)$ has special features that we review below from Lods and Pistone (2015, Sects. 4 and 6). Note that in that reference the Young functions Φ and $\Psi = \Phi_*$ were explicitly denoted as $(\cosh - 1)$ and $(\cosh - 1)_*$, respectively.

Proposition 2 *1. The Orlicz space $L^\Phi(M)$ contains all polynomials of degree up to two.*
2. The Orlicz space $L^\Psi(M)$ contains all polynomials.
3. The entropy $H \colon \mathcal{E}(M) \ni p \mapsto -\mathbb{E}_p[\log p]$ is finite and Frechét differentiable with statistical gradient grad $H(p) = -(\log p + H(p))$.

Let us compute the action on a density $p \in \mathcal{E}(M)$ of our running example of partial differential operator in Eq. (4), assuming all the needed differentiability. We write $p = \mathrm{e}^{U - K_M(U)} \cdot M$, $U \in \mathcal{S}_M$, and use repeatedly the equality $\mathbb{E}_M\left[f \frac{\partial}{\partial x_j} g\right] = \mathbb{E}_M\left[(X_j f - \frac{\partial}{\partial x_j} f) g\right]$ to get the following:

$$\frac{\partial}{\partial x_j} p(x) = \frac{\partial}{\partial x_j}\left(\mathrm{e}^{U(x) - K_M(U)} M(x)\right) = \left(\frac{\partial}{\partial x_j} U(x) - x_j\right) p(x) . \tag{10}$$

$$\begin{aligned}\frac{\partial}{\partial x_i}\left(a_{ij}(x) \frac{\partial}{\partial x_j} p(x)\right) = \frac{\partial}{\partial x_i}\left(a_{ij}(x)\left(\frac{\partial}{\partial x_j} U(x) - x_j\right) p(x)\right) = \\ \frac{\partial}{\partial x_i}\left[a_{ij}(x)\left(\frac{\partial}{\partial x_j} U(x) - x_j\right)\right] p(x) + \\ a_{ij}(x)\left(\frac{\partial}{\partial x_i} U(x) - x_i\right)\left(\frac{\partial}{\partial x_j} U(x) - x_j\right) p(x)\end{aligned} \tag{11}$$

and

$$\begin{aligned}p^{-1}(x) \sum_{i,j} \frac{\partial}{\partial x_i}\left(a_{ij}(x) \frac{\partial}{\partial x_j} p(x)\right) = \\ \sum_{i,j} \frac{\partial}{\partial x_i}\left[a_{ij}(x)\left(\frac{\partial}{\partial x_j} U(x) - x_j\right)\right] + \\ \sum_{i,j} a_{ij}(x)\left(\frac{\partial}{\partial x_i} U(x) - x_i\right)\left(\frac{\partial}{\partial x_j} U(x) - x_j\right) .\end{aligned}$$

Note that the left hand side is a random variable whose expectation at $p = \mathrm{e}^{U - K_M(U)} \cdot M$ is zero. Hence the right hand side is a candidate to be the expression in a chart of a section of the statistical predual bundle of Definition 2(3).

Example 1 If $[a_{ij}] = I$, then the expression of the PDE is

$$\frac{\partial}{\partial t} U(x, t) = \Delta U(x) - d + |\nabla U(x) - x|^2 ,$$

and for $d = 1$

$$\frac{\partial}{\partial t} U(x, t) = U''(x) - 1 + (U'(x) - x)^2.$$

This provides a simple example of finite dimensionality. Assume there is a solution of the form $U(x, t) = \theta_0(t) + \theta_1(t)x + \theta_2(t)x^2$, that is $p(x, t)$ is Gaussian. It follows

$$\begin{aligned} U''(x) - 1 + (U'(x) - x)^2 = & \\ & 2\theta_2(t) + (\theta_1(t) + 2\theta_2(t)x - x)^2 = \\ & (\theta_1(t)^2 + 2\theta_2(t)) + 2\theta_1(t)(2\theta_2(t) - 1)x + (2\theta_2(t) - 1)^2 x^2 \end{aligned}$$

where the value of the constant $\theta_0(t)$ follows from the section condition $\mathbb{E}_{p(t)}\,[U(t)] = 0$.

In the one-dimensional case $d = 1$, we can generalize easily the density $M(x)$ to $M_{1,m}(x)$, with m positive even integer, defined as

$$M_{1,m}(x) \propto \exp\left(-\frac{1}{m} x^m\right). \tag{12}$$

We could keep the multivariate case but the combinatorial complexity would become quite challenging, so we explain our idea in the scalar case.

The density $M_{1,m}$, chosen as background density, allows one to have in the exponent of the densities monomial terms up to x^{m-1} without any integrability problem, or up to x^m with restriction on the parameters. Suppose, for example, that we need a family of densities flexible enough to include bimodal densities. A natural choice (see Brigo et al. 1999; Armstrong and Brigo 2016) would be $m = 4$ and an exponential family of densities

$$\propto \exp(\theta_1 x + \theta_2 x^2 + \theta_3 x^3 + \theta_4 x^4)$$

with parameters $\theta \in \Theta$, open conved domain. However, if θ_4 goes to zero or even positive then we are in troubles. To avoid this, we may choose as background density $M_{1,6}$, so that

$$\propto \exp(\theta_1 x + \theta_2 x^2 + \theta_3 x^3 + \theta_4 x^4) M_{1,6}(x) = \exp(\theta_1 x + \theta_2 x^2 + \theta_3 x^3 + \theta_4 x^4 - (1/6)x^6)$$

will be always well defined as a probability density, for all θ. We briefly mention that densities such as the above have a number of computational advantages when used to obtain finite dimensional approximations of infinite dimensional evolution equations such as Fokker–Planck or Kushner–Stratonovich or Zakai. These advantages are related to an algebraic ring structure, see Armstrong and Brigo (2016).

Let us discuss the action of differential operators of interest on a density $p \in \mathcal{E}(M)_{1,m}$, assuming moreover the differentiability where needed. Dropping the index $(1, m)$ from M for brevity, we write $p = \mathrm{e}^{u-K_M(u)} \cdot M, u \in \mathcal{S}_M$, to get the following

$$\frac{\partial}{\partial x} p(x) = \frac{\partial}{\partial x}\left(\mathrm{e}^{u(x)-K_M(u)} M(x)\right) = \left(\frac{\partial}{\partial x} u(x) - x^{m-1}\right) p(x) .$$

$$\frac{\partial}{\partial x}\left(a(x)\frac{\partial}{\partial x} p(x)\right) = \frac{\partial}{\partial x}\left(a(x)\left(\frac{\partial}{\partial x} u(x) - x^{m-1}\right) p(x)\right) =$$
$$= \frac{\partial}{\partial x}\left[a(x)\left(\frac{\partial}{\partial x} u(x) - x^{m-1}\right)\right] p(x) + a(x)\left(\frac{\partial}{\partial x} u(x) - x^{m-1}\right)^2 p(x)$$

and

$$p^{-1}(x)\frac{\partial}{\partial x}\left(a(x)\frac{\partial}{\partial x} p(x)\right) = \frac{\partial}{\partial x}\left[a(x)\left(\frac{\partial}{\partial x} u(x) - x^{m-1}\right)\right] +$$
$$a(x)\left(\frac{\partial}{\partial x} u(x) - x^{m-1}\right)^2$$

Example 2 If $a = 1$, which in case $d = 1$ is usually obtained from a general diffusion via the Lamperti transform, then the previous equation becomes

$$p^{-1}(x)\Delta p(x) = \Delta u(x) - (m-1)x^{m-2} + \left|\nabla u(x) - x^{m-1}\right|^2$$

An important feature of the statistical bundles $S\mathcal{E}(M)$ and ${}^*S\mathcal{E}(M)$ is the possibility to define Orlicz–Sobolev spaces (see e.g. Musielak 1983) for the fibers and use this setup in the study of partial differential equations, cf. Lods and Pistone (2015, Sect. 6).

Definition 4 1. The exponential Orlicz–Sobolev spaces of $\mathcal{E}(M)$ are the vector spaces

$$W^1_\Phi = \left\{f \in L^\Phi(M) \middle| \partial_j f \in L^\Phi(M), j = 1, \ldots, d\right\}$$
$$W^1_\Psi = \left\{f \in L^\Psi(M) \middle| \partial_j f \in L^\Psi(M), j = 1, \ldots, d\right\}$$

where ∂_j is the derivative in the sense of distributions. These spaces become Banach spaces when endowed with the graph norm. The spaces defined with respect to any $p \in \mathcal{E}(M)$ are equal as vector space and isomorphic as Banach spaces.

2. The W^1_Φ-exponential family at M is

$$\mathcal{E}_1(M) = \left\{\mathrm{e}^{u-K_M(U)} \cdot M \middle| U \in \mathcal{S}_M \cap W^1_\Phi\right\}$$

The set $\mathcal{S}_M^1 = \mathcal{S}_M \cap W_\Phi^1$ is a convex open set

$$\mathcal{S}_M^1 \subset B_M^1 = \left\{U \in W_\Phi^1 \middle| \mathbb{E}_M\left[U\right] = 0\right\}$$

It contains all coordinate functions X_i and polynomials of order two, cf Pistone (2014).

The following proposition shows the regularity of the densities in the W_Φ^1-exponential family $\mathcal{E}_1\left(M\right)$ and the Stein's identity in the Orlicz–Sobolev setup, cf. Lods and Pistone (2015, Sect. 6). It should be noted that these properties were actually needed above in the derivation of the expression of the running example of PDE.

Proposition 3 *Assume* $U \in \mathcal{S}_M^1$, $p = e^{U-K_M(U)} \cdot M \in \mathcal{E}_1\left(M\right)$, *and* $f \in W_\Phi^1$.

1. *It follows* $fe^{U-K_p(U)} \in W_{\Phi_*}^1$ *and* $fe^{U-K_p(U)} \cdot M = fp \in W_{\Phi_*}^1$.
2. $\nabla e^{U-K_p(U)} = \nabla U e^{U-K_p(U)}$ *and* $\nabla(e^{U-K_p(U)}M) = (\nabla U - X)e^{U-K_p(U)}M$.
3. (Multiplication operator) *If* $f \in W_\Psi^1$, *then* $X_j f \in L^\Psi\left(M\right)$.
4. (Stein's identity) *If* $f \in W_\Psi^1$ *and* $g \in W_\Phi^1(M)$, *then*

$$\left\langle f, \partial_j g\right\rangle_M = \left\langle X_j f - \partial_j f, g\right\rangle_M.$$

We now define a differentiable version of the statistical bundles.

Definition 5 1. The (statistical) *differentiable exponential bundle* is the manifold defined on the set

$$S\mathcal{E}_1\left(M\right) = \left\{(p, V)\middle| p \in \mathcal{E}_1\left(M\right), V \in B_p^1\right\}$$

by the affine atlas of global charts

$$\sigma_p \colon (q, V) \mapsto \left(s_p(q), {}^e\mathbb{U}_q^p V\right) \in B_p^1 \times B_p^1, \quad p \in \mathcal{E}_1\left(M\right)$$

2. The (statistial) *differentiable predual bundle* is the manifold defined on the set of fibers

$${}^*S\mathcal{E}_1\left(M\right) = \left\{(p, V)\middle| p \in \mathcal{E}_1\left(M\right), V \in {}^*B_p^1\right\}$$

by the affine atlas of global charts

$${}^*\sigma_p \colon {}^*S\mathcal{E}_1\left(M\right) \ni (q, V) \mapsto \left(s_p(q), {}^m\mathbb{U}_q^p V\right) \in B_p^1 \times {}^*B_p^1,$$

We have given a setup such that we can look at a parabolic equation $\frac{\partial}{\partial t}p(x, t) = \mathcal{L}p(x, t)$ as the equation $p(x, t)^{-1}\frac{\partial}{\partial t}p(x, t) = p(x, t)^{-1}\mathcal{L}p(x, t)$, where the left hand side is the score of the solution curve $t \mapsto p(t)$ and the right hand side is a section of an appropriate statistical bundle. This type of equation requires the development of a full theory. We here restrict to finite dimensional cases, where the section is actually a section of a finite dimensional submodel.

4 Submodels and Submanifolds

Before turning to the main topic of this paper, namely finite dimensional approximations, requiring finite dimensional subspaces structures to be introduced, we study more general subspaces structures that can still be infinite dimensional in general. In particular, this will lead to a first definition of exponential and mixture families associated to subspaces. We will see that while this general exponential family subspace will be similar to the finite dimensional case we will use for the approximation later, the mixture case is subtler, as there are two different notions of mixture family that may however coincide in special cases.

We first consider the following adaptation of the standard definition of submanifold, as it is for example given in the monograph Lang (1995) or that by Abraham et al. (1988). Our definition is tentative and it is intended to go along with the special features of the exponential manifold $\mathcal{E}$, namely the duality between the pre-fibers ${}^*B_p$ and the fibers B_p, $p \in \mathcal{E}$. We shall consider two types of substructure, that we call respectively sub-model and sub-manifold.

Definition 6 (*Sub-model, sub-manifold*) Let $\mathcal{N}$ be a subset of the maximal exponential family $\mathcal{E}$ and, for each density $p \in \mathcal{N}$, let V_p^1 be a closed subspace of B_p and V_p^2 a closed subspace of ${}^*B_p$, such that $V_p^1 \cap V_p^2 = \{0\}$ with continuous immersions $B_p \hookrightarrow V_p^1 \oplus V_p^2 \hookrightarrow {}^*B_p$. Let σ be a diffeomorphism of a neighborhood $\mathcal{W}_p$ of p onto the product of two open sets $\mathcal{V}_p^1 \times \mathcal{V}_p^2$ of $V_p^1 \times V_p^2$ that maps $\mathcal{N} \cap \mathcal{W}_p$ onto $\mathcal{V}_p^1 \times \{0\}$. Assume there exists an atlas Σ of such mappings σ that covers $\mathcal{N}$.

1. It follows that $\mathcal{N}$ is a manifold with charts $\sigma_{|\mathcal{N}}$, $\sigma \in \Sigma$, with tangent spaces $T_p\mathcal{N}$ isomorphic to V_p, $p \in \mathcal{N}$. We say that such a manifold is a *sub-model* of $\mathcal{E}$.
2. If the space V_p^2 is a closed subspace of B_p, that is V_p^1 *splits* in B_p, then $\mathcal{N}$ is a *sub-manifold* of $\mathcal{E}$.

It should be noted that the splitting condition in Item 2 above is quite restrictive in our context. In fact, while a closed subspace of an Hilbert space always splits with its orthogonal complement, the same is not generally true in our Orlicz spaces. It is generally true only in the finite state space case. However, in the applications we are looking for, either the space $V_p^1 \subset B_p$ or the space $V_p^2 \in {}^*B_p$ is finite dimensional. Each one of these assumptions allows for a special treatment, as it is shown in the following sections.

The submanifold issue was originally discussed in Pistone and Rogantin (1999). In particular, it was observed there that each p-conditional expectation provides a splitting in B_p, because $U \mapsto \mathbb{E}_p[U|\mathcal{Y}]$ is an idempotent continuous linear mapping on B_p. The complementary space is the kernel of the conditional expectation. It follows, for example, that each marginalization is a submersion of the exponential manifolds.

The classical theory of parametric exponential families (see Brown 1986) uses a special splitting of the parameter's space which is called mixed parameterization. Our approach actually mimics the same approach in a more abstract and functional language. In fact, if V_p^1 is a closed subset of the space B_p, its orthogonal space or annihilator is actually a subspace of the predual space $^*B_p$, so that $(V_p^1)^\perp \subset {}^*B_p$. For this reason we have slightly modified the classical definition of sub-manifold in order to accommodate this special structure of interest.

4.1 Exponential Family and Mixture (-Closed) Family Submodels

Our basic example of sub-model is an exponential family in the maximal exponential family $\mathcal{E}$.

Definition 7 (*Exponential family* EF (V_p)) Let V_p be a closed subspace of B_p and define

$$\mathrm{EF}\left(V_p\right) = \left\{q \in \mathcal{E}\left(p\right) \middle| s_p(q) \in V_p\right\} .$$

That is, each $q \in \mathrm{EF}\left(V_p\right)$ is of the form $q = \mathrm{e}^{u - K_p(u)} \cdot p$ with $u \in V_p \cap \mathcal{S}_p$.

Recall the exponential transport $^e\mathbb{U}_p^q \colon B_p \to B_q$, $p, q \in \mathcal{E}$ is defined by $^e\mathbb{U}_p^q U = U - \mathbb{E}_q\left[U\right]$. We define the family of parallel spaces $V_q =^e \mathbb{U}_p^q V_p$, $q \in \mathcal{E}$. The exponential families of two parallel spaces, EF $\left(V_p\right)$ and EF $\left(V_q\right)$, are either equal or disjoint. If fact, if $q \in \mathrm{EF}\left(V_p\right)$ then $q = \exp\left(\bar{U} - K_p(\bar{U})\right) \cdot p$ and for each $U \in V_p$ it holds

$$\begin{aligned}\exp\left(U - K_q(U)\right) \cdot p = \exp\left(U - K_p(U) - \bar{U} + K_p(\bar{U})\right) \cdot q = \\ \exp\left({}^e\mathbb{U}_p^q(U - \bar{U}) - \mathbb{E}_q\left[U - \bar{U}\right] + K_p(U) + K_p(\bar{U})\right) \cdot q = \\ \exp\left(V - K_q(V)\right) \cdot q\end{aligned}$$

with $V = {}^e\mathbb{U}_p^q\left(U - \bar{U}\right) \in V_q$. If $q \notin \mathrm{EF}\left(V_p\right)$ then there is no common part otherwise the previous computation would show equality.

The exponential families based on the transport of a subspace V_p form a partition in a covering of statistical models. The next notion of mixture family provides a way to choose a representative in each class.

The mixture family and the complementary spaces are defined as follows.

Definition 8 (*Mixture-closed family*)

1. For each closed subspace $V_p \subset B_p$ define its orthogonal space to be its annihilator $V_p^\perp \subset {}^*B_p$, that is $V_p^\perp = \left\{v \in {}^*B_p \middle| \langle v, u\rangle_p = 0, u \in V_p\right\}$.
2. The *mixture-closed family* (or mixture family shortly) of V_p, is the set of densities MF $\left(V_p\right) \subset \mathcal{E}$ with zero expectation on V_p,

$$\mathrm{MF}\left(V_p\right) = \left\{q \in \mathcal{E} \middle| \mathbb{E}_q\left[U\right] = 0, U \in V_p\right\} .$$

Equivalently, the set of its mixture coordinates centered at p belongs to $V_p^{\perp}$,

$$\eta_p\left(\mathrm{MF}\left(V_p\right)\right) = \left\{ \frac{q}{p} - 1 \middle| q \in \mathcal{M}(V_p) \right\} = V_p^{\perp} \cap \eta_p\left(\mathcal{E}\right).$$

Remark 1 The mixture family $\mathrm{MF}\left(V_p\right)$ is convex and deserves its name because it is closed under mixtures, that is convex combinations. However, this name could be misleading as this set in not closed topologically, since we assumed it to be a subset of the maximal exponential family $\mathcal{E}\left(p\right)$. In general, our mixture families will not contain any extremal point nor will they be generated by a mixture of extremal points. Hence "closed" is to be understood in the convex combination sense and not topologically. We will come back to this distinction in the finite dimensional case below. The general problem of mixtures in a maximal exponential family has been discussed in Santacroce et al. (2015)

As we defined the family of subspaces parallel to V_p to be $V_q = {}^{e}\mathbb{U}_p^q V_p$, $q \in \mathcal{E}$, similarly we have the parallel family of orthogonal spaces $V_q^{\perp} = {}^{m}\mathbb{U}_p^q V_p^{\perp}$, where the mixture transport ${}^{m}\mathbb{U}_p^q \colon {}^{*}B_p \to {}^{*}B_q$ is defined by ${}^{m}\mathbb{U}_p^q V = \frac{p}{q} V_p$. In fact, $\left\langle {}^{e}\mathbb{U}_p^q U, V \right\rangle_q = \left\langle U, {}^{m}\mathbb{U}_p^q V \right\rangle_p$. The mixture families $\mathrm{MF}\left(V_q\right)$, $q \in \mathcal{E}$, are either equal or disjoint. In fact, if $q \in \mathrm{MF}\left(V_p\right)$, then

$$\begin{aligned}\mathrm{MF}\left(V_q\right) = &\left\{r \in \mathcal{E} \middle| \mathbb{E}_r\left[V\right] = 0,\, V \in V_q\right\} = \left\{r \in \mathcal{E} \middle| \mathbb{E}_r\left[{}^{e}\mathbb{U}_p^q U\right] = 0,\, U \in V_p\right\} = \\ &\left\{r \in \mathcal{E} \middle| \mathbb{E}_r\left[U\right] = \mathbb{E}_q\left[U\right],\, U \in V_p\right\} = \mathrm{MF}\left(V_p\right).\end{aligned}$$

The following proposition clarifies the relative position of $\mathrm{EF}\left(V_p\right)$ and $\mathrm{MF}\left(V_p\right)$.

Proposition 4 *1. The unique intersection of* $\mathrm{EF}\left(V_p\right)$ *and* $\mathrm{MF}\left(V_p\right)$ *is p.*

2. *The space of scores at q of regular curves in* $\mathrm{EF}\left(V_p\right)$ *is V_q.*
3. *If a regular curve through r is contained in* $\mathrm{MF}\left(V_p\right)$*, then its score at r is contained in $V_r^{\perp}$.*
4. *Assume V_p^1 splits in B_p with complementary space V_p^2. Then both* $\mathrm{EF}\left(V_p^1\right)$ *and* $\mathrm{EF}\left(V_p^2\right)$ *are sub-manifolds of $\mathcal{E}$ with tangent spaces at p respectively V_p^2 and V_p^2.*
5. *Assume V_p^1 splits in B_p with complementary space V_p^2, $u = \Pi_1(u) + \Pi_2(u)$ and assume the mapping*

$$q \mapsto u = s_p(q) \mapsto (\Pi_1(u), (\nabla K_p)^{-1} \circ \Pi_2(u))$$

is a diffeomorphism around p. Then $\mathrm{MF}\left(V_p\right)$ *is a sub-manifold of $\mathcal{E}$ with tangent spaces at p equal to V_p^2.*

Proof 1. First, $p = \mathrm{e}^{0-K_p(0)} \cdot p$ and $\mathbb{E}_p\left[U\right] = 0$, if $U \in V_p$. Second, assume $q \in \mathrm{EF}\left(V_p\right) \cap \mathrm{MF}\left(V_p\right)$. It follows that $q = \mathrm{e}^{U-K_p(U)} \cdot p$ and $\mathbb{E}_q\left[U\right] = 0$ for a $U \in V_p$. Hence $0 \geq \mathrm{D}\left(q \,\|p\right) = \mathbb{E}_q\left[U - K_p(U)\right] = \mathbb{E}_q\left[U\right] - K_p(U) = -K_p(u) \leq 0$, hence $U = 0$ and $q = p$.

2. Follows easily from the definition of exponential family.
3. For $r(t) = \mathrm{e}^{u(t)-K_r(u(t))} \cdot r \in \mathrm{MF}(V_r)$ and $u \in V_r^1$ we have

$$0 = \frac{d}{dt}\mathbb{E}_{r(t)}[u]\bigg|_{t=0} = \mathrm{Cov}_{r(t)}(u, \dot{u}(t))\big|_{t=0} = \langle u, \dot{u}(0)\rangle_r \ .$$

4. Let Π_i, $i = 1, 2, \ldots$ be the projections induced by the splitting and let S be the open convex set such that both $u_1, u_2 \in \mathcal{S}_p$, namely $S = \Pi_1^{-1}(\mathcal{S}_p) \cap \Pi_2^{-1}(\mathcal{S}_p)$. The mapping $q \mapsto (u_1, u_2)$ satisfies Definition 6(2).
5. As the main assumption in Definition 6(2) is now an assumption, we have only to check the image of $U \mapsto (0, (\nabla K_p)^{-1}(U_2))$. In fact, $q = \mathrm{e}_p(\nabla K_p)^{-1}(U_2)$ satisfies

$$\mathbb{E}_q[V] = dK_p \circ (\nabla K_p)^{-1}(U_2)[V] = \\ \left\langle (\nabla K_p) \circ (\nabla K_p)^{-1}(U_2), V \right\rangle_p = \langle U_2, V\rangle_p = 0, \quad V \in V_p^1 \ .$$

□

In conclusion, each $p \in \mathcal{E}$ is at the intersection of an exponential and a mixture family and such families can be sub-models or sub-manifolds under proper conditions. This provides a special type of coordinate system namely a mixed system, partly exponential and partly mixture, see Fig. 1. The following proposition summarizes basic facts from the literature and relates the splitting we are looking for with the classical characterization of exponential families, cf e.g. I. Csiszar's paper (Csiszár 1975) and the monograph (Brown 1986). Special cases of interest will be discussed in the following sections.

Proposition 5 *Let be given $p \in \mathcal{E}$ and $V_p \hookrightarrow B_p$, so that the families* $\mathrm{EF}(V_p)$ *and* $\mathrm{MF}(V_p)$ *are defined.*

1. *Assume that $q \in \mathcal{E}$ is such that the intersection of* $\mathrm{EF}(V_q)$ *and* $\mathrm{MF}(V_p)$ *is non empty and contains $\hat{q}$. The triple of densities $q, \hat{q}, r$, $r \in \mathrm{MF}(V_p)$ satisfies the*

Fig. 1 Mixed charts

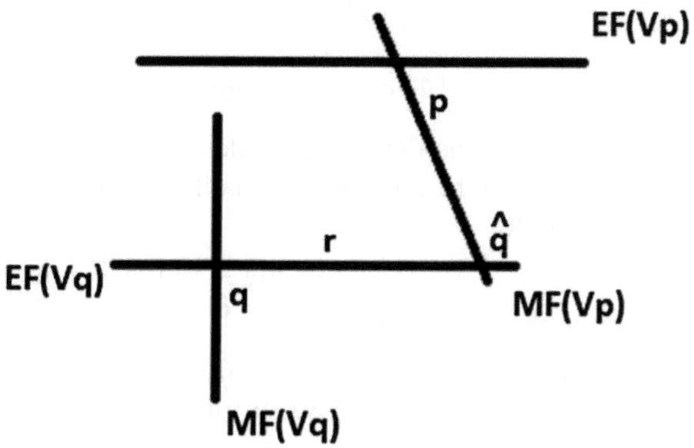

Pythagorean identity

$$\mathrm{D}\left(r \,\|\hat{q}\right) + \mathrm{D}\left(\hat{q} \,\|q\right) = \mathrm{D}\left(r \,\|q\right)$$

and the equivalent equation

$$\mathbb{E}_r\left[\log\frac{\hat{q}}{q}\right] = \mathbb{E}_{\hat{q}}\left[\log\frac{\hat{q}}{q}\right]$$

2. *It follows that any such intersection $\hat{q}$ strictly minimizes the divergence of* $\mathrm{MF}\left(V_p\right)$ *with respect to q, namely*

$$\mathrm{D}\left(\hat{q} \,\|q\right) \leq \mathrm{D}\left(r \,\|q\right), \quad r \in \mathrm{MF}\left(V_p\right) ,$$

with equality only if $r = \hat{q}$.
3. *Then such intersection $\hat{q}$ is unique and moreover* $\mathrm{EF}\left(V_q\right) = \mathrm{EF}\left(V_{\hat{q}}\right)$ *and* $\mathrm{MF}\left(V_p\right) = \mathrm{MF}\left(V_{\hat{q}}\right)$.
4. *Assume there is an open neighborhood $\mathcal{W}_p$ of $p \in \mathcal{E}$ such that for each $q \in \mathcal{W}_p$ there exist the intersection $\hat{q} =$ of* $\mathrm{EF}\left(V_q\right)$ *and* $\mathrm{MF}\left(V_p\right)$. *We can uniquely write $q = e^{\hat{u}-K_{\hat{q}}(\hat{u})} \cdot \hat{q}$ with $\hat{u} \in V_{\hat{q}}^1$ and $\hat{q} \in$* $\mathrm{MF}\left(V_p\right)$, *The map*

$$\mathcal{W}_p \ni q \mapsto \left(\hat{u} - \mathbb{E}_p\left[\hat{u}\right], \frac{\hat{q}}{p} - 1\right) \in V_p \times V_p^{\perp}$$

is injective and separates $\mathrm{EF}\left(V_p\right)$ *and* $\mathrm{MF}\left(V_p\right)$.

Proof 1. Let us write $\hat{q} \in \mathrm{EF}\left(V_q\right) \cap \mathrm{MF}\left(V_p\right)$ and $r \in \mathrm{MF}\left(V_p\right)$ in the chart centered at q as $\hat{q} = \mathrm{e}^{\hat{u}-K_q(\hat{u})} \cdot q$ and $r = \mathrm{e}^{v-K_q(v)} \cdot q$.

$$\begin{aligned}
&\mathrm{D}\left(r \,\|q\right) - \mathrm{D}\left(r \,\|\hat{q}\right) - \mathrm{D}\left(\hat{q} \,\|q\right) = \\
&\quad \mathbb{E}_r\left[v - K_q(v)\right] - \mathbb{E}_r\left[v - K_q(v) - \hat{u} + K_q(\hat{u})\right] - \mathbb{E}_{\hat{q}}\left[\hat{u} - K_q(\hat{u})\right] = \\
&\qquad \mathbb{E}_r\left[\hat{u}\right] - \mathbb{E}_{\hat{q}}\left[\hat{u}\right] = \mathbb{E}_r\left[\hat{u} - \mathbb{E}_p\left[\hat{u}\right]\right] - \mathbb{E}_{\hat{q}}\left[\hat{u} - \mathbb{E}_p\left[\hat{u}\right]\right] = 0 ,
\end{aligned}$$

because $\hat{u} - \mathbb{E}_p\left[\hat{u}\right] \in V_p$ and both $\hat{q}, r \in \mathrm{MF}\left(V_p\right)$.

2. Follows from the Pythagorean Identity and properties of the divergence.
3. Follows from the previous inequality and the definition of the families.
4. Let us write

$$\log\frac{q}{p} = \log\frac{q}{\hat{q}} + \log\frac{\hat{q}}{p}$$

with: $q = \mathrm{e}^{u-K_p(u)} \cdot p$, $u \in B_p$; $\hat{q} = \mathrm{e}^{v-K_p(v)} \cdot p$, $v \in B_p$ and $\mathbb{E}_{\hat{q}}\left[v\right] = 0$ if $v \in V_p$; $q = \mathrm{e}^{\hat{u}-K_{\hat{q}}(\hat{u})} \cdot \hat{q}$, $\hat{u} \in V_{\hat{q}}$. It follows

$$u - K_p(u) = \hat{u} - K_{\hat{q}}(\hat{u}) + v - K_p(v).$$

The p-expectation on both sides gives

$$-K_p(u) = \mathbb{E}_p\left[\hat{u}\right] - K_{\hat{q}}(\hat{u}) - K_p(v),$$

so that the equality becomes

$$u = \hat{u} - \mathbb{E}_p\left[\hat{u}\right] + v.$$

This splitting is unique, because $0 = \hat{u} - \mathbb{E}_p\left[\hat{u}\right] + v$ implies $v \in V_p$, hence $\hat{q} = \mathrm{e}^{v - K_p(v)} \cdot p \in \mathrm{EF}\left(V_p\right) \cap \mathrm{MF}\left(V_p\right)$, so that $\hat{u} = 0$ and $v = 0$.

5 Finite Dimensional Families

The most important practical applications of dimensionality reduction for infinite dimensional problems aim at transforming an infinite dimensional problem into a finite dimensional one. This is because, in order to be able to implement a numerical method in a machine, one needs a finite dimensional approximation. It is therefore particularly important to study finite dimensional submanifolds of the statistical manifold on which we might wish to approximate the full, infinite dimensional solution of a problem.

5.1 Finite Dimensional Exponential Family EF(c)

Our first special case is the parametric exponential family associated to a finite family of random variable $c = (c_1, \ldots, c_n)$.

$$\begin{aligned} \mathrm{EF}\,(c) &= \{p(\cdot, \theta), \theta \in \Theta\}, \\ p(\cdot, \theta) &= p_\theta = \exp\left(\theta^T c(\cdot) - \psi(\theta)\right), \end{aligned} \tag{13}$$

where Θ is a maximal convex open set in $\mathbb{R}^n$.

From the definition it is clear that all densities in the exponential family are connected by an open exponential arc. It follows that the exponential family is a subset of the maximal exponential family containing any of its elements, say $\mathcal{E} = \mathcal{E}\,(p)$, for some $p \in \mathrm{EF}\,(c)$. In fact, it is a special case of Definition 7. Precisely, the expression of each $p_\theta \in \mathrm{EF}\,(c)$ in the chart s_p is given by

$$\begin{aligned} p(\cdot, \theta) &= \exp\left(\theta^T c(\cdot) - \psi(\theta)\right) \\ &= \exp\left((\theta - \theta_0)^T c(\cdot) - (\psi(\theta) - \psi(\theta_0))\right) \cdot p \\ &= \exp\left((\theta - \theta_0)^T (c(\cdot) - \mathbb{E}_{p_0}[c]) - (\psi(\theta) - \psi(\theta_0) - (\theta - \theta_0)^T \mathbb{E}_{p_0}[c])\right) \cdot p \\ &= \exp\left(U(\theta) - K_{p_0}(U(\theta))\right) \cdot p \end{aligned}$$

with

$$U(\theta) = (\theta - \theta_0)^T (c(\cdot) - \mathbb{E}_{p_0}[c]) \in B_p$$
$$K_{p_0}(U(\theta)) = \psi(\theta) - \psi(\theta_0) - (\theta - \theta_0)^T \mathbb{E}_p[c]$$

For each $\theta \in \Theta$ let us define the subspace V_θ^1 of B_{p_θ} given by

$$\begin{aligned} V_\theta^1 = V_{p_\theta}^1 &= \mathrm{Span}\left(c_1 - \mathbb{E}_{p_\theta}[c_1], \ldots, c_n - \mathbb{E}_{p_\theta}[c_n]\right) \\ &= \mathrm{Span}\left(c_j - \frac{\partial}{\partial\theta_j}\psi(\theta)\middle| j = 1, \ldots, n\right) \end{aligned} \tag{14}$$

and let $\Pi_\theta \colon B_{p_\theta} \to V_\theta^1$ be the orthogonal projector. The orthogonal projection is well defined because $B_{p_\theta} \hookrightarrow L_0^2(p_\theta)$ and V_p^1 is a closed subspace of $L_0^2(p_\theta)$. If $g(\theta) = [\mathrm{Cov}_{p_\theta}(c_i, c_j)]_{i,j=1}^n = \mathrm{Hess}\,\psi(\theta)$ is the Fisher Information matrix of the exponential family and $[g^{ij}]_{i,j=1}^n = g^{-1}(\theta)$ denotes its inverse, then for all $U \in B_{p_\theta}$.

$$\Pi_\theta U = \sum_{j=1}^n \sum_{i=1}^n g^{ij}(\theta)\,\mathrm{Cov}_{p_\theta}(U, c_i)\,(c_j - \mathbb{E}_{p_\theta}[c_j]). \tag{15}$$

The mapping

$$B_{p_\theta} \ni U \mapsto (\Pi_\theta U, (I - \Pi_\theta)U) \in V_\theta^1 \times V_\theta^2,$$

with

$$V_\theta^2 = (I - \Pi_\theta)B_{p_\theta} = \left\{V \in B_{p_\theta} \middle| \langle V, U\rangle_{p_\theta} = 0, U \in V_\theta^1\right\} \hookrightarrow (V_{p_\theta}^1)^\perp,$$

is a splitting because the decomposition is unique and the spaces are both closed.

Here Definition 6(2) applies and splitting chart at p_θ is defined on the open domain where the projection is feasible, namely $\left\{p = \mathrm{e}_{p_\theta}(U) \in \mathcal{E} \middle| \Pi_\theta U \in \mathcal{S}_{p_\theta}\right\}$, by

$$\begin{aligned} p \mapsto U = s_{p_\theta}(p) \mapsto (U^1 = \Pi_\theta U, U^2 = U - \Pi_\theta U) \mapsto &(\mathrm{e}_{p_\theta}(U^1), U^2) \\ &\in \mathrm{EF}(c) \times (\mathcal{S}_{p_\theta} \cap \ker \Pi_\theta) \end{aligned}$$

Note that this splitting chart does provide an immersion of the exponential family into the maximal exponential family, together with a complementary model given by the infinite dimensional exponential family $\mathcal{E}_{\ker \Pi_\theta}(p_\theta) = \left\{\mathrm{e}_{p_\theta}(U^2) \middle| \Pi_\theta(U^2) = 0\right\}$, but it does not provide directly a complementary submanifold in the form of a mixture model. However, a different approach is usually taken to describe the complementary manifold, namely Propositions 4 and 5.

Let us fix $p_0 = p_{\theta_0} \in \mathrm{EF}(c)$ with associated vector space of centered statistics $V_0^1 \subset B_{p_0}$. Consider the vector space $V_0^2 = \left\{U^2 \in {}^*B_{p_0} \middle| \langle U^2, U^1\rangle_{p_{\theta_0}} = 0, U^1 \in V_0^1\right\}$, and observe that the mapping $\eta_{p_0}\colon \mathcal{S}_{p_0} \ni U \mapsto dK_{p_0}(U) \in B^*_{p_0}$, defined by $\left\langle V, \eta_{p_0}(U)\right\rangle_{p_0} = dK_{p_0}(U)[V]$, $V \in B_{p_0}$, is one-to-one because of the strict convexity of the cumulant generating functional $U \mapsto K_{p_0}(U)$.

Assume now $U \in \mathcal{S}_{p_0}$ and moreover $\eta_{p_0}(U) \in V_0^2$. It follows that the corresponding density $\mathrm{e}_{p_0}(U) \in \mathcal{E}$ is such that

$$\mathbb{E}_{\mathrm{e}_{p_0}(U)}\left[U^1\right] = dK_{p_0}(U)[U^1] = \left\langle \eta_{p_0}(U), U^1\right\rangle_{p_0} = 0, \quad U^1 \in V_0^1 .$$

Let $\mathrm{EF}(c) = \mathrm{EF}(c_1, \ldots, c_n)$ be an exponential family in the maximal exponential family $\mathcal{E}$, and let $V^1 = \mathrm{Span}\,(c_1, \ldots, c_n)$. Let us define the linear family

$$\mathcal{L}(c; \alpha) = \left\{q \in \mathcal{E} \middle| \mathbb{E}_q\left[c\right] = \alpha\right\} ,$$

where the expected value is meant to be applied componentwise.

Proposition 6 *1. Given $q \in \mathcal{E}$, compute the expected value of the c's statistics, $\mathbb{E}_q\left[c\right] = \alpha$, so that q belongs to the linear family $\mathcal{L}(c; \alpha)$. Assume there is a nonempty intersection $p \in \mathrm{EF}\,(c) \cap \mathcal{L}(c; \alpha)$, namely $p \in \mathrm{EF}\,(c)$ such that $\mathbb{E}_p\left[c\right] = \mathbb{E}_q\left[c\right]$. Then such a p is unique.*

2. *Let us express q in the chart centered at p, $q = \mathrm{e}_p(U^2)$. Then $\eta_p(U^2)$ is orthogonal to V_p^1.*
3. *p is the* information-projection *of any element $\bar{p}$ of the exponential family* $\mathrm{EF}\,(c)$ *on $\mathcal{L}(c; \mathbb{E}_q\left[c\right])$, that is*

$$\mathrm{D}\left(p \,\|\, \bar{p}\right) \leq \mathrm{D}\left(r \,\|\, \bar{p}\right), \quad r \in \mathcal{L}(c; \mathbb{E}_q\left[c\right]),\ \bar{p} \in \mathrm{EF}\,(c) ,$$

and the Pytagorean equality holds

$$\mathrm{D}\left(q \,\|\, p\right) + \mathrm{D}\left(p \,\|\, \bar{p}\right) = \mathrm{D}\left(q \,\|\, \bar{p}\right)$$

4. *p is the* reverse information-projection *of q on the exponential family* $\mathrm{EF}\,(c)$, *that is*

$$\mathrm{D}\left(q \,\|\, p\right) \leq \mathrm{D}\left(q \,\|\, \bar{p}\right), \quad \bar{p} \in \mathrm{EF}\,(c)\,,\ p \in \mathrm{EF}\,(c) \cap \mathcal{L}(c, \mathbb{E}_q\left[c\right]) .$$

Proof 1. Follows from the strict convexity of the cumulant generating function $\theta \mapsto \psi(\theta)$ and $\mathbb{E}_{p_\theta}\left[c_j\right] = \partial_j\psi(\theta)$, $j = 1, \ldots, n$ and $\theta \in \Theta$. If $\partial_j\psi(\theta_1) = \partial_j\psi(\theta_2)$, $j = 1, \ldots, n$, then $\sum_{j=1}^n(\partial_j\psi(\theta_1) - \partial_j(\theta_2))(\theta_{1j} - \theta_{2j}) = 0$, which implies $\theta_1 = \theta_2$ because of $\nabla\psi$ strict monotonicity.

2. The defining equality is equivalent to $\mathbb{E}_q\left[c_j - \mathbb{E}_p\left[c_j\right]\right] = 0$, $j = 1, \ldots, n$, hence $\mathbb{E}_q\left[V\right] = 0$ if $V \in V_p^1$. It follows $0 = dK_p(U^2)[V] = \left\langle U^2, V\right\rangle_p$.
3. Let us express r and $\bar{p}$ in the chart centered at p, namely $r = \mathrm{e}_p(U^2)$ and $\bar{p} = \mathrm{e}_p(U^1)$, so that $\mathbb{E}_r\left[U^1\right] = \mathbb{E}_p\left[U^1\right] = 0$. It follows that

$$\begin{aligned}
&\mathrm{D}\left(r \,\|\, \bar{p}\right) - \mathrm{D}\left(p \,\|\, \bar{p}\right) \\
&\quad = \mathbb{E}_r\left[U^2 - U^1 - K_p(U^2) + K_p(U^1)\right] - \mathbb{E}_p\left[-U^1 + K_p(U^1)\right] \\
&\quad = \mathbb{E}_r\left[U^2\right] - K_p(U^2) \\
&\quad = \mathrm{D}\left(r \,\|\, p\right)
\end{aligned}$$

The Pythagorean equality is proved by expressing each density in the chart centered at p.

4. By expressing $\bar{p}$ in the chart centered at p, namely $\bar{p} = \mathrm{e}_p(U^1)$, $U^1 \in V_p^1$, we have

$$\begin{aligned}
\mathrm{D}\left(q \,\|\, \bar{p}\right) - \mathrm{D}\left(q \,\|\, p\right) &= \mathbb{E}_q\left[\log \frac{q}{\bar{p}}\right] - \mathbb{E}_q\left[\log \frac{q}{p}\right] \\
&= \mathbb{E}_q\left[\log \frac{p}{\bar{p}}\right] \\
&= -\mathbb{E}_q\left[U^1\right] + K_p(U^1) = K_p(U^1)
\end{aligned}$$

which is minimized at $U^1 = 0$

Remark 2 1. For each $q \in \mathcal{E}$ such that there exists $p \in \mathrm{EF}\left(c\right)$ satisfying the previous proposition, there is a splitting parameterization $q \mapsto (p, \mathrm{e}_p(q)) \in \mathrm{EF}\left(c\right) \times V_p^{\perp}$. The critical issue is the closure of $V_p^{\perp}$ into B_p.

2. Item 4 suggests to characterize the feasible set for the splitting by considering the minimum of the mapping

$$q \mapsto \operatorname{argmin}\left\{\mathrm{D}\left(q \,\|\, \bar{p}\right) | \bar{p} \in \mathrm{EF}\left(c\right)\right\} .$$

Let us assume (without restriction) that the entropy $H\left(q\right) = -\mathbb{E}_q\left[\log q\right]$ is finite, so that $\mathrm{D}\left(q \,\|\, \bar{p}\right) = -H(q) + \mathbb{E}_q\left[\log \bar{p}\right] = -H(q) + \sum_{j=1}^n \theta_j \mathbb{E}_q\left[c_j\right] - \psi(\theta)$. we have

$$\inf \mathrm{D}\left(q \,\|\, \bar{p}\right) = -H(q) + \max \theta' \mathbb{E}_q\left[c\right] - \psi(\theta) = -H(q) + \psi_*(\mathbb{E}_q\left[c\right])$$

It follows that the feasible set for the splitting is the open set

$$\left\{q \in \mathcal{E} \middle| \mathbb{E}_q\left[c\right] \in \mathrm{Dom}\left(\psi_*\right)^{\circ}\right\} .$$

5.2 *Finite Dimensional Mixture(-Generated) Family MG(q)*

The basic splitting we have used in the previous sections consists of a closed subspace $V_p^1 \subset B_p$ together with its pre-dual annihilator $V_p^2 \subset {}^*B_p$. As the model space B_p is not an Hilbert space unless the base space is finite, there is no identification of $V_p^1 \times V_p^2$ within B_p, but we only have the immersion $B_p \hookrightarrow V_p^1 \oplus V_p^2$. However, the technicalities are somehow easier to control if one of the two splitting spaces is finite dimensional, as it was the case for V_p^1 in the previous section.

We have defined a mixture-closed (by convex combinations) family $\mathrm{MF}\left(V_p\right)$ in Definition 8. Here, we first define a family as the mixture generated by a given family through convex combinations and later we show how this is related with the mixture-closed family. Suppose we are given $n+1$ fixed probability densities, say $q = [q_1, q_2, \ldots, q_{n+1}]^T$. Consider the convex hull of q, generated by all possible convex combinations of q elements, which we term "mixture generated family" (MG)

$$\mathrm{MG}\,(q) = \left\{\theta^T q \middle| \theta \in \Delta(n)\right\} ,$$

were $\Delta(n) = \left\{\theta \in \mathbb{R}_+^{n+1} \middle| \sum_{i=1}^{n+1} \theta_i = 1\right\}$ is the standard simplex.

We now state a proposition giving conditions under which the two different notions of mixture family coincide in the finite dimensional case, namely we give conditions under which MF = MG.

Proposition 7 *1. If all q_i belong to the same maximal exponential family $\mathcal{E}(p)$, then* $\mathrm{MG}\,(q) \subset \mathcal{E}(p)$. *In particular, we can choose $p \in \mathrm{MG}\,(q)$.*

2. *In such a case, let $V_p^1 = \left\{U \in B_p \middle| \mathbb{E}_{q_j}[U] = 0, j = 1, \ldots, n+1\right\}$. Then this space is closed in B_p and* $\mathrm{MG}\,(q) \subset \mathrm{MF}\left(V_p^1\right)$.
3. *If moreover $\hat{q} = \sum_{i=1}^{n=1} \alpha_i q_i$ with $\sum_{i=1}^{n=1} \alpha_i = 1$ is a positive density only if $\alpha_i \geq 0$, $i = 1, \ldots, n+1$, then* $\mathrm{MG}\,(q) = \mathrm{MF}\left(V_p^1\right)$.

Proof 1. (Cf. Santacroce et al. 2015) We use Portmanteu Theorem 1.6. Given $q_1, q_2 \in \mathcal{E}(p)$, $q_1 = \mathrm{e}_p(U_1)$ and $q_2 = \mathrm{e}_p(U_2)$ consider the convex combination $q_\theta = (1-\theta)q_1 + \theta q_2$, $0 < \theta < 1$. From the convexity of $x \mapsto x^{1+\epsilon}$ we derive

$$\begin{aligned}\int \left(\frac{q_\theta}{p}\right)^{1+\epsilon} p &= \int \left(\frac{(1-\theta)q_1 + \theta q_2}{p}\right)^{1+\epsilon} p \\ &\leq (1-\theta)\int \left(\frac{q_1}{p}\right)^{1+\epsilon} p + \theta \int \left(\frac{q_2}{p}\right)^{1+\epsilon} p ,\end{aligned}$$

where both integrals are finite for some $\epsilon > 0$.
From the convexity of $x \mapsto x^{-\epsilon}$ we derive

$$\begin{aligned}\int \left(\frac{p}{q_\theta}\right)^{1+\epsilon} q_\theta &= \int \left(\frac{p}{(1-\theta)q_1 + \theta q_2}\right)^{1+\epsilon} ((1-\theta)q_1 + \theta q_2) \\ &= \int p^{1+\epsilon}((1-\theta)q_1 + \theta q_2)^{-\epsilon}\end{aligned}$$

$$\leq (1-\theta)\int p^{1+\epsilon}q_1^{-\epsilon} + \theta\int p^{1+\epsilon}q_2^{-\epsilon}$$
$$= (1-\theta)\int \left(\frac{p}{q_1}\right)^{1+\epsilon} q_1 + \theta\int \left(\frac{p}{q_2}\right)^{1+\epsilon} q_2\,,$$

where both integrals are finite for some $\epsilon > 0$.

2. Consider the vector space V_p^2 generated in ${}^*B_p$ by $\frac{q_i}{p} - 1$, $i = 1, \ldots, n+1$. As $V_p^1 = (V_p^2)^\perp$, we have $(V_p^1)^\perp = V_p^2$ so that $\mathrm{MF}\left(V_p\right) = V_p^2 \cap \mathcal{E}$. A generic $v \in V_p^2$ is a linear combination $v = \sum_{i=1}^{n+1} \alpha_j (\frac{q_j}{p} - 1)$, and $v = \frac{\bar{q}}{p} - 1$ for a density $\bar{q}$ if $\sum_{i=1}^{n=1} \alpha_j = 1$. In particular this is true for each $\bar{q} \in \mathrm{MG}\,(q)$.
3. If the assumption holds true, all α_i's that produce a density are nonnegative.

□

The exponential transport ${}^e\mathbb{U}_p^{\bar{q}} U = U - \mathbb{E}_{\bar{q}}[U]$, $\bar{q} \in \mathrm{MG}\,(q)$ acts on V_p^1 as $U - \sum_{j=1}^{n+1} \mathbb{E}_{q_j}[U] = U$, so that

$$ {}^e\mathbb{U}_p^{\bar{q}} V_p^1 = \left\{ {}^e\mathbb{U}_p^{\bar{q}} U \middle| U \in B_p, \mathbb{E}_{q_i}[U] = 0, i = 1, \ldots, n+1 \right\} = \left\{ V \in B_{\bar{q}} \middle| \mathbb{E}_{q_i}[V] = 0, i = 1, \ldots, n+1 \right\} = V_q^1$$

We define the *exponential family orthogonal to* $\mathrm{MG}\,(q)$ to be $\mathrm{EF}\left(V_{\bar{q}}^1\right) = \left\{ \mathrm{e}_{\bar{q}}(U) \middle| U \in V_{\bar{q}}^1 \right\}$ for any $\bar{q} \in \mathrm{MG}\,(q)$. Note that the same exponential family can be expressed at any p, in which case the base space is

$$V_p^1 =^e \mathbb{U}_{\bar{q}}^p V_{\bar{q}}^1 = \left\{ {}^e\mathbb{U}_{\bar{q}}^p U \middle| U \in B_{\bar{q}}, \mathbb{E}_{q_i}[U] = 0, i = 1, \ldots, n+1 \right\} = \left\{ V \in B_p \middle| U \in \mathbb{E}_{q_i}[V] = \mathbb{E}_{\bar{q}}[V], i = 1, \ldots, n+1 \right\}.$$

The families $\mathrm{EF}\left(V_p^1\right)$, $\mathrm{MF}\left(V_p^1\right)$ described above form a couple as discussed in Sect. 4.

6 Finite Dimensional Approximations by Projection

We now have all the tools we need to derive finite dimensional approximations of infinite dimensional evolution equations for probability measures, such as the ones we have highlighted in Sect. 2 from probability theory, signal processing, social sciences, physics and quantum theory. This can be done with the rigorous infinite dimensional manifold structure from G. Pistone and co-authors we have summarized in the previous sections.

As we have mentioned in the introduction, this has been done in the past by D. Brigo and co-authors in Brigo et al. (1998, 1999), Armstrong and Brigo (2016) for the filtering problem and in Brigo (1997, 1999) for the Fokker–Planck equation, but using the whole L^2 space as superstructure, without specifically investigating the geometric structures at play in the infinite-dimensional environment, except for the enveloping exponential manifold discussion in Brigo et al. (1999).

Here we will develop the case of the Fokker–Planck PDE since, as we explained in Sect. 2, this is really the element that brings about infinite dimensionality even in the more complex cases of signal processing and quantum theory stochastic PDEs. The Fokker–Planck equation is thus the ideal benchmark case where one can study dimensionality reduction at the crossroad of different areas.

We should also mention briefly that the SPDE case we do not treat here involves infinite-dimensional evolution equations driven by noise and rough paths. The driving rough paths motivate possibly different types of projections related to stochastic differential geometry and introduce different notions of optimality of the projection of the equation solution. We do not have this problem here, since our Fokker–Planck benchmark case will simply be a PDE and will not be driven by noise, but for the general case see the forthcoming paper by Armstrong and Brigo (2015) in this same volume.

Before turning to the Fokker–Planck equation, however, we first consider our running example of Sect. 2.5.

6.1 Finite Dimensional Approximation for the Heat Equation

With the notations of Definition 4, let p be a density in the W^1_Φ-exponential family, $p \in \mathcal{E}_1(M)$, that is $p = \mathrm{e}^{U-K_M(U)} \cdot M$ and $U \in \mathcal{S}^1_M = \mathcal{S}_M \cap B_M \cap W^1_\Phi$.

Let $\mathcal{A}p$ be the non-linear differential operator $p^{-1}\mathcal{L}^* p$ where $\mathcal{L}^*$ is the differential operator for our running example equation of Sect. 2.5, where we assume bounded and uniformly positive definite matrix of coefficients $[a_{ij}]$. Namely, we are considering the anisotropic heath equation.

$$\mathcal{A}p(x) = p(x)^{-1} \sum_{i,j=1}^{d} \frac{\partial}{\partial x_i}\left(a_{ij}(x)\frac{\partial}{\partial x_j}p(x)\right), \quad x \in \mathbb{R}^d .$$

Conditions on the coefficients $[a_{ij}]$ are to be given in order to show that the operator on a sufficiently large domain $\mathcal{D}$ is a section of the differentiable mixture bundle, namely $\mathcal{A}(p) \in {}^*B^1_p$, $p \in \mathcal{D} \subset \mathcal{E}_1(M)$. We do not want to discuss here such conditions. It was done in Lods and Pistone (2015) for the special case of the Laplacian, and we assume this property from now on. Note that the zero expectation condition is trivially verified by

$$\mathbb{E}_p\left[\mathcal{A}p(x)\right] = \int \mathcal{L}^* p(x)\, dx = \int p(x)\mathcal{L}1\, dx = 0.$$

Recall that the differentiable predual bundle has an affine atlas of charts, see Definition 5(2). The chart centered at p is

$$^*\sigma_p \colon {}^*S\mathcal{E}_1\,(M) \ni (q, V) \mapsto \left(s_p(q), {}^m\mathbb{U}_q^p V\right) \in B_p^1 \times {}^*B_p^1.$$

where the exponential chart is $s_p(q) = \log\frac{q}{p} - \mathbb{E}_p\left[\log\frac{q}{p}\right]$ and the linear transport ${}^m\mathbb{U}_q^p \colon {}^*B_q^1 \to {}^*B_p^1$ is defined by $V \mapsto \frac{q}{p}V$.

Example 3 In the chart centered at M,

$$^*\sigma_M(\mathrm{e}^{U-K_M(U)}\cdot M, V) = \left(U, \mathrm{e}^{U-K_M(U)}V\right) \in B_M^1 \times {}^*B_M^1.$$

It follows that the expression of the operator $\mathcal{A}$ in the charts centered at M is of the form

$$U \mapsto \widehat{\mathcal{A}}_M(U) = \mathrm{e}^{U-K_M(U)}\mathcal{A}(\mathrm{e}^{U-K_M(U)}\cdot M) = \frac{\mathrm{e}^{U-K_M(U)}}{\mathrm{e}^{U-K_M(U)}\cdot M}\mathcal{L}^*(\mathrm{e}^{U-K_M(U)}\cdot M) = M^{-1}\mathcal{L}^*(\mathrm{e}^{U-K_M(U)}\cdot M)$$

The computation in Eq. (11) gives

$$M^{-1}\mathcal{L}^*(\mathrm{e}^{U-K_M(U)}\cdot M) = \mathrm{e}^{U-K_M(U)}\sum_{i,j=1}^{d}\frac{\partial}{\partial x_i}\left[a_{ij}(x)\left(\frac{\partial}{\partial x_j}U(x) - x_j\right)\right] + \mathrm{e}^{U-K_M(U)}\sum_{i,j=1}^{d}a_{ij}(x)\left(\frac{\partial}{\partial x_i}U(x) - x_i\right)\left(\frac{\partial}{\partial x_j}U(x) - x_j\right).$$

We want now to consider the weak form of the operator, which is defined for each $V \in B_p^1$ by

$$\begin{aligned}\langle \mathcal{A}p, V\rangle_p &= \int p(x)dx\; p(x)^{-1}\sum_{i,j=1}^{d}\frac{\partial}{\partial x_i}a_{ij}(x)\frac{\partial}{\partial x_j}p(x)\;V(x)\\ &= \sum_{i,j=1}^{d}\int dx\;\frac{\partial}{\partial x_i}a_{ij}(x)\frac{\partial}{\partial x_j}p(x)\;V(x)\\ &= -\sum_{i,j=1}^{d}\int dx\;a_{ij}(x)\frac{\partial}{\partial x_j}p(x)\frac{\partial}{\partial x_i}V(x).\end{aligned}$$

Note that the weak form we have defined at each p is just the usual weak form of the operator $\mathcal{L}^*$, so that it is negative definite. If we proceed with the exponential charts and Eq. (10) we get

$$\begin{aligned}\langle \mathcal{A}p, V\rangle_p &= -\sum_{i,j=1}^{d}\int p(x)dx\, a_{ij}(x)\left(\frac{\partial}{\partial x_j}U(x) - x_j\right)\frac{\partial}{\partial x_i}V(x) \\ &= \sum_{i,j=1}^{d}\left\langle a_{ij}(X)(X_j - \partial_j U), \partial_i V\right\rangle_p \\ &= \sum_{i,j=1}^{d}\left\langle a_{ij}(X)X_j, \partial_i V\right\rangle_p - \sum_{i,j=1}^{d}\left\langle a_{ij}(X)\partial_j U, \partial_i V\right\rangle_p .\end{aligned}$$

Note that U belongs to B^1_M, so that X_j and $\partial_j U$ both belong to $L^{\Phi}(M)$. It is sufficient to assume $[a_{ij}]$ uniformly bounded. Weaker conditions are allowed, as we actually need to assume that the multiplication operator $W \mapsto a_{ij}(X)W$ maps $L^{\Psi}(p)$ into itself for all p.

To define a Galerkin-style projection, we want finite dimensional subspaces $V_n(p)$ of the fibers B^1_p. Such subspaces are obtained from a reference one $V_n(M)$ via the application of the exponential parallel transport. Assume $V_n \in B^1_M$ is a vector space of dimension n and take $U \in V_n$ and $V \in V_n(p) = {}^e\mathbb{U}^p_M V_n$. As the exponential transport has no effect on the partial derivatives, we have for $U, V \in B^1_M$

$$\begin{aligned}\left\langle \mathcal{A}p, {}^e\mathbb{U}^p_M V\right\rangle_p &= \sum_{i,j=1}^{d}\left\langle a_{ij}(X)(X_j - \partial_j)U, \partial_i V\right\rangle_p \\ &= \sum_{i,j=1}^{d}\left\langle a_{ij}(X)X_j, \partial_i V\right\rangle_p - \sum_{i,j=1}^{d}\left\langle a_{ij}(x)\partial_j U, \partial_i V\right\rangle_p\end{aligned}$$

Let $(W_1, \ldots, W_n)$ be a basis of V_n, so that $(W_1 - \mathbb{E}_p[W_1], \ldots, W_n - \mathbb{E}_p[W_n])$ is a basis of $V_n(p)$. We can write

$$\begin{aligned}U &= \sum_{h=1}^{n}\theta_h W_h \\ V &= \sum_{k=1}^{n}\alpha_k W_k\end{aligned}$$

and

$$\left\langle \mathcal{A}p, {}^e\mathbb{U}^p_M V\right\rangle_p = \sum_{h,k=1}^{n}\theta_h\alpha_k\sum_{i,j=1}^{d}\left\langle a_{ij}(X)(X_j - \partial_j)W_h, \partial_i W_k\right\rangle_p$$

Equivalently,

$$\left\langle \mathcal{A}p, {}^{e}\mathbb{U}_{M}^{p} W_k \right\rangle_p = \sum_{h=1}^{n} \theta_h \sum_{i,j=1}^{d} \left\langle a_{ij}(X)(X_j - \partial_j) W_h, \partial_i W_k \right\rangle_p, \quad k = 1, \dots, n$$

In the exponential family of densities of the form

$$p = \exp\left(\sum_{h=1}^{n} \theta_k W_k - \psi(\theta) \right) \cdot M$$

we look for a curve $t \mapsto p(t)$ whose score $Dp(t)$ is such that

$$\left\langle Dp(t) - \mathcal{A}p(t), {}^{e}\mathbb{U}_{M}^{p(t)} W_k \right\rangle_{p(t)} = 0, \quad k = 1, \dots, n. \tag{16}$$

In fact, the curve $t \mapsto (p(t), Dp(t) - \mathcal{A}p(t))$ belongs to a statistical bundle, hence has to be checked against a moving frame. The score can be written in the moving frame as

$$Dp(t) = \frac{\dot{p}(t)}{p(t)} = \sum_{h=1}^{n} \dot{\theta}_h(t) {}^{e}\mathbb{U}_{M}^{p(t)} W_h$$

so that

$$\left\langle Dp(t), {}^{e}\mathbb{U}_{M}^{p} W_k \right\rangle_{p(t)} = \sum_{h=1}^{n} \dot{\theta}_h(t) \left\langle {}^{e}\mathbb{U}_{M}^{p(t)} W_h, {}^{e}\mathbb{U}_{M}^{p} W_k \right\rangle_{p_{\theta(t)}} = \sum_{h=1}^{n} g_{hk}(t) \dot{\theta}_h(t),$$

where we have used the Fisher matrix

$$g(\theta) = [\left\langle {}^{e}\mathbb{U}_{M}^{p_\theta} W_h, {}^{e}\mathbb{U}_{M}^{p_\theta} W_k \right\rangle_{p_\theta}]_{h,k} = [\mathrm{Cov}_{p_\theta}(W_h, W_k)]_{h,k} = \mathrm{Hess}\, \psi(\theta).$$

Equation (16) becomes

$$\sum_{h=1}^{n} g_{kh}(\theta(t)) \dot{\theta}_h(t) = \sum_{h=1}^{n} \sum_{i,j=1}^{d} \left\langle a_{ij}(X)(X_j - \partial_j) W_h, \partial_i W_k \right\rangle_{p_{\theta(t)}} \theta_h(t), \tag{17}$$

for all $k = 1, \dots, n$.

If the inverse Fisher matrix is $g(\theta)^{-1} = [g^{lk}(\theta)]$, we can multiply the equation by $g^{lk}(\theta(t))$ and sum over k to get the system of non linear differential differential equations:

$$\dot{\theta}_l(t) = \sum_{h=1}^{n} \sum_{i,j=1}^{d} \left\langle a_{ij}(X)(X_j - \partial_j)W_h, \partial_i \sum_{k=1}^{d} g^{lk}(\theta(t))W_k \right\rangle_{p_{\theta(t)}} \theta_h(t), \tag{18}$$

for all $l = 1, \ldots, n$.

We have shown that it is possible, at least in principle, to derive Galerkin-type approximations of our running example. To proceed to a practical implementation it would be necessary to choose a suitable basis $(W_1, \ldots, W_n)$ for which the Galekin equation (18) is computable.

We now turn to examine from a different perspective a second example, the Fokker–Plank equation.

6.2 Fokker–Planck Equation in Statistical Manifold Coordinates

We could apply the same techniques we used in the running example pari passu to the Fokker–Planck equation (2), keeping in mind the definition of the related operators $\mathcal{L}$ and $\mathcal{L}^*$. However, we will proceed at a low pace given the more complicated nature of (2) compared to our running example. We proceed step by step by showing how the specific structure of (2) is dealt with in the statistical manifold context of this paper.

We may want to avoid using necessarily the Gaussian density M as background density, so for simplicity in this section we work in a single chart and assume the equation is written until the first exit time from the manifold. For example, again in the case $c_1(x) = x$, $c_2(x) = x^2, \ldots, c_n(x) = x^n$, n even natural number, this would correspond to the first exit time from $\{\theta_n < 0\}$. We might avoid the exit time by introducing a suitable background density, for example $M_{1,n+2}$, but for simplicity we do not assume a background density in the derivation. We will discuss again the possible use of a background density when considering the $\mathcal{L}$ eigenfunctions later.

Now we rewrite Eq. (2) in exponential coordinates. Consider as local reference density the solution p_t of FPE at time t. We are now working around p_t. Consider a curve around p_t corresponding to the solution of FPE around time t expressed in B_{p_t} coordinates:

$$h \mapsto s_{p_t}(p_{t+h}) =: u_h.$$

The function u_h represents the expression in coordinates of the density

$$p_{t+h} = \exp[u_h - K_{p_t}(u_h)]p_t =: e_h p_t. \tag{19}$$

Now consider FPE around t, i.e.

$$\frac{\partial p_{t+h}}{\partial h} = \mathcal{L}^*_{t+h} p_{t+h}.$$

Substitute (19) in this last equation in order to obtain

$$\frac{\partial e_h p_t}{\partial h} = \mathcal{L}^*_{t+h}(e_h p_t).$$

Write

$$\frac{\partial e_h}{\partial h} = \frac{\mathcal{L}^*_{t+h}(e_h p_t)}{p_t}$$

and set $h = 0$, since we are concerned with the behavior in t. Notice that $e_0 = \exp[u_0 - K_{p_t}(u_0)] = \exp(0) = 1$, and that

$$\left.\frac{\partial e_h}{\partial h}\right|_{h=0} = \{e_h \frac{\partial [u_h - K_{p_t}(u_h)]}{\partial h}\}|_{h=0} = \frac{\partial [u_h - K_{p_t}(u_h)]}{\partial h}|_{h=0}.$$

Moreover, by straightforward computations (write explicitly the map K_{p_t}, use $u_h = s_{p_t}(p_{t+h})$ and differentiate wrt h under the expectation E_{p_t}) one verifies

$$\left.\frac{\partial K_{p_t}(u_h)}{\partial h}\right|_{h=0} = 0,$$

so that

$$\left.\frac{\partial u_h}{\partial h}\right|_{h=0} = \frac{\mathcal{L}^*_t p_t}{p_t} \tag{20}$$

is the formal representation in exponential coordinates of the vector in the statistical exponential (vector) bundle $S\mathcal{E}$ at p_t. Notice that, again by straightforward computations, and omitting the time arguments in f and a for brevity,

$$\begin{aligned}\alpha_t := \alpha_t(p) = \frac{\mathcal{L}^*_t p}{p} &= -\sum_{i=1}^N \left(f_i \frac{\partial}{\partial x_i}(\log p) + \frac{\partial f_i}{\partial x_i} \right) + \\ &+ \frac{1}{2} \sum_{i,j=1}^N \left[a_{ij} \frac{\partial^2}{\partial x_i \partial x_j}(\log p) + a_{ij} \frac{\partial}{\partial x_i}(\log p) \frac{\partial}{\partial x_j}(\log p) + \right. \\ &\left. + 2 \frac{\partial a_{ij}}{\partial x_j} \frac{\partial}{\partial x_i}(\log p) + \frac{\partial^2 a_{ij}}{\partial x_i \partial x_j} \right].\end{aligned} \tag{21}$$

Summarizing: consider the curve expressing FPE around p_t in B_{p_t} coordinates. Its tangent vector/fiber in the statistical exponential bundle $S\mathcal{E}$ at p_t is given by α_t. Under suitable assumptions on the coefficients f_t and a_t the function α_t belongs to B_{p_t}, according to the convention that locally identifies the tangent bundle of a normed space with the normed space itself. To render the computation not only formal we need α_t to be really a tangent vector/fiber for our bundle structure. This in turn requires the curve $t \mapsto p_t$ to be differentiable in the proper sense. Below we give a regularity result expressing a condition under which this happens and whose proof is immediate. Moreover, we give a condition which can be used to check whether the evolution stays in a given submanifold.

Proposition 8 (Regularity and finite dimensionality of the solution of FPE)

(i) *If the map $t \mapsto p_t$ is differentiable in the manifold $\mathcal{E}$ then α_t given in Eq. (21) is a tangent vector.*
(ii) *If the map $t \mapsto \alpha_t$ is continuous at t_0 into L^Φ, then $t \mapsto p_t$ is differentiable at t_0 as a map into $\mathcal{E}$.*
(iii) *Let be given a submanifold $\mathcal{N}$ such that $p_0 \in \mathcal{N}$. If the previous condition is satisfied and*

$$\frac{\mathcal{L}_t^* p}{p}$$

is tangent to $\mathcal{N}$ at p for all $p \in \mathcal{N}$, then p_t evolves in $\mathcal{N}$.

Sufficient conditions under which condition (ii) in the proposition holds are related to boundedness for all possible $T > 0$ and i, j of $f, \partial_{x_i} f, a, \partial_{x_i} a, \partial^2_{x_i x_j} a$ in $[0\ T] \times \mathbb{R}$ plus classical assumptions ensuring (D). This follows from the fact that if $\alpha_t(x)$ is continuous and bounded in both t and x, then it is continuous as a map $t \mapsto \alpha_t$ from $[0\ T]$ to L^Φ.

6.3 Examples of Finite Dimensional Fokker–Planck

In the following we give examples where Proposition 8 applies in the special case $N = 1$. Some of them are obtained from Brigo (1997) (see also Brigo 2000) where the detailed derivation is given.

Example 4 (Linear case) If $f_t(x) = F_t x$ for all $t \geq 0,\ x \in \mathbb{R}$ (f linear in x) and if $a_t(x) = A_t$ for all $t \geq 0,\ x \in \mathbb{R}$ (a does not depend on x) and if finally $p_0 \sim \mathcal{N}(m_0, Q_0)$ then it is known that $p_t \sim \mathcal{N}(m_t, Q_t)$ where $m_t = m_0 \exp \int_0^t F_s ds$ and Q_t is the (unique) positive solution of the (scalar Lyapunov) equation

$$\dot{Q}_t = 2F_t Q_t + A_t,$$

with initial condition Q_0 given. Consider now a generic Gaussian density $p \sim \mathcal{N}(m, Q)$ and compute

$$\left(\frac{\mathcal{L}_t^* p}{p}\right)(x) = \left(\frac{F_t}{Q} + \frac{A_t}{2Q^2}\right)x^2 - \left(\frac{F_t m}{Q} + \frac{A_t m}{Q^2}\right)x + \frac{A_t m^2}{2Q^2} - F_t - \frac{A_t}{2Q}. \tag{22}$$

When applied to p_t, the previous formula yields α_t:

$$\alpha_t = \left(\frac{F_t}{Q_t} + \frac{A_t}{2Q_t^2}\right)x^2 - \left(\frac{F_t m_t}{Q_t} + \frac{A_t m_t}{Q_t^2}\right)x + \frac{A_t m_t^2}{2Q_t^2} - F_t - \frac{A_t}{2Q_t},$$

where m_t and Q_t have been defined above.

In this case the previous proposition applies. First, one sees that $t \mapsto \alpha_t$ is indeed continuous at any t_0 in L^{Φ}. Secondly, one can deduce already from (22) *without solving the Fokker–Planck equation* that the solution will have a Gaussian density. Indeed, one can easily check that the tangent space to the Gaussian submanifold of $\mathcal{E}$ expressed in B coordinates contains the function space span$\{1, x, x^2\}$. Since by expression (22) we see that $(\mathcal{L}_t^* p)/p$ lies in span$\{1, x, x^2\}$ for all p in the Gaussian submanifold, we deduce that the solution of the Fokker–Planck equation will evolve in the Gaussian submanifold.

Example 5 (*Nonlinear diffusions with unit variance Gaussian law*) Let be given a diffusion coefficient $\sigma_t(x)$ satisfying assumptions (B) and assumption (C) when the drift vanishes, i.e. when $f = 0$ (we set as usual $a := \sigma^2$). In Brigo (1997) it is shown that by defining the drift

$$f_t(x) := \frac{1}{2}\frac{\partial a_t}{\partial x}(x) + \frac{1}{2}a_t(x)[kt - x] + k,$$

the Fokker–Planck equation for the density of the solution of the stochastic differential equation

$$dX_t = f_t(X_t)dt + \sigma_t(X_t)dW_t \quad X_0 \sim \mathcal{N}(0, 1),$$

is solved by $p_t \sim \mathcal{N}(kt, 1)$ for all possible diffusion coefficients $\sigma_t(x)$. Here the solution of the Fokker–Planck equation evolves in a submanifold of $\mathcal{E}$ given by Gaussian densities with unit variance. Actually, the mean of p_t evolves linearly in time and the variance is fixed to one. Note that in this case

$$\alpha_t = \partial_t \log p_t = k(x - kt),$$

and the curve $t \mapsto \alpha_t$ is clearly continuous at any t_0 in L^{Φ}. One might check a priori that if a given p belongs to the submanifold of Gaussian densities with unit variance, then $(\mathcal{L}_t^* p)/p$ belongs to the tangent space of this submanifold if the mean is given by kt. Indeed, we are considering the family

$$p(x, \theta) = \frac{1}{\sqrt{2\pi}}\exp\left[-\frac{1}{2}(x - \theta)^2\right] \sim \mathcal{N}(\theta, 1), \quad \theta \in \mathbb{R},$$

and its tangent space expressed in $B_{p(\cdot,\theta)}$ coordinates, span$\{x - \theta\}$. Let us compute

$$\alpha_{t,\theta}(x) = \frac{1}{2}(\partial_x a_t(x))\,(kt - \theta) + \frac{1}{2}a_t(x)(x - \theta)(kt - \theta) \;+\; k(x - \theta).$$

Under reasonable assumptions on a, this function belongs to the tangent space span$\{x - \theta\}$ if and only if $\theta = kt$. We have been able to check that the density of the diffusion X evolves according to $p_t \sim \mathcal{N}(kt, 1)$ without solving the Fokker–Planck equation.

From examples given in Brigo (1997, 2000) one can construct other nonlinear cases where the above proposition applies.

6.4 *Projection of the Infinite Dimensional Fokker–Planck Equation*

In reaching Eq. (20) we assumed implicitly a few facts. We are assuming that there always exists a neighborhood of $h = 0$ such that in this neighborhood $p_{t+h} \in \mathcal{E}(p_t)$. Conditions under which this happens will be examined in the future. We only remark that when projecting on a finite dimensional exponential manifold, these conditions are not necessary for the projected equation to exist and make sense, see below. Neither we need Eq. (20) to have a solution to obtain existence of the solutions of the projected equation. Now we shall project this equation on a finite dimensional parametrized exponential manifold EF (c). We will assume the following on the family EF (c) (see Brigo et al. 1999 for other more specific assumptions):

$$(E) \qquad \text{We assume} \quad c \in C^2.$$

A rapid projection computation based on Formula (15) and involving integration by parts between $\mathcal{L}$ and $\mathcal{L}^*$ and standard results on the normalization constant $\psi(\theta)$ of exponential families (such as $\partial_{\theta_i}\psi(\theta) = E_\theta c_i$) yields

$$\mathcal{P}_{t,\theta} := \Pi_\theta\left[\frac{\mathcal{L}_t^* p(\cdot,\theta)}{p(\cdot,\theta)}\right] = E_\theta[\mathcal{L}_t c]^T\; g^{-1}(\theta)\;[c(\cdot) - E_\theta c],$$

where integrals of vector functions are meant to be applied to their components. Note that this map is regular in θ under reasonable assumptions on f, a and c. At this point we project Eq. (20) via this projection. By remembering expression (14) for tangent vectors and the above formula for the projection we obtain the following (n–dimensional) ordinary differential equation (in vector form) in the coordinates of the manifold EF (c):

$$\dot{\theta}_t = g^{-1}(\theta_t)\; E_{\theta_t}\{\mathcal{L}_t\; c\}. \tag{23}$$

Notice that, as anticipated above, Eq. (23) is well defined and admits locally a unique solution if the following condition (ensuring existence of the norm of $\alpha_t(p(\cdot, \theta_t))$ associated to the inner product $\mathrm{Cov}_{p_{\theta_t}}(\cdot, \cdot)$) holds:

$$(F) \qquad E_\theta\{\alpha_{t,\theta}^2\} < \infty \ \ \forall \theta \in \Theta, \ \forall t \geq 0, \tag{24}$$

$$\alpha_{t,\theta} := \frac{\mathcal{L}_t^* p(\cdot, \theta)}{p(\cdot, \theta)} = -\sum_{i=1}^N \left(f_i \frac{\partial}{\partial x_i}(\theta^T c) + \frac{\partial f_i}{\partial x_i} \right) +$$
$$+\frac{1}{2} \sum_{i,j=1}^N \left[a_{ij} \frac{\partial^2}{\partial x_i \partial x_j}(\theta^T c) + a_{ij} \frac{\partial}{\partial x_i}(\theta^T c) \frac{\partial}{\partial x_j}(\theta^T c) + \right.$$
$$\left. + 2 \frac{\partial a_{ij}}{\partial x_j} \frac{\partial}{\partial x_i}(\theta^T c) + \frac{\partial^2 a_{ij}}{\partial x_i \partial x_j} \right].$$

We will assume such condition to hold in the following. Sufficient explicit conditions for (F) to hold for EF (c) can be easily given. For example, (F) holds if f and its first derivatives with respect to x, a and its first and second derivatives with respect to x, and c and its first and second derivatives have at most polynomial growth, and if densities in EF (c) integrate any polynomial, see for example Brigo et al. (1999).

We have thus proven the following

Proposition 9 (Projected evolution of the density of an Itô diffusion) *Assume assumptions (A), (B), (C), (E) and (F) on the coefficients f, a, on the initial condition X_0 of the Itô diffusion X, and on the sufficient statistics $c_1, \ldots, c_n$ of the exponential family* EF (c) *are satisfied. Then the projection of Fokker–Planck equation describing the evolution of $p_t = p_{X_t}$ onto* EF (c) *reads, in B_{p_t} coordinates:*

$$[c(\cdot) - E_{\theta_t} c]^T \dot{\theta}_t = E_{\theta_t}[\mathcal{L}c]^T \, g^{-1}(\theta_t) \, [c(\cdot) - E_{\theta_t} c], \tag{25}$$

and the differential equation describing the evolution of the parameters for the projected density-evolution is

$$\dot{\theta}_t = g^{-1}(\theta_t) \, E_{\theta_t}\{\mathcal{L}_t \, c\}.$$

Notice that the projected equations exist under conditions which are more general than conditions for existence of the solution of the original Fokker–Planck equation. For more details see Brigo (1997). Notice also that this equation is substantially the same we had derived in the running example with a Galerkin-inspired approach: Compare (23) with (18) after viewing the right hand side of (18) as coming from an integration by parts.

6.5 *Interpretation of the Projected Density as Density of a Different Diffusion*

In this section we shortly expose a problem which was treated in Brigo (1997), see also Brigo (2000). Consider the projected density $p(\cdot, \theta_t)$, expressing the projection of the density-evolution of the one dimensional diffusion X onto the finite dimensional exponential manifold EF (c). The question is: Can we define a diffusion Y_t whose density is the projected density $p(\cdot, \theta_t)$? If the answer is yes, Y_t is a diffusion whose density evolves in a finite dimensional exponential manifold assigned a priori (for example Gaussian). For simplicity, we treat the case $N = 1$. In order to proceed, define a diffusion

$$dY_t = u_t(Y_t)dt + \sigma_t(Y_t)dW_t, \quad Y_0 = X_0, \tag{26}$$

with the same diffusion coefficient as X_t. We shall try to define the drift u in such a way that the density-evolution of Y_t coincides with $p(\cdot, \theta_t)$. Call $\mathcal{T}_t$ the backward differential operator of Y_t:

$$\mathcal{T}_t = u_t \frac{\partial}{\partial x} + \frac{1}{2} a_t \frac{\partial^2}{\partial x^2}.$$

Consider the right hand sides of (20) and (25). Clearly, the density of Y_t coincides with $p(\cdot, \theta_t)$ if

$$\frac{\mathcal{T}^* p(\cdot, \theta_t)}{p(\cdot, \theta_t)} = E_{\theta_t}[\mathcal{L}c]^T \, g^{-1}(\theta_t) \, [c(\cdot) - E_{\theta_t} c]$$

which we can rewrite as

$$\mathcal{T}^* p(\cdot, \theta_t) = \mathcal{P}_{t,\theta_t} \, p(\cdot, \theta_t).$$

By simple calculations one can rewrite the above equation as the following PDE for u, where we do not expand the second partial derivative of $a_t p(\cdot, \theta)$:

$$\frac{\partial u_t}{\partial x} + \theta_t^T \frac{\partial c}{\partial x} u_t = \frac{1}{2 \, p(\cdot, \theta_t)} \frac{\partial^2}{\partial x^2} (a_t p(\cdot, \theta_t)) - \mathcal{P}_{t,\theta_t}$$

Call $\mathcal{B}_{t,\theta_t}$ the right hand side of such equation. A solution is given by

$$u_t^*(x) := \exp[-\theta_t^T c(x)] \int_{-\infty}^{x} \mathcal{B}_{t,\theta_t}(y) \exp[\theta_t^T c(y)] dy,$$

as one can verify immediately by substitution. Straightforward calculations yield

$$u_t^*(x) := \frac{1}{p(x,\theta_t)} \int_{-\infty}^{x} \left[\frac{\partial^2_{xx}(a_t(y)p(y,\theta_t))}{p(y,\theta_t)} - \Pi_{\theta_t} \left\{ \frac{\partial^2_{xx}(a_t(y)p(y,\theta_t))}{p(y,\theta_t)} \right\} \right.$$

$$\left. + \Pi_{\theta_t} \left\{ \frac{\partial_x(f_t(y)p(y,\theta_t))}{p(y,\theta_t)} \right\} \right] p(y,\theta_t)dy \tag{27}$$

$$= \frac{1}{2}\frac{\partial a_t}{\partial x}(x) + \frac{1}{2} a_t(x)\theta_t^T \frac{\partial c}{\partial x}(x)$$

$$- E_{\theta_t}\{\mathcal{L}_t c\}^T g^{-1}(\theta_t) \int_{-\infty}^{x} (c(y) - E_{\theta_t} c) \, \exp[\theta_t^T (c(y) - c(x))] dy.$$

From this last equation one sees that under condition (24) and under the assumption that densities of EF (c) are integrable, the above integral always exists.

We have thus proven the following

Proposition 10 (Interpretation of the projected density-evolution) *Assume assumptions (A), (B), (C), (E) and (F) on the coefficients f, a and on the initial condition X_0 of the Itô diffusion X and on the sufficient statistics c of the exponential family* EF (c) *are satisfied. Let $p(\cdot,\theta_t)$ be the projected density evolution, according to Proposition 9. Define*

$$dY_t = u_t^*(Y_t)dt + \sigma_t(Y_t)dW_t,$$

$$u_t^*(x) := \frac{1}{2}\frac{\partial a_t}{\partial x}(x) + \frac{1}{2} a_t(x)\theta_t^T \frac{\partial c}{\partial x}(x) +$$

$$- E_{\theta_t}\{\mathcal{L}_t c\}^T g^{-1}(\theta_t) \int_{-\infty}^{x} (c(y) - E_{\theta_t} c) \, \exp[\theta_t^T (c(y) - c(x))] dy.$$

Then Y is an Itô diffusion whose density-evolution coincides with the projected density-evolution $p(\cdot,\theta_t)$ of X_t onto EF (c).

6.6 *Quality of the Finite Dimensional Approximation*

In order to assess how good the projection is locally, and to have a measure for how far the projected evolution is, locally, from the original one, we now define a local projection residual as the duality-based norm of the Fokker Planck infinite dimensional vector field minus its finite-dimensional orthogonal projection. Define the vector field minus its projection as

$$\varepsilon_t(\theta) := \frac{\mathcal{L}_t^* p(\cdot, \theta)}{p(\cdot, \theta)} - \Pi_\theta \left[\frac{\mathcal{L}_t^* p(\cdot, \theta)}{p(\cdot, \theta)} \right].$$

Then the projection residual R_t is defined as

$$R_t^2 := \mathrm{Cov}_{p_\theta} (\varepsilon_t(\theta), \varepsilon_t(\theta)) = \langle \varepsilon_t(\theta), \varepsilon_t(\theta) \rangle_{p(\cdot,\theta)}$$

and can be computed jointly with the projected equation evolution (23) to have a local measure of the goodness of the approximation involved in the projection.

Monitoring the projection residual and its peaks can be helpful in tracking the projection method performance, see also Brigo et al. (1998, 1999) for examples of L^2-based projection residuals in the more complex case of the Kushner–Stratonovich equations of nonlinear filtering. However, the projection residual only allows for a local approximation error numerical analysis. To have an idea of how good the approximation is we need to relate it to the global approximation error.

We could define the global approximation error as follows. Rather than projecting the Fokker Planck equation vector field instant by instant, we could project the true solution as a point onto the exponential family EF (c). To appreciate the difference with what we have done so far, let us recap the method we have followed so far, which we call "vector field projection". We denote time steps with $0, 1, 2, \ldots$ for simplicity but in the real equation they correspond to infinitesimal time steps. To make the point, we are artificially separating projection and propagation and the local and global errors. This is not completely precise but allows us to make an important point on our method.

- Assume at time 0 we have $p_0(x) = p(x; \theta_0)$, so we start from the family.
- Now the vector field of Fokker Planck $\frac{\mathcal{L}^* p(\cdot,\theta_0)}{p(\cdot,\theta_0)}$ is not in the tangent space of EF (c) in general and therefore would bring us out of the exponential family at time 1. To stay in the exponential family, we project this vector field onto the tangent space of EF (c) and follow the projected vector for the evolution, moving on the tangent space to time 1. By doing this, we get a new $p(\cdot, \theta_1)$ on the manifold.
- Now we start again. We apply the vector field of the Fokker Planck equation to $p(\cdot, \theta_1)$. Note that this is not right if comparing with the true evolution. We are applying the vector field to the wrong point at time 1, because $p(\cdot, \theta_1)$ is not the true p_1, and now we are not applying the vector field to p_1 but to $p(\cdot, \theta_1)$. But even starting from $p(\cdot, \theta_1)$, the vector field $\frac{\mathcal{L}^* p(\cdot,\theta_1)}{p(\cdot,\theta_1)}$ is not in the tangent space of EF (c) in general and therefore would bring us out of EF. To stay in EF, we project this vector field onto the tangent space of the exponential family and follow the projected vector for the evolution, moving on the tangent space. By doing this, we get a new $p(\cdot, \theta_2)$ at time 2 on the manifold.
- We continue like this and obtain an evolution of the manifold, but none of the projections was based on projecting the vector field starting from the true solution, except for the first step.

This method has two types of approximations, so to speak: on one hand, we approximate the true equation vector field with a projection. On the other hand, we apply the true equation vector field not to the true solution but already to an approximated solution coming from the previous steps. The two steps are related in the limit, clearly, and with some very sophisticated analysis one might be able to bound the global error based on the local one. However, let us continue with the artificial setting with separate steps. We can say that while it is possible to measure locally the error in the first type of approximation, for example via R_t above, it is difficult to measure the effect of the second one, unless one obtains a very precise approximation of the true solutions by some other method and then compares the outputs. But if one has the true solution to a very good precision already, there is clearly no point in finding a finite dimensional approximation.

If we leave the global approximation error analysis aside for a minute, the big advantage of the above method is that it does not require us to know the true solution of the Fokker Planck equation to be implemented. Indeed, Eq. (23) works perfectly well without knowing the true solution p_t.

As we mentioned above, to study the global error, we now introduce a second projection method. This one will require us to know the true solution, so as an approximation method it will be pointless. However, it will help us with the global error analysis, and a modification of the method based on the assumed density approximation will allow us to find an algorithm that does not require the true solution.

This method works as follows.

- Assume at time 0 we have $p_0(x) = p(x; \theta_0)$, so we start from the family.
- Now the vector field of Fokker Planck $\frac{\mathcal{L}^* p(\cdot,\theta_0)}{p(\cdot,\theta_0)}$ is not in the tangent space of EF (c) in general and therefore would bring us out of the exponential family at time 1. We accept this, follow it, and move to p_1 outside $EF(c)$. To go back to EF, we project p_1 onto the exponential family by minimizing the divergence, or Kullback Leibler information of p_1 with respect to EF (c), finding the orthogonal projection of p_1 on EF. It is well known that the orthogonal projection in Kullback Leibler divergence is obtained by matching the sufficient statistics expectations of the true density. Namely, the projection is the particular exponential density of EF (c) with c-expectations
$$\eta_1 = E_{p_1}[c].$$
See for example Brigo (1998) for a quick proof and an application to filtering in discrete time. We know that EF (c), besides θ, admits another important coordinate system, the expectation parameters η. If one defines
$$\eta(\theta) = E_{p(\theta)}[c]$$
then $d\eta(\theta) = g(\theta) d\theta$ where g is the Fisher metric. Thus, we can take the η_1 above coming from the true density p_1 and look for the exponential density $p(\cdot; \eta_1)$ sharing these c-expectations. This will be the closest in Kullback Leibler to p_1 in EF (c).
- Now from p_1 we keep following the true vector field of the Fokker Planck equation, and in general we start from outside the manifold EF (c) and we stay outside. We

reach p_2. Now again we project p_2 onto the exponential family in Kullback Leibler, finding $\eta_2 = E_{p_2}[c]$ and the projection is the exponential density $p(\cdot;\eta_2)$.
- We continue like this

The advantage of this method compared to the previous vector field based one is that we find at every time the best possible approximation ("maximum likelihood") of the true solution in EF. The disadvantage is that in order to compute the projection at every time, such as for example $\eta_1 = E_{p_1}[c]$, we need to know the true solution p_1 at that time. Clearly if we know the true solution there is no point in developing an approximation by projection in the first place.

However it turns out that we can somewhat combine the two ideas and analyze the error if we invoke the assumed density approximation. This works as follows.

6.7 Maximum Likelihood Estimation and $\mathcal{L}$ Eigenfunctions

Consider the second type of projection, namely

$$\eta_t = E_{p_t}[c].$$

Differentiate both sides (d_t here denotes differentiation with respect to time) to obtain

$$d_t\eta_t = d_t \int c(x)p_t(x)dx = \int c(x)d_t p_t(x)dx = \int c(x)\mathcal{L}_t^* p_t(x)dx = E_{p_t}[\mathcal{L}c]dt$$

so that

$$d_t\eta_t = E_{p_t}[\mathcal{L}c]dt.$$

This last equation is not a closed equation, since p_t in the right hand side is not characterized by η. Thus, to be solved this equation should be coupled with the original Fokker Planck for p_t. Again, this makes this equation useless as an approximation. However, at this point we can close the equation by invoking the assumed density approximation (see Brigo et al. 1999): we replace p_t with the exponential density $p(\cdot,\eta_t)$. We obtain

$$d_t\tilde{\eta}_t = E_{p(\cdot,\tilde{\eta}_t)}[\mathcal{L}c]dt.$$

This is now a finite dimensional ODE for the expectation parameters. There is more: if we use $d\eta = g(\theta)d\theta$ and substitute, in the θ coordinates this last equation is the same as our earlier vector field based projected Eq. (23).

Theorem 2 *Closing the evolution equation for the Kullback Leibler projection of the Fokker Planck solution onto* EF (c) *by forcing an exponential density on the right hand side is equivalent to the approximation based on the vector field projection in Fisher metric.*

We can now attempt an analysis of the error between the best possible projection η_t and the vector field based (or equivalently assumed density approximation based) projection $\tilde{\eta}$. To do this, write

$$\epsilon_t := \eta_t - \tilde{\eta}_t,$$

expressing the difference between the best possible approximation and the vector field projection/assumed density one, in expectation coordinates. Differentiating we see easily that

$$d\epsilon_t = (E_{p_t}[\mathcal{L}c] - E_{p(\tilde{\eta}_t)}[\mathcal{L}c])dt.$$

Now suppose that the c statistics in EF (c) are chosen among the eigenfunctions of the operator $\mathcal{L}$, so that

$$\mathcal{L}c = -\Lambda c$$

where Λ is a $n \times n$ diagonal matrix with the eigenvalues corresponding to the chosen eigenfunctions. Substituing, we obtain

$$d\epsilon_t = -\Lambda(E_{p_t}[c] - E_{p(\tilde{\eta}_t)}[c])dt$$

or

$$d\epsilon_t = -\Lambda\epsilon_t dt$$

from which

$$\epsilon_t = \exp(-\Lambda t)\epsilon_0$$

so that if we start from the manifold the error is zero, meaning that the vector field projection gives us the best possible approximation. If we don't start from the manifold, i.e. if p_0 is outside EF (c), then the difference between the vector field approach and the best possible approximation dies out exponentially fast in time provided we have negative eigenvalues for the chosen eigenfunctions.

Theorem 3 (Maximum Likelihood Estimator for the Fokker Planck Equation and Fisher–Rao projection) *The vector field projection approach leading to* (23) *provides the best possible approximation of the Fokker Planck equation solution in Kullback Leibler in the family* EF (c), *provided that the sufficient statistics* c *are chosen among the eigenfunctions of the adjoint operator* $\mathcal{L}$ *of the original Fokker Planck equation, and provided that* EF (c) *is an exponential family when using such eigenfunctions. In other words, under such conditions the Fisher Rao projected Eq.* (23) *provides the exact maximum likelihood estimator for the solution of the Fokker Planck equation in the related exponential family.*

The choice or availability of suitable eigenfunctions is not always straightforward, except in a few simple cases. See Pavliotis (2014) for a discussion on eigenfunctions for the Fokker Planck equation. For example, in the one dimensional case $N = 1$ where the diffusion is on a bounded domain $[\ell, r]$ with reflecting boundaries and strictly positive diffusion coefficient σ then the spectrum of the operator $\mathcal{L}$ is discrete, there is a stationary density and eigenfunctions can be expressed with respect to this stationary density. In our framework it would be natural to use the stationary density as background density replacing $M(x)$ and then use the eigenfunctions and the related negative real eigenvalues to study the approximation of the Fokker Planck equation.

For the case $N > 1$ only special types of SDEs allow for a specific eigenfunctions/eigenvalue analysis, see for example the Ornstein Uhlenbeck case and SDEs with constant diffusion matrices and drifts associated to potentials in Pavliotis (2014). Further research is needed to explore the eigenfunctions approach in connection with maximum likelihood.

6.8 The Direct L^2 Metric Projection

As we mentioned at the end of Sect. 3.1, the L^2 structure based on square roots of densities (Hellinger distance) and the exponential statistical manifold lead to the same finite dimensional metric on any finite dimensional manifold p_θ (not just EF (c)), but the direct L^2 metric based on densities rather than their square roots leads to a different finite dimensional metric. Under a background measure μ, by generalizing straightforwardly (15) and the related derivation to a general family p_θ we see that the statistical manifold induces on finite dimensional families the inner product

$$\mathrm{Cov}_{p_\theta}\left(\frac{\partial \log p_\theta}{\partial \theta_i}, \frac{\partial \log p_\theta}{\partial \theta_j}\right) = \left\langle \frac{\partial \log p_\theta}{\partial \theta_i}, \frac{\partial \log p_\theta}{\partial \theta_j} \right\rangle_{p_\theta} = g_{i,j}(\theta)$$

and the $L^2(\mu)$ based Hellinger distance leads to

$$\left\langle \frac{\partial \sqrt{p_\theta}}{\partial \theta_i}, \frac{\partial \sqrt{p_\theta}}{\partial \theta_j} \right\rangle_{\mu} = \frac{1}{4} g_{i,j}(\theta),$$

essentially giving the same Fisher–Rao metric on the finite dimensional manifold. However, the direct metric yields

$$\left\langle \frac{\partial p_\theta}{\partial \theta_i}, \frac{\partial p_\theta}{\partial \theta_j} \right\rangle_{\mu} = \gamma_{i,j}(\theta) \neq g_{i,j}(\theta).$$

This means that the direct metric leads to a different finite dimensional metric γ, different from the Fisher Rao g given by the Hellinger distance or the statistical manifold structure. This finite dimensional geometry related to γ works quite well when projecting infinite dimensional evolution equations on subspaces MG (q) generated by mixtures of a given finite set of densities q, see Brigo (2011), Armstrong and Brigo (2016), and coincides with traditional Galerkin methods based on L^2 bases for p directly. The g metric works well when projecting on finite dimensional exponential families such as EF (c). The direct metric approach to dimensionality reduction with MG (q) mixtures will not be pursued further here given that its induced finite dimensional geometry is different from the statistical manifold induced geometry.

7 Conclusions and Further Work

We have proposed a dimensionality reduction method for infinite–dimensional measure–valued evolution equations such as the Fokker Planck equation or the Kushner–Stratonovich/Duncan Mortensen Zakai equations, with potential applications to signal processing, quantitative finance, heat flows and quantum theory. This dimensionality reduction method is based on a projection coming from a duality argument and allows one to design a finite dimensional approximation for the evolution equation that is optimal locally according to the statistical manifold structure by G. Pistone and co-authors. Clearly the choice of the finite dimensional manifold on which one should project the infinite dimensional equation is crucial, and we proposed finite dimensional exponential and mixture families as in previous works by D. Brigo and co-authors inspired by the L^2 structure instead.

Given the work of Newton (2012, 2013, 2015) on finding an infinite dimensional manifold structure on the space of measures that combines the exponential manifold structure of G. Pistone and co-authors and the L^2 full-space structure used by D. Brigo and co-authors, further work is to be done to see how dimensionality reduction based on Newton's framework would look like and would relate to this paper.

It would also be important to see how convergence works when the finite dimensional manifold dimension tends to infinity. Indeed, one further natural question is whether it is possible to prove that the finite dimensional approximated solution converges to the infinite dimensional solution when the dimension of the finite dimensional manifold tends to infinity. More precisely, suppose we are given a sequence of functions $(c_j)_{j\in\mathbb{N}}$. Call $c^m := \{c_1\ c_2\ \ldots c_m\}$, and assume that for an infinite subset $\mathbb{M} \subset \mathbb{N}$ and for $m \in \mathbb{M}$ the family EF (c^m) is a finite dimensional exponential manifold satisfying assumptions (E) and (F). For example, in the monomial case where $c_i(x) = x^i$, we could have that $\mathbb{M}$ is the set of natural even numbers. Call $p(\cdot, \theta_t^m)$ the density coming from projection of Fokker–Planck equation onto EF (c^m), $m \in \mathbb{M}$. It is conceivable that in case the infinite sequence c_k, $k \in \mathbb{N}$ is chosen carefully, one can prove that if $\mathbb{M} \ni m \to +\infty$ then $p(\cdot, \theta^m(t)) \to p_t$ where p_t is the original infinite dimensional density coming from the Fokker Planck equation being approximated. The way to approach this would be to treat the c_k as a basis of an infinite dimensional space and to use Sobolev spaces and weak convergence arguments. We will try to find the weakest possible conditions under which convergence is attained in future work.

Further work is also needed to explore the eigenfunctions approach. We have sketched a proof of the fact that if the sufficient statistics c of the exponential family EF (c) are chosen among the eigenfunctions of the operator $\mathcal{L}$ associated with the Fokker Planck equation then the Fisher metric projection gives us also the best maximum likelihood estimator of the exact solution. We need to identify SDEs for which the eigenfunction approach is feasible and to study the related approximation. We might be able to show that by including more and more eigenfunctions we could converge in some sense to the true solution.

In this paper we also tried to clarify how the finite dimensional and infinite dimensional terminology for exponential and mixture spaces are related, since the terms are often used with different meaning in different contexts. This has been clarified to some extent but not completely, and further work remains to be done.

Further work is needed to clarify the L^2 direct metric projection in terms of statistical manifolds. The projection based on the L^2 structure on densities rather than their square roots, and the related metric, have been used in Armstrong and Brigo (2016) to work with projection of infinite dimensional evolution equations on finite dimensional mixture families such as the MG (q) above. In further work we would like to relate this projection to the statistical and mixture manifold structures based on Orlicz spaces given here rather than in terms of the blunt whole L^2 space.

We would also like to study in the statistical manifold framework the different projections suggested in Armstrong and Brigo (2015) for evolution equations driven by rough paths. For such equations there is more than one possible projection, depending on the notion of optimality one chooses, which is related to the rough paths properties. This would combine geometry in the space of probability laws with geometry in the state space.

Finally, we would like to examine different measure evolution equations than the few we worked with here. This too will be investigated in further work.

Acknowledgements The authors are grateful to the organizers and participants of the conference *Computational information geometry for image and signal processing*, held at the ICMS in Edinburgh on September 21–25, 2015. They are also grateful to Frank Nielsen for feedback on this preprint and to an anonynous referee for suggesting investigating the approximation error, as this prompted us to derive the MLE theorem. G. Pistone is supported by deCastro Statistics, Collegio Carlo Alberto, Moncalieri, and he is a member of GNAFA-INDAM.

References

Abraham, R., Marsden, J. E., & Ratiu, T. (1988). *Manifolds, tensor analysis, and applications*. Applied mathematical sciences (2nd ed., Vol. 75). New York: Springer.

Amari, S. (1987). Dual connections on the Hilbert bundles of statistical models. *Geometrization of statistical theory (Lancaster, 1987)* (pp. 123–151). Lancaster: ULDM Publ.

Amari, S., & Nagaoka, H. (2000). *Methods of information geometry*. Providence: American Mathematical Society. (Translated from the 1993 Japanese original by Daishi Harada).

Armstrong, J., & Brigo, D. (2015). *Extrinsic projection of Itô SDEs on submanifolds with applications to non-linear filtering*. To appear in the same volume of this paper.

Armstrong, J., & Brigo, D. (2016). Nonlinear filtering via stochastic PDE projection on mixture manifolds in L^2 direct metric. *Mathematics of Control, Signals and Systems*, *28*(1), Art.5, p. 33.

Ay, N., Jost, J., Lê, H.V., & Schwachhöfer, L. (2016). *Parametrized measure models*. arXiv:1510.07305.

Brezis, H. (2011). *Functional analysis, Sobolev spaces and partial differential equations*, Universitext. New York: Springer.

Brigo, D. (1997). On nonlinear SDEs whose densities evolve in a finite–dimensional family. *Stochastic differential and difference equations, Progress in systems and control theory* (Vol. 23, pp. 11–19). Boston: Birkhäuser.

Brigo, D. (1998). On some filtering problems arising in mathematical finance. *Insurance: Mathematics and Economics, 22*(1), 53–64.

Brigo, D. (1999). Diffusion processes, manifolds of exponential densities, and nonlinear filtering. In O. E. Barndorff-Nielsen, et al. (Eds.), *Geometry in present day science. Proceedings of the Conference, Aarhus, Denmark, January 16–18, 1997* (pp. 75–96). Singapore: World Scientific.

Brigo, D. (2000). On SDEs with marginal laws evolving in finite-dimensional exponential families. *Statistics & Probability Letters, 49*(2), 127–134.

Brigo, D. (2011). The direct L2 geometric structure on a manifold of probability densities with applications to Filtering. arXiv:1111.6801.

Brigo, D., & Pistone, G. (1996). Projecting the Fokker-Planck equation onto a finite dimensional exponential family. Preprint 4/1996, Department of Mathematics, University of Padua, posted in 2009 on arXiv:0901.1308.

Brigo, D., Hanzon, B., & Le Gland, F. (1998). A differential geometric approach to nonlinear filtering: the projection filter. *IEEE Transactions on Automatic Control, 43*(2), 247–252.

Brigo, D., Hanzon, B., & Le Gland, F. (1999). Approximate nonlinear filtering by projection on exponential manifolds of densities. *Bernoulli, 5*(3), 495–534.

Brown, L. D. (1986). *Fundamentals of statistical exponential families with applications in statistical decision theory*. IMS Lecture Notes–Monograph Series (Vol. 9). Hayward, CA: IMS.

Cena, A., & Pistone, G. (2007). Exponential statistical manifold. *Annals of the Institute of Statistical Mathematics, 59*(1), 27–56.

Csiszár, I. (1975). *I*-divergence geometry of probability distributions and minimization problems. *Annals of Probability, 3*, 146–158.

Friedman, A. (1975). *Stochastic differential equations and applications* (Vol. I). New York: Academic Press.

Gibilisco, P., & Pistone, G. (1998). Connections on non-parametric statistical manifolds by Orlicz space geometry. *IDAQP, 1*(2), 325–347.

Hanzon, B. (1987). A dierential-geometric approach to approximate nonlinear ltering. In C. Dodson (Ed.), *Geometrization of statistical theory* (pp. 219–233). Lancaster: ULMD Publ.

Hazewinkel, M., Marcus, S., & Sussmann, H. (1983). Nonexistence of finite-dimensional filters for conditional statistics of the cubic sensor problem. *Systems & Control Letters, 3*(6), 331–340.

Lang, S. (1995). *Differential and Riemannian manifolds*. Graduate texts in mathematics (2nd ed., Vol. 160). New York: Springer.

Lods, B., & Pistone, G. (2015). Information geometry formalism for the spatially homogeneous Boltzmann equation. *Entropy, 17*(6), 4323–4363.

Mitter, S. K. (1979). On the analogy between mathematical problems of non-linear filtering theory and quantum physics. *Ricerche di Automatica, 10*(2), 163–216.

Musielak, J. (1983). *Orlicz spaces and modular spaces*, Lecture Notes in Mathematics (Vol. 1034). Berlin: Springer.

Naudts, J. (2011). *Generalised thermostatistics*. London: Springer London Ltd.

Newton, N. J. (2012). An infinite-dimensional statistical manifold modelled on Hilbert space. *Journal of Functional Analysis, 263*(6), 1661–1681.

Newton, N. J. (2013). Infinite-dimensional manifolds of finite-entropy probability measures. In F. Barbaresco & F. Nielsen (Eds.), *Geometric science of information*, Springer LNCS (Vol. 8085, pp. 713–720). Berlin: Springer.

Newton, N. J. (2015). Information geometric nonlinear filtering. *Infinite Dimensional Analysis Quantum Probability And Related Topics, 18*(2), 1550014, 24.

Pavliotis, G. A. (2014). *Stochastic processes and applications: Diffusion processes, the Fokker-Planck and Langevin equations*. New York: Springer.

Pistone, G. (2013). Examples of the application of nonparametric information geometry to statistical physics. *Entropy, 15*(10), 4042–4065.

Pistone, G. (2014). A version of the geometry of the multivariate Gaussian model, with applications. In *XLVII Scientific Meeting SIS June 11–13*. Cagliari: Società Italiana di Statistica.

Pistone, G., & Rogantin, M. (1999). The exponential statistical manifold: mean parameters, orthogonality and space transformations. *Bernoulli*, *5*(4), 721–760.

Pistone, G., & Sempi, C. (1995). An infinite-dimensional geometric structure on the space of all the probability measures equivalent to a given one. *Annals of Statistics*, *23*(5), 1543–1561.

Santacroce, M., Siri, P., & Trivellato, B. (2015). New results on mixture and exponential models by Orlicz spaces. *Bernoulli*, *22*(3), 1431–1447.

Schwachhöfer, L., Ay, N., Jost, J., & Lê, H. V. (2015). Invariant geometric structures in statistical models. In F. Barbaresco & F. Nielsen (Eds.), *Geometric science of information*, Springer LNCS (Vol. 8085, pp. 713–720). Berlin: Springer.

Shima, H. (2007). *The geometry of Hessian structures*. Hackensack: World Scientific Publishing Co. Pte. Ltd.

Stroock, D. W., & Varadhan, S. R. S. (1979). *Multidimensional diffusion processes*. Berlin-New York: Springer.

van Handel, R., & Mabuchi, H. (2005). Quantum projection filter for a highly nonlinear model in cavity qed. *Journal of Optics B: Quantum and Semiclassical Optics*, *7*(10), S226.

Batch and Online Mixture Learning: A Review with Extensions

Christophe Saint-Jean and Frank Nielsen

1 Introduction

Mixture models $f(x;\theta)$ are a powerful and flexible tool to approximate any unknown *smooth* probability density function π as a finite convex combination of parametric density functions $g_j(x;\theta_j)$:

$$\pi(x) \approx f(x;\theta) = \sum_{j=1}^{K} w_j g_j(x;\theta_j), \text{ with } w_j > 0 \text{ and } \sum_{j=1}^{K} w_j = 1, \tag{1}$$

where $K \in \mathbb{N}$ denotes the number of components of the mixture. Fitting such a kind of semi-parametric model amounts to find a "good" candidate within a parametric family of distributions $\mathcal{F}_\theta$ defined by a set of parameters θ. Among all those distributions, the closest candidate in $\mathcal{F}_\theta$ to π will be denoted f^* (related to the approximation error). Figure 1 depicts the case of a density of a continuous random variable modeled as a mixture of three univariate normal components.

This mixture learning task received much attention in the literature since it is a core operation for both theoretical purposes, and it is widely used in many applications. Classically, one may distinguish two main approaches:

1. Maximum Likelihood Estimation (MLE), and
2. Bayesian Learning.

C. Saint-Jean (✉)
Mathématiques, Image, Applications (MIA), Université de La Rochelle, La Rochelle, France
e-mail: christophe.saint-jean@univ-lr.fr

F. Nielsen
LIX, École Polytechnique, Palaiseau, France
e-mail: Frank.Nielsen@acm.org

F. Nielsen
Sony CSL, Tokyo, Japan

F. Nielsen et al. (eds.), *Computational Information Geometry*,
Signals and Communication Technology, DOI 10.1007/978-3-319-47058-0_11

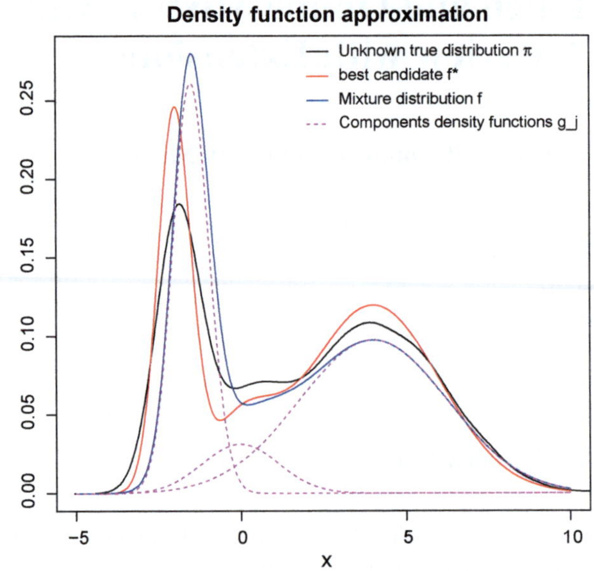

Fig. 1 Approximating an unknown distribution π (*black curve*) with a mixture distribution $f(x;\theta)$ (*blue curve*) of three normal component distributions ($K = 3$, *dashed magenta curves*) (color figure online)

While the former approach gives a *point estimate* of mixture parameters, the latter considers the *posterior distribution* of the parameters given a *prior distribution* on them. In this work, we restrict ourselves to the MLE approach since it is by far the most popular approach.

Consider a random sample $\chi = \{x_i\}_{i=1}^N$ of N independent and identically distributed (iid) observations from π. Under this assumption, the *joint probability* of set χ regarding a particular value for θ is simply $f(\chi;\theta) = \prod_i f(x_i;\theta)$. Viewing χ as a fixed set and θ as a parameter vector, the maximum likelihood estimator (MLE) $\hat{\theta}^{(N)}$ is defined as the maximizer of the likelihood, or equivalently of the average log-likelihood:

$$\bar{l}(\theta;\chi) = N^{-1}\sum_{i=1}^{N}\log f(x_i;\theta) = N^{-1}\sum_{i=1}^{N}\log\left(\sum_{j=1}^{K} w_j g_j(x_i;\theta_j)\right). \quad (2)$$

In the remainder of this chapter, we will discuss the case when sample χ is not fully known in a whole. That is, we shall consider that the observations x_i are available one after another (e.g. in the data stream model, useful for dealing with very large data sets). Thus online methods differ from batch methods, and ideally aim to get same convergence and efficiency properties as batch ones while having a single pass over the full dataset. This topic receives increasing attention due to the recent challenges associated to massive datasets.

2 Online Learning with Gradient-Based Methods

In this section, we recall the basics of gradient-based optimization and of stochastic approximation. Most of the content below comes from the paper (Bottou 1998; Amari 1997; Bottou and Bousquet 2011).

2.1 Gradient-Based Optimization

The maximization of $\bar{l}$ takes the form of a *sum-minimization* problem (M-estimation):

$$C_N(\theta) = N^{-1} \sum_{i=1}^{N} C(x_i, \theta)$$

for the loss function $C(x, \theta) = -\log f(x; \theta)$. The empirical risk $C_N(\theta)$ evaluated on sample χ of size N is an approximation of the expected risk

$$C(\theta) = \mathbb{E}_\pi[C(x, \theta)].$$

The iterative minimization of the empirical risk with a batch gradient descent (GD) takes the following form:

At iteration t:

$$\boxed{\hat{\theta}^{(t+1)} = \hat{\theta}^{(t)} - \alpha^{(t)} R^{-1}(\hat{\theta}^{(t)})\, \underbrace{N^{-1} \sum_{i=1}^{N} \nabla_\theta C(x_i, \hat{\theta}^{(t)})}_{\nabla C_N(\hat{\theta}^{(t)})}} \tag{3}$$

where $\hat{\theta}^{(t)}$ is the parameter estimates, $\alpha^{(t)}$ is the learning rate, and positive definite matrix $R \succ 0$ a rescaling matrix. When R is chosen as the identity matrix, this amounts to ordinary first-order gradient ascent. For $R = \nabla^2 C_N$ chosen as the hessian matrix of C_N, this defines the Newton–Raphson method for finding extrema. Since naïve versions of these methods involve costly operations at each iteration (computation of gradients, hessians for all observations and a matrix inversion), quasi-newton methods (e.g., L-BGFS) which approximate the inverse of hessians are generally preferred.

When the parameter space Θ is a Riemannian manifold with tensor metric G, the direction of the steepest descent at θ is given by the *natural gradient* (Amari 1997, 1998, 2016):

$$\tilde{\nabla}_\theta C_N(\theta) \doteq G^{-1}(\theta) \nabla_\theta C_N(\theta) \tag{4}$$

So, picking $R(\hat{\theta}^{(t)})$ as $G(\hat{\theta}^{(t)})$ defines the natural gradient descent method (Amari 1998).

In information geometry, a D-dimensional parametric exponential or mixture family has a dually flat structure (Amari 2016) induced by a convex potential function F with a canonical divergence a Bregman divergence (with convex generator defined modulo an affine term). The convex function induced two dual coordinate systems θ and η such that $\eta = \nabla F(\theta)$ and $\theta = \nabla F^*(\eta)$, where F^* is the Legendre convex conjugate (Amari 2016). In a dually flat space, the dual basis vectors $e^i = \partial_i = \frac{\partial}{\partial \eta_i}$ and $e_j = \partial^j = \frac{\partial}{\partial \theta^j}$ are orthogonal since $\langle e^i, e_j \rangle = \delta^i_j$ (with $\delta^i_j = 1$ iff $i = j$, and 0 otherwise). We can define a *mixed coordinate system* (Amari 2016, p. 144) ξ by choosing the first k components from the θ-coordinate system and the $D - k$ remaining coordinates from the η-coordinate system. Then then Riemannian metric G in this mixed coordinate system has a *block-diagonal* structure by construction:

$$G = \begin{bmatrix} g_{ij} & 0 \\ 0 & g^{lm} \end{bmatrix},$$

where $g_{ij} = \langle e_i, e_j \rangle$ and $g^{lm} = \langle e^l, e^m \rangle$.

It follows that when $D = 2$, the mixed coordinate systems always ensure a *diagonal* Riemannian (Fisher information) matrix (see Miura (2011) for an example of such parameter orthogonalization). Computing the inverse G^{-1} of a diagonal matrix $G = \text{diag}(a_{11}, \ldots, a_{DD})$ is fast since $G^{-1} = \text{diag}(a_{11}^{-1}, \ldots, a_{DD}^{-1})$, and the gradient-based optimization efficient. However,

Note that the ordinary gradient is obtained for $G = I$ (the identity matrix), and it makes sense to consider this natural gradient updating rule since $\Delta^{(t+1)} = \theta^{(t+1)} - \theta^{(t)}$ is a contravariant vector and ∇l is a covariant derivative. Therefore in the natural gradient, the factor G^{-1} converts a covariant to contravariant vector (Amari 1997).

2.2 Stochastic Gradient Descent Methods

While batch methods have good convergence properties (linear or quadratic), their costs in time and memory is prohibitive when the sample size increases. During the last decade, stochastic methods for optimization (especially those based on GD) have been proven to be very effective in the situation.

Following the seminal work of Robbins and Monro (1951), the observations $\nabla_\theta C(x_1, \theta), \nabla_\theta C(x_2, \theta), \ldots$ can be considered as "noise corrupted" ones of $\nabla_\theta C(\theta)$ for which a root θ^* is searched. Under the assumptions that learning rates $\alpha^{(t)}$ satisfy $\sum_{t \geq 0} \alpha^{(t)} = \infty$ (diverge) and $\sum_{t \geq 0} \alpha^{(t)^2} < \infty$ (converge), they proved that the sequence $\hat{\theta}^{(0)}, \hat{\theta}^{(1)}, \ldots$ in Eq. 5 converges almost surely to θ^*. This method is referred in the literature as the *Stochastic Gradient Descent* (SGD):

<u>At iteration t:</u>

$$\boxed{\hat{\theta}^{(t+1)} = \hat{\theta}^{(t)} - \alpha^{(t)} R(\hat{\theta}^{(t)})^{-1} \nabla_\theta C(x_t, \hat{\theta}^{(t)})} \quad (5)$$

Again, if the parameter space has a non-Euclidean Riemannian structure, it is preferable to use the *stochastic natural gradient descent* (SNGD).

At iteration t:

$$\boxed{\hat{\theta}^{(t+1)} = \hat{\theta}^{(t)} - \alpha^{(t)} \tilde{\nabla}_\theta C(x_t, \hat{\theta}^{(t)})} \tag{6}$$

One strength of the natural gradient descent for online learning besides its invariance under reparameterization is that is it provably Fisher efficient (Amari 1997, 2016), meaning that it meets asymptotically the Cramér-Rao lower bound Amari (2016).

There exist many extensions to this algorithm. In the sequel, we report some old and new heuristics:

(Minibatch SGD) In order the reduce the variance in the parameter update, the gradient of C may be estimated from a limited sample B_t (a.k.a. mini-batch, see Sculley 2010). Since this mini-batch is created at each iteration (successive picks in the stream or through the sampling without replacement from χ), the resulting estimate is also a noisy observation of $\nabla_\theta C(\hat{\theta}^{(t)})$.

At iteration t:

$$\boxed{\hat{\theta}^{(t+1)} = \hat{\theta}^{(t)} - \alpha^{(t)} |B_t|^{-1} \sum_{x \in B_t} \nabla_\theta C(x, \hat{\theta}^{(t)})} \tag{7}$$

(Momentum SGD) Another strategy for regularizing the parameter update consists in doing a convex combination between the previous update and the gradient.[1]

At iteration t:

$$\boxed{\begin{aligned} \Delta^{(t+1)} &= \epsilon^{(t)} \Delta^{(t)} - \alpha^{(t)} \nabla_\theta C(x_t, \hat{\theta}^{(t)}) \\ \hat{\theta}^{(t+1)} &= \hat{\theta}^{(t)} + \Delta^{(t+1)} \end{aligned}} \tag{8}$$

Doing such modification enforces velocity vector Δ to accumulate directions of steepest descent. Momentum coefficient $\epsilon^{(t)}$ is an additional hyper-parameter which has to be set in $[0, 1]$. A popular setting of ϵ consists in taking it around 0.5 in the warmup phase (initial learning) then to increase it towards 0.9 simultaneously to the iterations to enforce the stability of the update.

Better methods have been proposed when a sequence of gradients or parameters over iterations is used. This leads the following heuristics:

(Average SGD) Polyak–Ruppert averaging Polyak and Juditsky (1992) refers to a post-processing method where a second sequence $\bar{\theta}^{(0)}, \bar{\theta}^{(1)}, \ldots$ is generated by averaging estimates after t_0 iterations:

$$\bar{\theta}^{(t)} = \begin{cases} \hat{\theta}^{(t)} & t \leq t_0, \\ \frac{1}{t - t_0} \sum_{t'=t_0+1}^{t} \hat{\theta}^{(t')} & \text{otherwise} \end{cases} \tag{9}$$

[1]It is equivalent to an exponentially decaying moving average of past gradients.

In practice, recursive reformulations are always preferable since it avoids a significant memory cost.
At iteration t:

$$\boxed{\begin{aligned}\hat{\theta}^{(t+1)} &= \hat{\theta}^{(t)} - \alpha^{(t)} R(\hat{\theta}^{(t)})^{-1} \nabla_\theta C(x_t, \hat{\theta}^{(t)})\\ \bar{\theta}^{(t+1)} &= \left((t - t_0)\, \bar{\theta}^{(t)} + \hat{\theta}^{(t+1)}\right) / (t - t_0 + 1) \text{ if } t > t_0\end{aligned}} \tag{10}$$

There are many policies for setting $\alpha^{(t)}$. The original proposition in Robbins and Monro (1951) is to pick $\alpha^{(t)} = t^{-1}$ which meets the requirements mentioned before. Nowadays, a classical setting is $\alpha^{(t)} = \alpha^{(0)}(1 + Ct)^{-1}$ where $\alpha^{(0)}$ and C are prescribed constants. Because the convergence of the optimization depends strongly on these constants, several authors suggest to re-evaluate them periodically using a small validation dataset (different from the training set).

(Adam) This method (named after adaptive moment estimation) is a first-order method which use estimates of first and second moments of the gradient with respect to each parameter to estimate.
At iteration t:

$$\boxed{\begin{aligned}\text{for } j = 1, 2 \qquad m_j^{(t+1)} &= \beta_j\, m_j^{(t)} + (1 - \beta_j) \left(\nabla_\theta C(x_t, \hat{\theta}^{(t)})\right)^{\circ j}\\ \hat{m}_j^{(t+1)} &= m_j^{(t+1)} / (1 - \beta_j^t)\\ \hat{\theta}^{(t+1)} &= \hat{\theta}^{(t)} - \alpha\, \hat{m}_1^{(t+1)} / \left(\sqrt{\hat{m}_2^{(t+1)}} + \epsilon\right)\end{aligned}} \tag{11}$$

The first two steps consist in estimating moments of the gradient using exponential moving averages (the symbol $\circ j$ denotes the Hadamard power) then correct their biases. The biais correction are mandatory since the m_1 and m_2 are initialized as 0's vectors. Note that the learning rate is adapted for each parameter *independently*. One of the most appealing property of this method is that the magnitudes of parameter updates are invariant to rescaling of the gradient and are controlled by hyperparameter α (the term $\hat{m}_1 / \left(\sqrt{\hat{m}_2} + \epsilon\right)$ is unitless).

2.3 From Batch Learning to Online Learning

Observe that the previous methods (except Minibatch SGD) which do not need to remember previous observations are suitable for on-the-fly processing: the iteration number t becomes the observation number N. In such a case, since the examples are randomly drawn from the ground truth distribution π, the expected risk $C(\theta)$ is directly minimized. Note that the same methods applied to χ, a sample from π, lead to the minimization of $\mathbb{E}_{\pi_N}[C(x,\theta)]$ over an *empirical distribution* π_N with distribution:

$$\pi_N(x) = \frac{1}{N}\sum_{i=1}^{N}\delta_{x_i}(x),$$

where δ_x is the Dirac measure.

In order to prevent overfitting, the empirical risk is classically replaced by a *regularized risk* where a L_1 or L_2 penalty term is added. From the implementation perspective, a fixed dataset χ may be processed seamlessly as a data stream using function generators of modern programming languages (e.g. Python). Such kind of functions is able to yield an observation on demand by repeating infinitely χ (with shuffle). Also, let us mention that the parallelization of optimization techniques remains a very active research field leading to sophisticated hardware and software architectures.

3 Online Mixture Modelling

Before dealing with mixtures of multiple components, the simpler special case of a single component mixture is first discussed below.

3.1 Online Learning with a Single Component

Consider the case when $f = g_1$ is a (regular) exponential family (EF), that is f may be decomposed as

$$f(x;\theta) = \exp\{\langle\theta, s(x)\rangle + k(x) - F(\theta)\} \tag{12}$$

where θ, s, k, F are respectively the natural parameters, the sufficient statistics, the carrier measure, the log-partition function (see Nielsen and Garcia (2009) for further definitions). Most common distributions (but not the uniform, heavy-tailed Student t-, and Cauchy distributions) are regular exponential families: Gaussian, Dirichlet, Multinomial (including the categorical distribution), von Mises-Fisher, Wishart, Rayleigh, etc.

In case of EF, the loss function $C(x,\theta)$ takes the following expression:

$$C(x,\theta) = -\log f(x;\theta) = -\langle\theta, s(x)\rangle - k(x) + F(\theta) \tag{13}$$

The MLE $\hat{\theta}^{(N)}$ is given analytically by differentiating $C_N(\theta)$ with respect to θ:

$$\nabla F(\hat{\theta}^{(N)}) = \frac{1}{N}\sum_{i=1}^{N} s(x_i) \longrightarrow \hat{\theta}^{(N)} = (\nabla F)^{-1}\left(\frac{1}{N}\sum_{i=1}^{N} s(x_i)\right) \tag{14}$$

The functional reciprocal $(\nabla F)^{-1}$ of ∇F is generally available in an explicit formula for most (but not all) EF. It corresponds to the gradient of the dual Legendre convex conjugate (Nielsen and Garcia 2009): $(\nabla F)^{-1} = \nabla F^*$. The Fisher information matrix $I(\theta)$ of a regular exponential family is the Hessian of the log-normalizer:

$$I(\theta) = -E_\theta[\nabla^2 \log f(x;\theta)] = \nabla^2 F(\theta) \succ 0,$$

a positive-definite matrix for all $\theta \in \Theta$, where Θ denotes the natural parameter space. When switching to the online case, it is interesting to get an exact expression of the MLE by a *recursive formulation* of the average of the sufficient statistics. For that, it suffices to keep the sum of the previous sufficient statistics and update as:

$$\hat{\theta}^{(N+1)} = (\nabla F)^{-1}\left(\frac{\{\sum_{i=1}^{N} s(x_i)\} + s(x_{N+1})}{N+1}\right) \tag{15}$$

$$\text{or equivalently} \quad (\nabla F)^{-1}\left(\frac{N\nabla F(\hat{\theta}^{(N)}) + s(x_{N+1})}{N+1}\right) \tag{16}$$

The recursion in Eq. 5 appears more clearly when this formula is written in the Expectation parameter space H (Nielsen and Garcia 2009). Let $\eta = \nabla F(\theta)$. The recursive computation of the exact MLE is then given by[2]:

$$\hat{\eta}^{(N+1)} = \hat{\eta}^{(N)} + \{N+1\}^{-1}(s(x_{N+1}) - \hat{\eta}^{(N)}) \text{ and } \hat{\eta}^{(0)} = 0. \tag{17}$$

It is of interest to compare this expression to the one given by the SGD update (Eq. 5) on natural parameter space Θ:

$$\hat{\theta}^{(N+1)} = \hat{\theta}^{(N)} + \alpha^{(N+1)}\left(s(x_{N+1}) - \nabla_\theta F(\hat{\theta}^{(N)})\right) \tag{18}$$

For the same optimization but in the expectation space H, recall the bijection between exponential families and Bregman divergences (Banerjee et al. 2005):

$$\log f(x;\eta) = -B_{F^*}(s(x) : \eta) + F^*(s(x)) + k(x), \tag{19}$$

where B_{F^*} is the Bregman divergence associated with F^*, the convex conjugate of F. It follows that maximizing the loss function $C(\eta) = \mathbb{E}_\pi[-\log f(x;\eta)]$ leads to the following computation:

[2] When $(\nabla F)^{-1}$ is computed with numerical approximations, this may give a different result.

$$\begin{aligned} -\nabla_\eta \log f(x;\eta) &= \nabla_\eta B_{F^*}(s(x):\eta) \\ &= -\nabla_\eta F^*(\eta) - \nabla_\eta \langle s(x)-\eta, \nabla_\eta F^*(\eta)\rangle \\ &= -H(F^*)(\eta)(s(x)-\eta) \end{aligned} \tag{20}$$

where $H(F^*)(\eta)$ is the hessian of F^* at observed point η. Thus, the minimization of $C(\eta)$ with the stochastic gradient descent on H takes the following form:

$$\hat{\eta}^{(N+1)} = \hat{\eta}^{(N)} + \alpha^{(N+1)} H(F^*)(\hat{\eta}^{(N)})(s(x_{N+1}) - \hat{\eta}^{(N)}) \tag{21}$$

Section 4 of this chapter gives an empirical comparison of Eqs. 17, 18 and 21.

Algorithm 1: Exact Online MLE for regular exponential families

Input: a sequence $\mathcal{S} = x_1, x_2, \ldots$ of observations
Input: Functions s and $(\nabla F)^{-1}$ for some regular exponential family
Output: a sequence $\hat{\eta}^{(1)}, \hat{\eta}^{(2)}, \ldots$ of MLE where $\hat{\eta}^{(N)}$ is the exact MLE for the first N observations

$\hat{\eta}^{(0)} = 0;\quad N = 0;$
for $x_{N+1} \in \mathcal{S}$ **do**
  $\hat{\eta}^{(N+1)} = \hat{\eta}^{(N)} + \{N+1\}^{-1}(s(x_{N+1}) - \hat{\eta}^{(N)});$
  yield $\hat{\eta}^{(N+1)}$ or **yield** $\hat{\theta}^{(N+1)} = (\nabla F)^{-1}(\hat{\eta}^{(N+1)});$
  $N = N + 1;$

To conclude this part, recall that for a *regular exponential family*, the natural parameter space Θ is an open convex space, and F is strictly convex and differentiable function. It follows that f is a log-concave function and that $-\log f$ is a convex function. Since we consider data stream of many different observations, we are not concerned by the problem of existence of the MLE (see Bogdan and Bogdan (2000) for a rigorous treatment of that point) in Algorithm 1.

When f does not belong to an exponential family, its mathematical properties have to be studied (especially convexity, convex relaxations, etc) and numerical methods are often required (see previous section or Shalev-Shwartz 2011).

3.2 *Batch Mixture Learning with Multiple Components*

Before carrying on with details of online mixture learning methods, let us first recall the basics of Expectation-Maximization algorithm (EM) in the next subsection.

Batch mixture learning with EM For $K > 1$, the direct maximization of $\bar{l}$ is a difficult problem since $\log f$ is the logarithm of the sum of multiple terms ($-\log f$ is no more convex). However it can be made easier if we know the component, let say

z_i, which have generated x_i. This mechanism, called data augmentation, amounts to introduce a latent (unobservable) random variable.

Let Z_i be a categorical random variable over $1, \dots, K$ whose parameters are $\{w_j\}_j$, that is, $Z_i \sim \text{Cat}_K(\{w_j\}_j)$. Also, assuming that $X_i|Z_i = j \sim g_j(\cdot; \theta_j)$, the unconditional mixture distribution f in Eq. 1 is recovered by marginalizing their joint density p over Z_i (i.e. $f(x) = \sum_z p(x, z)$). Obviously, Z_i is a latent (unobservable) variable so that the realizations x_i of X_i (resp. (x_i, z_i) of (X_i, Z_i)) is often viewed as an incomplete (resp. complete) data observation. Alternatively, we may consider that Z_i is a random vector $[Z_{i,1}, Z_{i,2}, \dots, Z_{i,k}]$ where $Z_{i,j} = 1$ iff. X_i arises from the j-th component of the mixture and 0 otherwise. Thus, $Z_1, \dots, Z_N$ are unconditionally distributed according to the multinomial law $\mathcal{M}_K(1, \{w_j\}_j)$ which is an exponential family.

Similarly to Eq. 2, the average complete log-likelihood function can be written as:

$$\begin{aligned} \bar{l}_c(\theta; \chi_c) = \quad & N^{-1} \sum_{i=1}^{N} \log p(x_i, z_i; \theta) = N^{-1} \sum_{i=1}^{N} \log \prod_{j=1}^{K} \left(w_j g_j(x_i; \theta_j) \right)^{z_{i,j}} \\ = \quad & N^{-1} \sum_{i=1}^{N} \sum_{j=1}^{K} z_{i,j} \log(w_j g_j(x_i; \theta_j)) \end{aligned} \tag{22}$$

where $\chi_c = \{(x_i, z_i)\}_{i=1}^N$, is the set of complete data observations.

Here comes the EM algorithm (cf. Algorithm 2) which optimizes $\bar{l}(\theta; \chi)$ (proofs in Dempster et al. 1977; Robbins and Monro 1951; Titterington 1984; Amari 1997, 1998, 2016; Miura 2011; Cappé and Moulines 2009; Neal and Hinton 1999) by repeating two steps until convergence:

- Compute the conditional expectation of missing values

$$\begin{aligned} \mathcal{Q}(\theta; \hat{\theta}^{(t)}, \chi) &:= \mathbb{E}_{\hat{\theta}^{(t)}}[\bar{l}_c(\theta; \chi_c)|\chi] \\ &= N^{-1} \sum_{i=1}^{N} \sum_{j=1}^{K} \mathbb{E}_{\hat{\theta}^{(t)}}[Z_{i,j}|X_i = x_i] \log(w_j g_j(x_i; \theta_j)) \end{aligned}$$

- Maximize $\mathcal{Q}(\theta; \hat{\theta}^{(t)}, \chi)$ over θ.

Remark that while $\hat{w}_j^{(t+1)}$ is always known in closed-form whatever the chosen g_j, $\hat{\theta}_j^{(t+1)}$ are obtained by component-wise specific optimization involving **all** observations. More generally, the improvement of $\bar{l}(\theta; \chi)$ is guaranteed whatever the increase of $\mathcal{Q}$ is in the M-Step. This leads to the *Generalized EM* algorithm (GEM) when partial maximization (i.e., not necessarily global optimization) is performed.

Batch mixture learning with EM and EF Consider now the case when all the g_j's are exponential families (EF, e.g. gaussians densities or generalized gaussians densities). The joint density $p(x, z)$ may be decomposed as follows:

Algorithm 2: EM for fitting finite mixture models

Input: a set $\chi = x_1, x_2, \ldots, x_N$ of observations, $\hat{\theta}^{(0)} = \{\hat{w}_j^{(0)}, \hat{\theta}_j^{(0)}\}_j$ an initial parameter values where θ_j is the parameter of p.d.f. g_j.

Output: an estimate $\hat{\theta}$ of the mixture parameters

$t = 0$;

repeat

```
// E-Step : This step amounts to compute:
```

$$\hat{z}_{i,j}^{(t)} = \mathbb{E}_{\hat{\theta}^{(t)}}[Z_{i,j}|X_i = x_i] = \frac{\hat{w}_j^{(t)} g_j(x_i; \hat{\theta}_j^{(t)})}{\sum_{j'} \hat{w}_{j'}^{(t)} g_{j'}(x_i; \hat{\theta}_{j'}^{(t)})} \tag{23}$$

```
// M-Step: Separated maximization of {w_j}_j and {θ_j}_j
```

$$\hat{w}_j^{(t+1)} = \frac{\sum_{i=1}^N \hat{z}_{i,j}^{(t)}}{N}, \quad \hat{\theta}_j^{(t+1)} = \arg\max_{\theta_j \in \Theta_j} \sum_{i=1}^N \hat{z}_{i,j}^{(t)} \log\left(g_j(x_i; \theta_j)\right) \tag{24}$$

$t = t + 1$;

until *Convergence of* $\bar{l}_c(\theta; \chi_c)$;

return $\hat{\theta}^{(t)}$;

$$\begin{aligned}
\log p(x, z; \theta) &= \sum_{j=1}^K [z = j]\{\log(w_j) + \log g_j(x; \theta_j)\} \\
&= \sum_{j=1}^K [z = j]\{\log(w_j) + \langle \theta_j, s_j(x) \rangle + k_j(x) - F_j(\theta_j)\} \\
&= \sum_{j=1}^K \left\langle \begin{pmatrix} [z = j] \\ [z = j]\, s_j(x) \end{pmatrix}, \begin{pmatrix} \log w_j - F_j(\theta_j) \\ \theta_j \end{pmatrix} \right\rangle + \sum_{j=1}^K [z = j]\, k_j(x) \\
&= \langle s(x, z), \theta_c \rangle + \sum_{j=1}^K [z = j]\, k_j(x)
\end{aligned}$$

where $[z = j]$ denotes the Iverson's bracket,

$$s(x, z) := ([z = 1], [z = 1]\, s_1(x), \quad \ldots, [z = K], [z = K]\, s_K(x))^T \tag{25}$$

$$\theta_c := (\log w_1 - F_1(\theta_1), \theta_1, \quad \ldots, \log w_K - F_K(\theta_K), \theta_K)^T \tag{26}$$

Note that notation θ_c may be considered as ambiguous but in the paper the subscript j always refers to component-specific parameters. One can then recognize the form of an exponential family. Then, it follows very simple expressions for $\bar{l}_c$ and $\mathcal{Q}$:

$$\bar{l}_c(\theta; \chi_c) = N^{-1} \sum_{i=1}^{N} \langle s(x_i, z_i), \theta_c \rangle + N^{-1} \sum_{i=1}^{N} \sum_{j=1}^{K} z_{i,j}\, k_j(x_i) \tag{27}$$

$$\begin{aligned} \mathcal{Q}(\theta; \hat{\theta}^{(t)}, \chi) =& N^{-1} \sum_{i=1}^{N} \langle \mathbb{E}_{\hat{\theta}^{(t)}} \left[s(X_i, Z_i) | X_i = x_i \right], \theta_c \rangle + \\ & N^{-1} \sum_{i=1}^{N} \sum_{j=1}^{K} \mathbb{E}_{\hat{\theta}^{(t)}} \left[Z_{i,j} | X_i = x_i \right] k_j(x_i) \end{aligned} \tag{28}$$

Since the second term is irrelevant (i.e., a constant) to the maximization $\mathcal{Q}$, the EM algorithm for such distributions can be reformulated with sufficient statistics for complete data. The E-Step amounts to compute the vector $\hat{S}^{(t)}$, the empirical average of the conditional expectation of sufficient statistics for complete data (see Eq. 30). The M-Step consists in finding the value θ_c which maximizes the inner product with $\hat{S}^{(t)}$ (see Eq. 31). If this mapping is denoted by $\theta^{\dagger} : H \mapsto \Theta$, the EM algorithm for the mixture of EF can be written in one recurring formula:

$$\hat{S}^{(t+1)} = N^{-1} \sum_{i=1}^{N} \mathbb{E}_{\theta^{\dagger}(\hat{S}^{(t)})} [s(X_i, Z_i) | X_i = x_i] \tag{29}$$

where initial values $\hat{S}^{(0)}$ is given by $\hat{\theta}^{(0)}$ and Eq. 26.

Algorithm 3: EM for fitting finite mixture models of exponential families

---

Input: a set $\chi = x_1, x_2, \ldots, x_N$ of observations, $\hat{\theta}^{(0)} = \{\hat{w}_j^{(0)}, \hat{\theta}_j^{(0)}\}_j$ an initial parameter values where θ_j is the parameter of exponential family g_j.

Output: an estimate $\hat{\theta}$ of the mixture parameters

$t = 0$;

repeat

E-Step : $$\hat{S}^{(t)} = N^{-1} \sum_{i=1}^{N} \mathbb{E}_{\hat{\theta}^{(t)}} [s(X_i, Z_i) | X_i = x_i] \tag{30}$$

M-Step : $$\hat{w}_j^{(t+1)} = \hat{S}_{2j-1}^{(t)}, \quad \hat{\eta}_j^{(t+1)} = \nabla F_j(\hat{\theta}_j^{(t+1)}) = \hat{S}_{2j}^{(t)} / \hat{S}_{2j-1}^{(t)} \tag{31}$$

$t = t + 1$;

until *Convergence*;

return $\hat{\theta}^{(t)}$;

3.3 Online Mixture Learning with Multiple Components

The case of online mixture learning is discussed in the following. It is now more appropriate to denote $\hat{\theta}^{(N)}$ the current parameter estimate instead of $\hat{\theta}^{(t)}$.

Titterington's algorithm The first online algorithm, due to Titterington (1984), corresponds to the direct optimization of $\mathcal{Q}(\theta; \hat{\theta}^{(N)}, \chi)$ using a second-order stochastic gradient ascent:

$$\boxed{\hat{\theta}^{(N+1)} = \hat{\theta}^{(N)} + \alpha^{(N+1)} I_c^{-1}(\hat{\theta}^{(N)}) \nabla_\theta \log f(x_{N+1}; \hat{\theta}^{(N)})} \tag{32}$$

where $\{\alpha^{(N+1)}\}$ is a decreasing sequence of positive step sizes ($\alpha^{(N+1)} = N^{-1}$ in the original paper) and the hessian $\nabla^2 \mathcal{Q}$ of $\mathcal{Q}$ is approximated by the Fisher Information matrix I_c for the complete data:

$$I_c(\hat{\theta}^{(N)}) = -\mathbb{E}_{\hat{\theta}^{(N)}} \left[H(\log p(x, z; \theta)) \right],$$

where H denotes the hessian operator ∇^2.

The justification of this recursion relies on the Fisher's identity (see discussion in Dempster et al. 1977) for finite mixture models: for any value θ' for mixture parameters,

$$\begin{aligned}
\nabla_\theta \log f(x; \theta') &= f(x; \theta')^{-1} \nabla_\theta f(x; \theta') = f(x; \theta')^{-1} \sum_z \nabla_\theta p(x, z; \theta') \\
&= f(x; \theta')^{-1} \sum_z \left\{ p(x, z; \theta') \nabla_\theta \log p(x, z; \theta') \right\} \\
&= \sum_z \left\{ h(z|x; \theta') \nabla_\theta \log p(x, z; \theta') \right\} \\
&= \mathbb{E}_{\theta'} [\nabla_\theta \log p(X, Z; \theta') | X = x]
\end{aligned} \tag{33}$$

where $h(z|x; \theta)$ is the conditional density of z given x.

It follows that the gradient of $\mathcal{Q}$ at $\hat{\theta}^{(N)}$ (see Eq. 28) has a particular form especially when the model for the complete data is an exponential family:

$$\begin{aligned}
\nabla_\theta \mathcal{Q}(\hat{\theta}^{(N)}; \hat{\theta}^{(N)}, \chi) &= N^{-1} \sum_{i=1}^{N} \mathbb{E}_{\hat{\theta}^{(N)}} [\nabla_\theta \log p(X_i, Z_i; \hat{\theta}^{(N)}) | X_i = x_i] \\
&= N^{-1} \sum_{i=1}^{N} \nabla_\theta \log f(x_i; \hat{\theta}^{(N)})
\end{aligned}$$

In order to incorporate the constraint on weight components ($\sum_{j=1}^{K} w_j = 1$), the last component weight w_K is removed from the parameters to be optimized and set to

$w_K = 1 - \sum_{j=1}^{K-1} w_j$. Thus, if the θ-coordinate system is considered and parameters are ordered as $\theta = (w_1, \ldots, w_{K-1}, \theta_1, \ldots, \theta_K)$, we are able to further describe this algorithm for mixtures of exponential families (MEFs):

$$\text{For } j = 1, \ldots, K-1 \quad \frac{\partial \log f(x_{N+1}; \hat{\theta}^{(N)})}{\partial w_j} = \frac{g_j(x_{N+1}; \hat{\theta}_j^{(N)}) - g_K(x_{N+1}; \hat{\theta}_K^{(N)})}{f(x_{N+1}; \hat{\theta}^{(N)})} \tag{34}$$

$$\text{For } j = 1, \ldots, K \quad \frac{\partial \log f(x_{N+1}; \hat{\theta}^{(N)})}{\partial \theta_j} = \hat{z}_{N+1,j}^{(N)} \left(s_j(x_{N+1}) - \nabla_{\theta_j} F_j(\hat{\theta}_j^{(N)}) \right) \tag{35}$$

where $\hat{z}_{N+1,j}^{(N)} = w_j g_j(x_{N+1}; \hat{\theta}_j^{(N)}) / f(x_{N+1}; \hat{\theta}^{(N)})$.

Due to the chosen parametrization, the hessian of $\log p$ is a block diagonal matrix where the hessians $H(F_j)$ of all the F_j appear. It follows that the information matrix I_c is easier to compute:

$$I_c(\theta) = blockdiag\left(\left(diag(w_1^{-1}, \ldots, w_{K-1}^{-1}) - \frac{\mathbf{1}_{K-1}{}'\mathbf{1}_{K-1}}{w_K} \right), \right.$$
$$\left. w_1 H(F_1)(\theta_1), \ldots, w_K H(F_K)(\theta_K) \right) \tag{36}$$

The inverse of first block matrix is given by the Sherman-Morrison identity (see formula 160 in Petersen and Pedersen 2012). By plugging these results in Eq. 32, the update equations for a generic MEF are:

<u>For a new observation x_{N+1},</u>

$$\hat{w}_j^{(N+1)} = \hat{w}_j^{(N)} + \alpha^{(N+1)} (\hat{z}_{N+1,j}^{(N)} - \hat{w}_j^{(N)}) \text{ and } \hat{w}_K^{(N+1)} = 1 - \sum_{j=1}^{K-1} \hat{w}_j^{(N+1)} \tag{37}$$

$$\hat{\theta}_j^{(N+1)} = \hat{\theta}_j^{(N)} + \alpha^{(N+1)} \frac{\hat{z}_{N+1,j}^{(N)}}{\hat{w}_j^{(N)}} H(F_j)^{-1} (\hat{\theta}_j^{(N)}) \left(s_j(x_{N+1}) - \nabla_{\theta_j} F_j(\hat{\theta}_j^{(N)}) \right) \tag{38}$$

This recursive procedure does not necessarily take into account the constraints on the θ_j's (e.g. $\theta_j > 0$ for a mixture of Rayleigh distributions).

Online EM In a recent paper, Cappé and Moulines (2009) proposed to replace the E-Step by a *stochastic approximation* and leave the M-step unchanged. Here are the key ideas of their approach in the case of mixture of EFs.

When considering an infinite number of observations, the EM update given by an empirical average in Eq. 29 can be defined by the mapping $\mathcal{T} : H \mapsto H$ as follows:

$$\mathcal{T}(S) = \mathbb{E}_\pi \left[\mathbb{E}_{\theta^\dagger(S)}[s(X, Z) | X] \right] \tag{39}$$

Thus, the sequence $\hat{S}^{(0)}, \hat{S}^{(1)}, \hat{S}^{(2)}, \ldots$ converges to thc sequence $\hat{S}^{(0)}, \mathcal{T}(\hat{S}^{(0)}), \mathcal{T}(\mathcal{T}(\hat{S}^{(0)})), \ldots$ which depends only on $\hat{S}^{(0)}$. Finding the limit of this sequence amounts to find a fixed point of $\mathcal{T}$ or equivalently to look for a root of the function $C : H \mapsto H$:

$$C(S) = \mathcal{T}(S) - S = \mathbb{E}_{\pi}\left[\mathbb{E}_{\theta^{\dagger}(S)}[s(X, Z)|X] - S\right] \quad (40)$$

According again to the framework of Robbins-Monro, one can get the solution by iterating :

$$\tilde{S}^{(N+1)} = \tilde{S}^{(N)} + \alpha^{(N+1)}\left(\mathbb{E}_{\theta^{\dagger}(\tilde{S}^{(N)})}[s(x_{N+1}, z_{N+1})|x_{N+1}] - \tilde{S}^{(N)}\right) \quad (41)$$

The initial value for parameters $\hat{\theta}^{(0)}$ is transformed $\tilde{S}^{(0)}$ by Eq. 31. Obviously, this formula is comparable to the one for $K = 1$ (see Eq. 17).

This approach guarantees that parameter constraints are automatically respected solving a known problem for Titterington's approach. The authors have proved that two algorithms are asymptotically equivalent. The link between the two approaches will be discussed later on.

4 Experiments

4.1 *Online Learning for a Gaussian Distribution*

The aim of this first experiment is to test several methods of optimization for the simple case of the online learning of a single univariate gaussian distribution. This experiment may appear to be unnecessary since a recursive formulation for the MLE is known from Eq. 17. Hence, many properties of previous optimization methods can be exhibited from this case. This distribution is an exponential family for which the canonical decomposition is recalled in Appendix 6.1. In particular, Eqs. 54, 58 and 66 are needed for the different update formulas 17, 18 and 21.

The experiment consists in the recursive estimation of the parameters on an univariate gaussian $\mathcal{N}(\mu = 1, \sigma^2 = 4)$ from a continuous stream of its realizations. The dataset of size 60,000 is splitted on two parts, one for training (1/3) and the other for the validation (2/3). Two criteria are used to evaluate the estimates $(\hat{\mu}^{(N)}, \hat{\sigma}^{2(N)})$: the average log-likelihood on training and testing datasets, the Kullback–Leibler (KL) divergence (see. Eq. 68) between true parameters values and their estimates.

The results of the recursive estimation with exact formula are reported on Fig. 2. As expected, since the variance of the MLE for a N-sample is $\{NI(\lambda)\}^{-1}$ (see Eq. 69), a convergence is achieved quite quickly. This method does not depend on a particular initialization and one can remark that the average log-likelihood does not necessarily increase after incorporating a new observation. This property is common to many

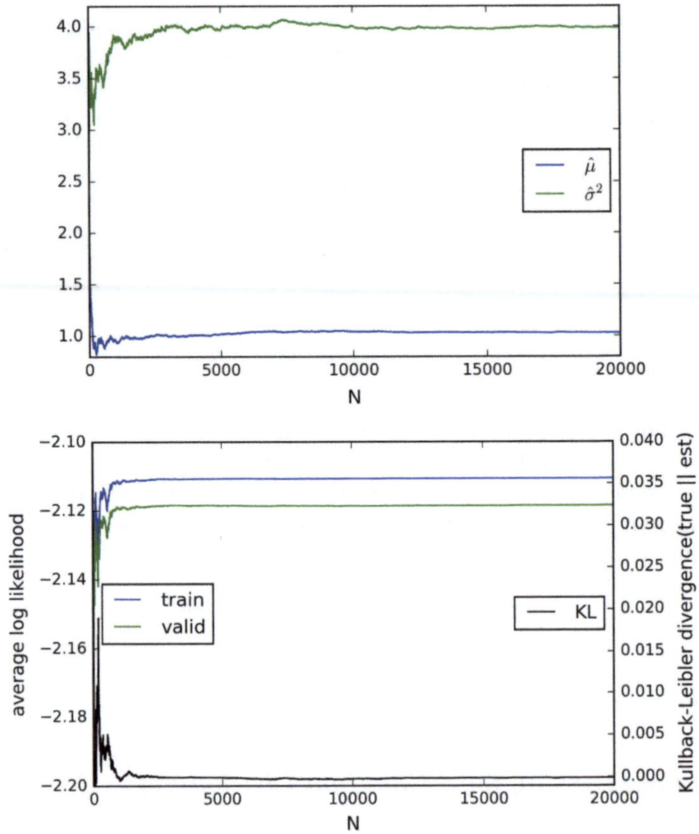

Fig. 2 Recursive estimation with exact formula: parameters estimates (*top*) - Average log-likelihood/KL divergence (*bottom*) w.r.t. the number of observations (color figure online)

recursive methods. On the right side, one may notice that green and blue curves are very similar, but the shift between them shows the training error is an optimistically biased criterion.

Figure 3 shows the estimates of μ and σ^2 through the iterations with various settings (space, fixed learning rate) but same initialization. We can immediately see that the speed convergence of SGD methods is highly dependent on the learning rate. For some good values (e.g. $\alpha = 0.0316$ for SGD on source parameters), the online method is quite competitive with recursive estimation. When, the learning rate is too low, parameter update and therefore the convergence is very slow. In contrast, when α is too large, the estimates oscillate around the global maximum. During these optimizations, the updates can lead to estimates that are outside the domain of admissible values for them. To cope with that, several strategies can be implemented: reject the update, project onto the set of admissibles values, Uzawa's method (Boyd and Vandenberghe 2004), etc.

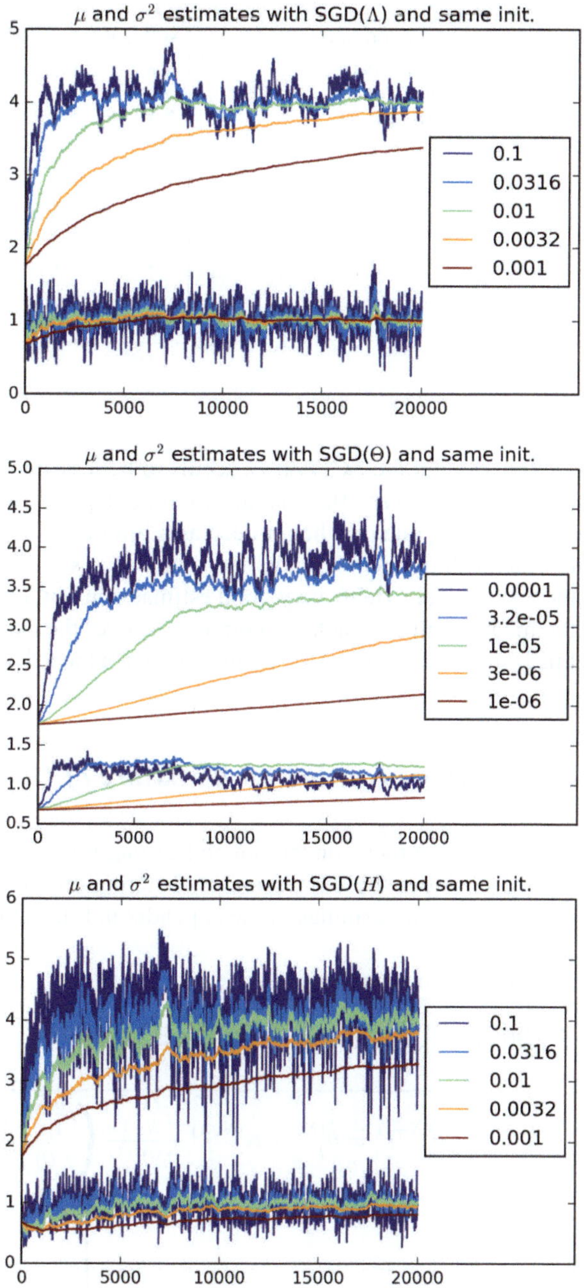

Fig. 3 Recursive estimation with SGD on source space (*left*), natural space (*middle*), expectation space (*right*) with same initialization and different $\alpha^{(N+1)}$

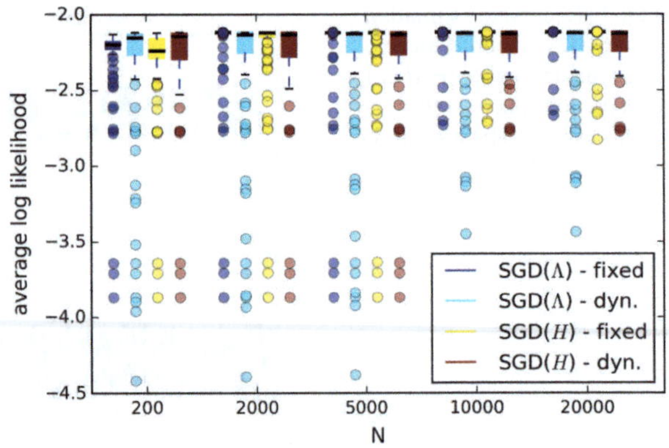

Fig. 4 Average log-likelihood over 100 runs with SGD on source space and on expectation space ($\alpha^{(N+1)} = 0.0316$ or $\alpha^{(N+1)} = n^{-0.85}$)

Remark that the SGD on H seems to be less stable. Figure 4 shows the average log-likelihood over 100 runs at different steps of the SGD on Λ and on H for a fixed learning rate or when it decreases after each iterations ($\alpha^{(N+1)} = n^{-0.85}$). The two algorithms seem to have more or less a similar behavior. Note that adapting the learning rate yields very good estimates in few iterations (black stroke indicates the median value) which are competitive over the exact MLE (cf. Fig. 2). However, this strategy seems to be too aggressive when the $\theta^{(0)}$ is far from the global optimum.

4.2 Online Learning for a Mixture of Gaussian Distributions

In this part, we focus on the online learning for a mixture of gaussian distributions. Firstly, consider the expression of Titterington's recursive EM for this case. By plugging several formulas of the appendix in Eq. 38, the following update equations are obtained:

For a new observation x_{N+1},

Estimate $\hat{z}^{(N)}_{N+1,j}$ and $\hat{w}_j^{(N+1)}$ using Eq. 38.

$$\hat{\theta}_j^{(N+1)} = \hat{\theta}_j^{(N)} + \alpha^{(N+1)} \frac{2\hat{z}^{(N)}_{N+1,j}}{\hat{w}_j^{(N+1)}} \begin{pmatrix} \hat{\theta}^{2(N)}_{1_j} + \hat{\theta}^{(N)}_{2_j} & \hat{\theta}^{(N)}_{1_j}\hat{\theta}^{(N)}_{2_j} \\ \hat{\theta}^{(N)}_{1_j}\hat{\theta}^{(N)}_{2_j} & \hat{\theta}^{2(N)}_{2_j} \end{pmatrix} \begin{pmatrix} x_{N+1} - \frac{\hat{\theta}^{(N)}_{1_j}}{2\hat{\theta}^{(N)}_{2_j}} \\ -x^2_{N+1} + \frac{\hat{\theta}^{2(N)}_{1_j}}{4\hat{\theta}^{2(N)}_{2_j}} + \frac{1}{2\hat{\theta}^{(N)}_{2_j}} \end{pmatrix} \tag{42}$$

This latter expression appears to be quite complicated. If this algorithm is applied on $\lambda = (\mu, \sigma^2)$-coordinates, the matrix I_c is almost diagonal:

$$I_c(\lambda) = blockdiag\left(\left(diag(w_1^{-1}, \ldots, w_{K-1}^{-1}) - \frac{\mathbf{1}_{K-1}{}^t\mathbf{1}_{K-1}}{w_K}\right), \right.$$
$$\left. w_1 I(\lambda_1), \ldots, w_K I(\lambda_K)\right) \tag{43}$$

where I represents in this case the Fisher information matrix on λ for the gaussian distribution (see Eq. 69). With this parametrization, the score vector given by Eq. 35 is partially composed by the following expressions:

$$\text{For } j = 1, \ldots, K \quad \frac{\partial \log f(x_{N+1}; \hat{\lambda}^{(N)})}{\partial \mu_j} = \hat{z}_{N+1,j}^{(N)} \frac{x_{N+1} - \hat{\mu}_j^{(N)}}{\hat{\sigma}_j^{2(N)}} \tag{44}$$

$$\frac{\partial \log f(x_{N+1}; \hat{\lambda}^{(N)})}{\partial \sigma_j^2} = \hat{z}_{N+1,j}^{(N)} \frac{(x_{N+1} - \hat{\mu}_j^{(N)})^2 - \hat{\sigma}_j^{2(N)}}{\hat{\sigma}_j^{4(N)}} \tag{45}$$

After few simplifications, the update equations in this coordinates system are:

$$\hat{\mu}_j^{(N+1)} = \hat{\mu}_j^{(N)} + \alpha^{(N+1)} \frac{\hat{z}_{N+1,j}^{(N)}}{\hat{w}_j} \left(x_{N+1} - \hat{\mu}_j^{(N)}\right) \tag{46}$$

$$\hat{\sigma}_j^{2(N+1)} = \hat{\sigma}_j^{2(N)} + \alpha^{(N+1)} \frac{\hat{z}_{N+1,j}^{(N)}}{\hat{w}_j} \left((x_{N+1} - \hat{\mu}_j^{(N)})^2 - \hat{\sigma}_j^{2(N)}\right) \tag{47}$$

Note the estimation of weight components remains unchanged.

In order to compare Titterington's algorithm with online EM, consider its formulation in the η-coordinates system. Recall that for a regular exponential family g_j:

$$\nabla_{\eta_j} \log g_j(x; \eta_j) = H(F_j^*)(\eta_j)(s_j(x) - \eta_j) \tag{48}$$

Moreover, since the matrix $I_c(\eta)$ is

$$I_c(\eta) = blockdiag\left(\left(diag(w_1^{-1}, \ldots, w_{K-1}^{-1}) - \frac{\mathbf{1}_{K-1}{}^t\mathbf{1}_{K-1}}{w_K}\right), \right.$$
$$\left. w_1 H(F_1^*)(\eta_1), \ldots, w_K H(F_K^*)(\eta_K)\right), \tag{49}$$

the recursion no longer requires to invert a matrix:

$$\hat{\eta}_j^{(N+1)} = \hat{\eta}_j^{(N)} + \alpha^{(N+1)} \frac{\hat{z}_{N+1,j}^{(N)}}{\hat{w}_j^{(N)}} \left(s_j(x_{N+1}) - \hat{\eta}_j^{(N)}\right) \tag{50}$$

Unfortunately, for all the above methods, the constraints on parameters ($\sigma_j^2 > 0$) have to be checked beforehand in order to accept the parameters update. Looking at equations Eqs. 31 and 41, we can conclude that the online EM differs in the way components parameters are updated:

$$\hat{\eta}_j^{(N+1)} = \frac{\hat{w}_j^{(N)}}{\hat{w}_j^{(N+1)}}\hat{\eta}_j^{(N)} + \alpha^{(N+1)}\left(\frac{\hat{z}_{N+1,j}^{(N)}}{\hat{w}_j^{(N+1)}}s_j(x_{N+1}) - \frac{\hat{w}_j^{(N)}}{\hat{w}_j^{(N+1)}}\hat{\eta}_j^{(N)}\right) \tag{51}$$

For more details about the convergence of these algorithms, the interested reader is referred to Cappé and Moulines (2009) for further information.

To illustrate these algorithms, two experiments on synthetic datasets are provided (see Fig. 5). Their respective parameters are:
Dataset 1 : $(w_1 = 0.5, \mu_1 = 0, \sigma_1^2 = 1), (w_2 = 0.5, \mu_2 = 4, \sigma_2^2 = 4)$
Dataset 2 : $(w_1 = 0.25, \mu_1 = 0.25, \sigma_1^2 = 0.15), (w_2 = 0.65, \mu_2 = -1, \sigma_2^2 = 0.4)$
$(w_3 = 0.1, \mu_3 = -0.5, \sigma_3^2 = 0.6)$
All the methods were initialized with same parameters values coming from the k-means algorithm.

The policy for the learning rate is also identical: $\alpha^{(N)} = \left(\frac{1}{N_0+N}\right)^{0.7}$ where N_0 is the number of observations used for k-means. The criteria used to evaluated the results are the average log-likelihood and the Kullback–Liebler divergence (KL). Since there is no closed form to evaluate this divergence, we rely on numerical integration which is reasonably fast and accurate in 1d. Figure 6 reports the results of all estimators on the two datasets. Additionally, Figs. 7 and 8 illustrates the best estimates KL resulting components.

As expected, since the dataset 1 contains two relatively separated components, the estimation converges very quickly for all methods except for the Recursive EM on θ-coordinates. For this experiment, we observe that most of the first updates are rejected due to the constraints on parameters ($\theta_2 > 0 \equiv \sigma^2 > 0$). Later in the recursion, the learning rate has decreased and the updates do not violate the constraints. Undoubtedly, and as also observed in first experiment, the choice of the learning rate policy should be different on natural parameter space. Also, one can observe that some methods are trapped in different local minima even if their initialization are

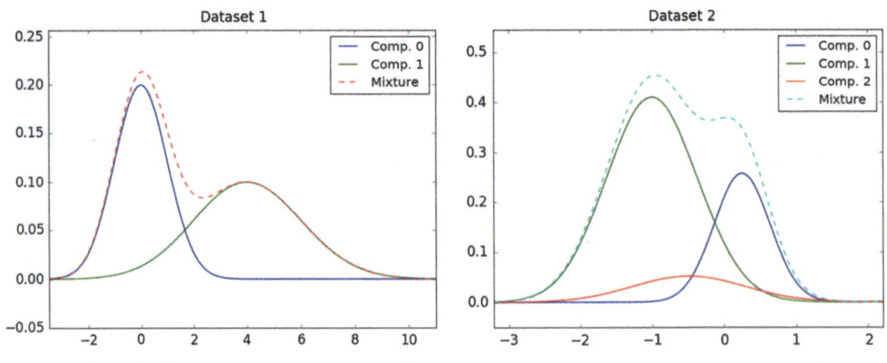

Fig. 5 Synthetic datasets

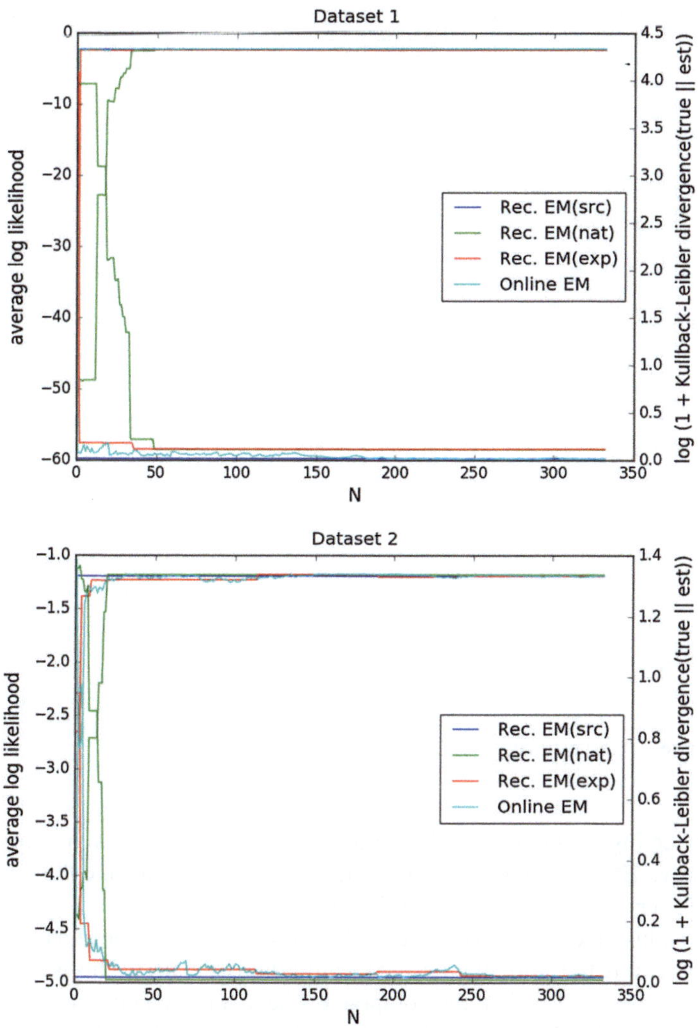

Fig. 6 Average log-likelihood and Kullback–Leibler for all estimators

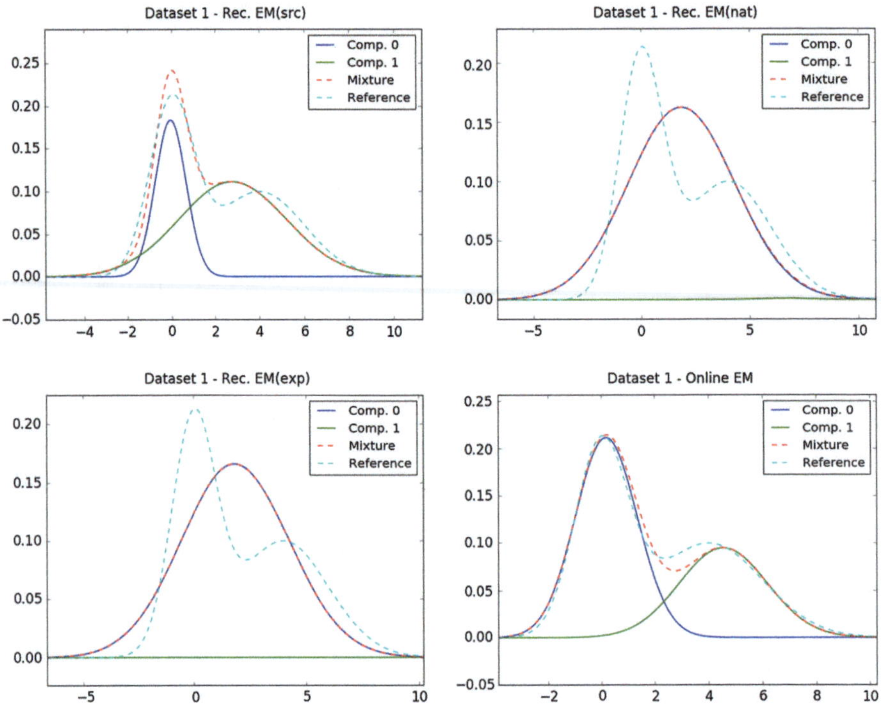

Fig. 7 Dataset 1: Best estimates w.r.t. to KL divergence

the same (see Fig. 7). For dataset 2, we remark the constraints prevents some updates for Recursive EM on natural and expectation parameters. Despite this, the mixture estimate is very good (see Recursive EM(natural) on Fig. 8).

As a conclusion of these experiments, Recursive EM on η-coordinates and online EM do not require the computation and the inversion of matrix. This is a very appealing property especially when the components have a more complicated parametric distribution (e.g. Wishart distributions Saint-Jean and Nielsen 2014). But in practice, this provides only easier to implement methods and does not guarantee better estimates. Since online EM makes a stochastic approximation of the E-Step of EM, the constraints on parameters are automatically guaranteed by the maximization step which is particularly efficient for exponential families.

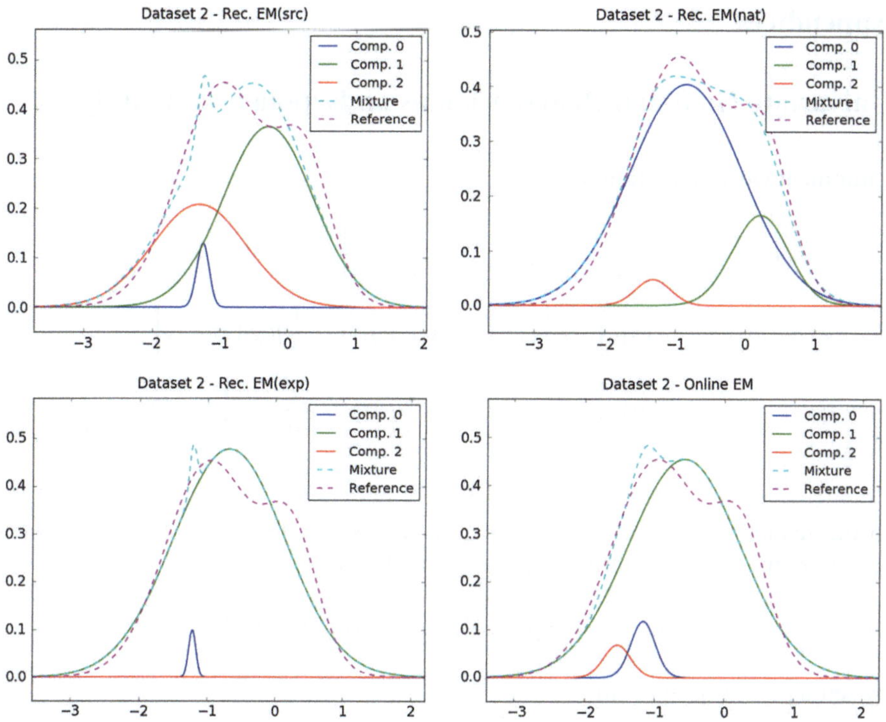

Fig. 8 Dataset 2: Best estimates w.r.t. to KL divergence

5 Conclusion

This paper addresses the problem of online learning of finite statistical mixtures with a special focus on distribution components belonging to the exponential families. Many details to compare Recursive EM and online EM from the practical point of view are given. The presented methods are fast since they require only one pass over the data stream. However, there is still room for improvement, especially for the Recursive EM method which is roughly a classical second-order stochastic gradient ascent. More recent optimization methods are described in the paper and leads to overcome the difficulty to choose an adequate policy for the learning rate. We might have also mentioned the incremental EM by Neal and Hinton (1999) which shares many properties with the online EM (partial E-Step). Further speed increase may be achieved by using distributed computing on a cluster of machines by aggregating partial sums of sufficient statistics (see Liu and Ihler 2014) since the statistical estimation is a decomposable problem.

Appendices

Univariate Gaussian Distribution as an Exponential Family

Canonical Decomposition and *F*

$$
\begin{aligned}
f(x;\mu,\sigma^2) &= \frac{1}{(2\pi\sigma^2)^{1/2}} \exp\left\{-\frac{(x-\mu)^2}{2\sigma^2}\right\}\\
&= \exp\left\{-\frac{1}{2\sigma^2}(x^2-2x\mu+\mu^2) - \frac{1}{2}\log\left(2\pi\sigma^2\right)\right\}\\
&= \exp\left\{\langle\frac{1}{2\sigma^2}, -x^2\rangle + \langle\frac{\mu}{\sigma^2}, x\rangle - \frac{\mu^2}{2\sigma^2} - \frac{1}{2}\log\left(2\pi\sigma^2\right)\right\}
\end{aligned}
$$

In the sequel, the vector of source parameters is denoted $\lambda = (\mu, \sigma^2)$. One may recognize the canonical form of an exponential family

$$f(x;\theta) = \exp\{< \theta, s(x) > + k(x) - F(\theta)\}$$

by setting $\theta = (\theta_1, \theta_2)$ with

$$\theta_1 = \frac{\mu}{\sigma^2} \Longleftrightarrow \mu = \frac{\theta_1}{2\theta_2} \tag{52}$$

$$\theta_2 = \frac{1}{2\sigma^2} \Longleftrightarrow \sigma^2 = \frac{1}{2\theta_2} \tag{53}$$

$$s(x) = (x, -x^2) \tag{54}$$

$$k(x) = 0 \tag{55}$$

$$
\begin{aligned}
f(x;\theta_1,\theta_2) &= \exp\left\{\langle\theta_2, -x^2\rangle + \langle\theta_1, x\rangle - \frac{1}{2}\frac{(\theta_1/2\theta_2)^2}{1/2\theta_2} - \frac{1}{2}\log(2\pi/2\theta_2)\right\}\\
&= \exp\left\{\langle\theta_2, -x^2\rangle + \langle\theta_1, x\rangle - \frac{\theta_1^2}{4\theta_2} - \frac{1}{2}\log(\pi) + \frac{1}{2}\log\theta_2\right\}
\end{aligned}
$$

with the log normalizer F as

$$F(\theta_1,\theta_2) = \frac{\theta_1^2}{4\theta_2} + \frac{1}{2}\log(\pi) - \frac{1}{2}\log\theta_2 \tag{56}$$

Gradient of the Log-Normalizer

The gradient of the log-normalizer is given by:

$$\frac{\partial F}{\partial \theta_1}(\theta_1, \theta_2) = \frac{\theta_1}{2\theta_2} \tag{57}$$

$$\frac{\partial F}{\partial \theta_2}(\theta_1, \theta_2) = -\frac{\theta_1^2}{4\theta_2^2} - \frac{1}{2\theta_2} \tag{58}$$

In order to get the dual coordinate system $\eta = (\eta_1, \eta_2)$, the following set of equations has to be inverted:

$$\eta_1 = \frac{\theta_1}{2\theta_2} \tag{59}$$

$$\eta_2 = -\frac{\theta_1^2}{4\theta_2^2} - \frac{1}{2\theta_2} \tag{60}$$

By plugging the first equation into the second one, it follows:

$$\eta_2 = -\eta_1^2 - \frac{1}{2\theta_2} \Longleftrightarrow \theta_2 = -\frac{1}{2(\eta_1^2 + \eta_2)} \qquad = \frac{\partial F^*}{\partial \eta_2}(\eta_1, \eta_2) \tag{61}$$

$$\theta_1 = 2\theta_2\eta_1 = -\frac{\eta_1}{(\eta_1^2 + \eta_2)} \qquad = \frac{\partial F^*}{\partial \eta_1}(\eta_1, \eta_2) \tag{62}$$

Formulas are even simpler regarding the source parameters since we know that

$$\eta_1 = \mathbb{E}[X] = \mu \Longleftrightarrow \qquad \mu = \eta_1 \tag{63}$$

$$\eta_2 = \mathbb{E}[-X^2] = -\left\{\mu^2 + \sigma^2\right\} \Longleftrightarrow \qquad \sigma^2 = -\left\{\eta_1^2 + \eta_2\right\} \tag{64}$$

In order to compute F^*, we simply have to reuse our previous results in

$$F^*(H) = \langle(\nabla F)^{-1}(H), H\rangle - F((\nabla F)^{-1}(H))$$

and obtain the following expression

$$F^*(\eta_1, \eta_2) = \left\langle -\frac{\eta_1}{(\eta_1^2 + \eta_2)}, \eta_1 \right\rangle + \left\langle -\frac{1}{2(\eta_1^2 + \eta_2)}, \eta_2 \right\rangle - \left\{ \frac{\left(-\frac{\eta_1}{(\eta_1^2+\eta_2)}\right)^2}{4\left(-\frac{1}{2(\eta_1^2+\eta_2)}\right)} + \frac{1}{2}\log(\pi) - \frac{1}{2}\log\left(-\frac{1}{2(\eta_1^2 + \eta_2)}\right) \right\}$$

$$= -\frac{\eta_1^2}{(\eta_1^2+\eta_2)} - \frac{\eta_2}{2(\eta_1^2+\eta_2)} + \frac{\frac{\eta_1^2}{(\eta_1^2+\eta_2)^2}}{\frac{2}{(\eta_1^2+\eta_2)}}$$

$$- \frac{1}{2}\log(\pi) + \frac{1}{2}\log((-2(\eta_1^2+\eta_2))^{-1})$$

$$= -\frac{\eta_1^2}{2(\eta_1^2+\eta_2)} - \frac{\eta_2}{2(\eta_1^2+\eta_2)} - \frac{1}{2}\log(\pi) - \frac{1}{2}\log(-2(\eta_1^2+\eta_2))$$

$$= -\frac{1}{2} - \frac{1}{2}\log(\pi) - \frac{1}{2}\log(-2(\eta_1^2+\eta_2))$$

$$= -\frac{1}{2}\log(e\pi) - \frac{1}{2}\log(-2(\eta_1^2+\eta_2))$$

The hessians $H(F)$, $H(F^*)$ of respectively F and F^* are

$$H(F)(\theta_1,\theta_2) = \begin{pmatrix} \frac{1}{2\theta_2} & -\frac{\theta_1}{2\theta_2^2} \\ -\frac{\theta_1}{2\theta_2^2} & \frac{\theta_1^2+\theta_2}{2\theta_2^3} \end{pmatrix} \tag{65}$$

$$H(F^*)(\eta_1,\eta_2) = \begin{pmatrix} \frac{\eta_1^2-\eta_2}{(\eta_1^2+\eta_2)^2} & \frac{\eta_1}{(\eta_1^2+\eta_2)^2} \\ \frac{\eta_1}{(\eta_1^2+\eta_2)^2} & \frac{1}{2(\eta_1^2+\eta_2)^2} \end{pmatrix} \tag{66}$$

Since the univariate normal distribution is an exponential family, the Kullback–Leibler divergence is a Bregman divergence for F^* on expectation parameters:

$$\begin{aligned} KL(\mathcal{N}(\mu_p,\sigma_p^2)||\mathcal{N}(\mu_q,\sigma_q^2)) &= B_{F^*}(\eta_p:\eta_q) \\ &= F^*(\eta_p) - F^*(\eta_q) - \langle \eta_p-\eta_q, \nabla F^*(\eta_q)\rangle \end{aligned}$$

After calculations, it follows:

$$B_F^*(\eta_p:\eta_q) = \frac{1}{2}\left(\log\left(\frac{\eta_{1_q}^2+\eta_{2_q}}{\eta_{1_p}^2+\eta_{2_p}}\right) + \frac{2(\eta_{1_p}-\eta_{1_q})\eta_{1_q}}{(\eta_{1_q}^2+\eta_{2_q})} + \frac{\eta_{2_p}-\eta_{2_q}}{(\eta_{1_q}^2+\eta_{2_q})}\right) \tag{67}$$

A simple rewrite of it with the source parameters leads to the known closed form:

$$\frac{1}{2}\left(\log\left(\frac{\eta_{1_q}^2+\eta_{2_q}}{\eta_{1_p}^2+\eta_{2_p}}\right) + \frac{2(\eta_{1_p}-\eta_{1_q})\eta_{1_q}}{(\eta_{1_q}^2+\eta_{2_q})} + \frac{\eta_{2_p}-\eta_{2_q}}{(\eta_{1_q}^2+\eta_{2_q})}\right) =$$

$$\frac{1}{2}\left(\log\left(\frac{\eta_{1_q}^2+\eta_{2_q}}{\eta_{1_p}^2+\eta_{2_p}}\right) + \frac{(\eta_{1_p}^2+\eta_{2_p}) - (\eta_{1_p}-\eta_{1_q})^2 - (\eta_{1_q}^2+\eta_{2_q})}{(\eta_{1_q}^2+\eta_{2_q})}\right) =$$

$$\frac{1}{2}\left(\log\left(\frac{\sigma_q^2}{\sigma_p^2}\right) + \frac{\sigma_p^2}{\sigma_q^2} + \frac{(\mu_p-\mu_q)^2}{\sigma_q^2} - 1\right) \tag{68}$$

The Fisher information matrix $I(\lambda)$ is obtained by computing the expectation of the product of Fisher score and its transposition:

$$\begin{aligned}
I(\lambda) &\overset{def}{=} \mathbb{E}\left[\nabla_\lambda \log f(x;\lambda).\nabla_\lambda \log f(x;\lambda)^T\right] \\
&= \mathbb{E}\left[\begin{pmatrix} \frac{x-\mu}{\sigma^2} \\ \frac{(x-\mu)^2-\sigma^2}{2\sigma^4} \end{pmatrix} \cdot \begin{pmatrix} \frac{x-\mu}{\sigma^2} & \frac{(x-\mu)^2-\sigma^2}{2\sigma^4} \end{pmatrix}\right] \\
&= \begin{pmatrix} \frac{1}{\sigma^2} & 0 \\ 0 & \frac{1}{2\sigma^4} \end{pmatrix}.
\end{aligned} \tag{69}$$

By change in coordinates or direct computation, the Fisher information matrix is also:

$$I(\theta) = H(F)(\theta) = \begin{pmatrix} \frac{1}{2\theta_2} & -\frac{\theta_1}{2\theta_2^2} \\ -\frac{\theta_1}{2\theta_2^2} & \frac{\theta_1^2+\theta_2}{2\theta_2^3} \end{pmatrix} \text{ and } I(\eta) = \frac{1}{(\eta_1^2+\eta_2)^2}\begin{pmatrix} (\eta_1^2-\eta_2) & \eta_1 \\ \eta_1 & \frac{1}{2} \end{pmatrix} \tag{70}$$

Multivariate Gaussian Distribution as an Exponential Family

Canonical Decomposition and *F*

$$\begin{aligned}
f(x;\mu,\Sigma) &= \frac{1}{(2\pi)^{d/2}|\Sigma|^{1/2}} \exp\left\{-\frac{{}^t(x-\mu)\Sigma^{-1}(x-\mu)}{2}\right\} \\
&= \exp\left\{-\frac{{}^tx\Sigma^{-1}x - {}^t\mu\Sigma^{-1}x - {}^tx\Sigma^{-1}\mu + {}^t\mu\Sigma^{-1}\mu}{2} - \log\left((2\pi)^{d/2}|\Sigma|^{1/2}\right)\right\} \\
&= \exp\left\{-\frac{tr({}^tx\Sigma^{-1}x) - \langle {}^t\Sigma^{-1}\mu, x\rangle - \langle x, \Sigma^{-1}\mu\rangle + \langle {}^t\Sigma^{-1}\mu, \Sigma\Sigma^{-1}\mu\rangle}{2} - \log\left(\pi^{d/2}|2\Sigma|^{1/2}\right)\right\}
\end{aligned}$$

Due to the cyclic property of the trace and to the symmetry of Σ^{-1}, it follows:

$$\begin{aligned}
f(x;\mu,\Sigma) &= \exp\left\{tr\left({}^t\left(\frac{1}{2}\Sigma^{-1}\right)(-x^tx)\right) + \langle\Sigma^{-1}\mu, x\rangle - \frac{1}{2}\langle\Sigma^{-1}\mu, \Sigma\Sigma^{-1}\mu\rangle - \frac{d}{2}\log(\pi) - \frac{1}{2}\log|2\Sigma|\right\} \\
&= \exp\left\{\langle\frac{1}{2}\Sigma^{-1}, -x^tx\rangle_F + \langle\Sigma^{-1}\mu, x\rangle - \frac{1}{4}{}^t(\Sigma^{-1}\mu)2\Sigma(\Sigma^{-1}\mu) - \frac{d}{2}\log(\pi) - \frac{1}{2}\log|2\Sigma|\right\}
\end{aligned}$$

where $\langle\cdot,\cdot\rangle_F$ is the Frobenius scalar product. One may recognize the canonical form of an exponential family

$$f(x;\Theta) = \exp\{< \Theta, t(x) > + k(x) - F(\Theta)\}$$

by setting:

$$\Theta = (\theta_1, \theta_2)$$

$$\theta_1 = \Sigma^{-1}\mu \Longleftrightarrow \mu = \frac{1}{2}\theta_2^{-1}\theta_1 \tag{71}$$

$$\theta_2 = \frac{1}{2}\Sigma^{-1} \Longleftrightarrow \Sigma = \frac{1}{2}\theta_2^{-1} \tag{72}$$

$$t(x) = (x, -x^t x) \tag{73}$$

$$k(x) = 0 \tag{74}$$

$$f(x;\theta_1,\theta_2) = \exp\left\{\langle\theta_2, -x^t x\rangle_F + \langle\theta_1, x\rangle - \frac{1}{4}{}^t\theta_1\theta_2^{-1}\theta_1 - \frac{d}{2}\log(\pi) + \frac{1}{2}\log|\theta_2|\right\} \tag{75}$$

with the log normalizer F:

$$F(\theta_1,\theta_2) = \frac{1}{4}{}^t\theta_1\theta_2^{-1}\theta_1 + \frac{d}{2}\log(\pi) - \frac{1}{2}\log|\theta_2| \tag{76}$$

Gradient of the Log-Normalizer

By applying the following formulas from the matrix cookbook (Petersen and Pedersen 2012)

identity 57

$$\frac{\partial \log|X|}{\partial X} = ({}^tX)^{-1} = {}^t(X^{-1})$$

identity 61

$$\frac{\partial^t a X^{-1} b}{\partial X} = -{}^tX^{-1}a^t b X^{-1}$$

identity 81

$$\frac{\partial^t x B x}{\partial x} = (B + {}^tB)x$$

the gradient of the log-normalizer is given by:

$$\frac{\partial F}{\partial \theta_1}(\theta_1,\theta_2) = \frac{1}{4}(\theta_2^{-1} + {}^t\theta_2^{-1})\theta_1 = \frac{1}{2}\theta_2^{-1}\theta_1 \tag{77}$$

$$\frac{\partial F}{\partial \theta_2}(\theta_1,\theta_2) = -\frac{1}{4}{}^t\theta_2^{-1}\theta_1{}^t\theta_1\theta_2^{-1} - \frac{1}{2}{}^t\theta_2^{-1} = -\left(\frac{1}{2}\theta_2^{-1}\theta_1\right)^t\left(\frac{1}{2}\theta_2^{-1}\theta_1\right) - \frac{1}{2}\theta_2^{-1} \tag{78}$$

In order to emphasize the coherence of these formulas, recall that the gradient of the log-normalizer corresponds the expectation of the sufficient statistics:

$$\mathbb{E}[x] = \mu \qquad \equiv \frac{1}{2}\theta_2^{-1}\theta_1 \tag{79}$$

$$\mathbb{E}[-x^t x] = -\mathbb{E}[x^t x] = -\mu^t \mu - \Sigma \qquad \equiv -\left(\frac{1}{2}\theta_2^{-1}\theta_1\right)^t \left(\frac{1}{2}\theta_2^{-1}\theta_1\right) - \frac{1}{2}\theta_2^{-1} \tag{80}$$

Last equation comes from the expansion of $\mathbb{E}[(x-\mu)^t(x-\mu)]$.

Convex Conjugate G of F and Its Gradient

In order to get the dual coordinate system $H = (\eta_1, \eta_2)$, the following set of equations has to be inverted:

$$\eta_1 = \frac{1}{2}\theta_2^{-1}\theta_1 \tag{81}$$

$$\eta_2 = -\left(\frac{1}{2}\theta_2^{-1}\theta_1\right)^t \left(\frac{1}{2}\theta_2^{-1}\theta_1\right) - \frac{1}{2}\theta_2^{-1} \tag{82}$$

By plugging the first equation into the second one, it follows

$$\eta_2 = -\eta_1{}^t\eta_1 - \frac{1}{2}\theta_2^{-1} \Longleftrightarrow \theta_2 = \frac{1}{2}(-\eta_1{}^t\eta_1 - \eta_2)^{-1} = \frac{\partial G}{\partial \eta_2}(\eta_1, \eta_2) \tag{83}$$

and

$$\theta_1 = 2\theta_2\eta_1 = (-\eta_1{}^t\eta_1 - \eta_2)^{-1}\eta_1 = \frac{\partial G}{\partial \eta_1}(\eta_1, \eta_2) \tag{84}$$

Formulas are even simpler regarding the source parameters since we know from Eqs. 79 and 80 that

$$\eta_1 = \mu \Longleftrightarrow \qquad \mu = \eta_1 \tag{85}$$

$$\eta_2 = -\mu^t\mu - \Sigma \Longleftrightarrow \qquad \Sigma = -\eta_1{}^t\eta_1 - \eta_2 \tag{86}$$

In order to compute $G := F^*$, we simply have to reuse our previous results in

$$G(H) = \langle (\nabla F)^{-1}(H), H \rangle - F((\nabla F)^{-1}(H))$$

and obtain the following expression

$$G(\eta_1, \eta_2) = \langle (-\eta_1{}^t\eta_1 - \eta_2)^{-1}\eta_1, \eta_1 \rangle + \langle \frac{1}{2}(-\eta_1{}^t\eta_1 - \eta_2)^{-1}, \eta_2 \rangle_F$$
$$- \frac{1}{4}{}^t((-\eta_1{}^t\eta_1 - \eta_2)^{-1}\eta_1)2(-\eta_1{}^t\eta_1 - \eta_2)(-\eta_1{}^t\eta_1 - \eta_2)^{-1}\eta_1$$

$$
\begin{aligned}
&-\frac{d}{2}\log(\pi)+\frac{1}{2}\log|\frac{1}{2}(-\eta_1{}^t\eta_1-\eta_2)^{-1}|\\
&={}^t\eta_1(-\eta_1{}^t\eta_1-\eta_2)^{-1}\eta_1+\frac{1}{2}tr({}^t(-\eta_1{}^t\eta_1-\eta_2)^{-1}\eta_2)\\
&-\frac{1}{2}{}^t\eta_1{}^t(-\eta_1{}^t\eta_1-\eta_2)^{-1}\eta_1\\
&-\frac{d}{2}\log(\pi)+\frac{1}{2}\log|(2(-\eta_1{}^t\eta_1-\eta_2))^{-1}|\\
&=\frac{1}{2}{}^t\eta_1(-\eta_1{}^t\eta_1-\eta_2)^{-1}\eta_1+\frac{1}{2}tr((-\eta_1{}^t\eta_1-\eta_2)^{-1}\eta_2)\\
&-\frac{d}{2}\log(\pi)-\frac{1}{2}\log|2(-\eta_1{}^t\eta_1-\eta_2)|\\
&=\frac{1}{2}\left(tr((-\eta_1{}^t\eta_1-\eta_2)^{-1}\eta_1{}^t\eta_1)+tr((-\eta_1{}^t\eta_1-\eta_2)^{-1}\eta_2)\right)\\
&-\frac{d}{2}\log(\pi)-\frac{1}{2}\log|2(-\eta_1{}^t\eta_1-\eta_2)|\\
&=-\frac{1}{2}tr((-\eta_1{}^t\eta_1-\eta_2)^{-1}(-\eta_1{}^t\eta_1-\eta_2))-\frac{d}{2}\log(\pi)\\
&-\frac{1}{2}\log|2(-\eta_1{}^t\eta_1-\eta_2)|\\
&=-\frac{1}{2}tr(I_d)-\frac{d}{2}\log(\pi)-\frac{1}{2}\log|2(-\eta_1{}^t\eta_1-\eta_2)|\\
&=-\frac{d}{2}\log(e\pi)-\frac{1}{2}\log|2(-\eta_1{}^t\eta_1-\eta_2)|
\end{aligned}
$$

Let us rewrite this expression with source parameters:

$$G(\mu,\Sigma)=-\frac{d}{2}\log(e\pi)-\frac{1}{2}\log|2\Sigma| \tag{87}$$

Kullback–Leibler Divergence

First recall that the Kullback–Leibler divergence between two p.d.f. p and q is

$$KL(p||q)=\int p(x)\log\frac{p(x)}{q(x)}dx$$

For two multivariate normal distributions, it is known in closed form

$$KL(\mathcal{N}(\mu_p,\Sigma_p)||\mathcal{N}(\mu_q,\Sigma_q))=\frac{1}{2}\left(\log\left(\frac{|\Sigma_q|}{|\Sigma_p|}\right)+tr(\Sigma_q^{-1}\Sigma_p)+{}^t(\mu_q-\mu_p)\Sigma_q^{-1}(\mu_q-\mu_p)-d\right) \tag{88}$$

Since the multivariate normal distribution is an E.F., the same result must be obtained using the bregman divergence for G on expectation parameters H_p and H_q:

$$KL(\mathcal{N}(\mu_p, \Sigma_p)||\mathcal{N}(\mu_q, \Sigma_q)) = B_G(H_p||H_q) = G(H_p) - G(H_q) - \langle H_p - H_q, \nabla G(H_q)\rangle$$

$$\begin{aligned}
G(H_p) - G(H_q) &= -\frac{d}{2}\log(e\pi) - \frac{1}{2}\log|-2({\eta_1}_p{}^t{\eta_1}_p + {\eta_2}_p)| \\
&+ \frac{d}{2}\log(e\pi) + \frac{1}{2}\log|-2({\eta_1}_q{}^t{\eta_1}_q + {\eta_2}_q)| \\
&= \frac{1}{2}\log\frac{|-({\eta_1}_q{}^t{\eta_1}_q + {\eta_2}_q)|}{|-({\eta_1}_p{}^t{\eta_1}_p + {\eta_2}_p)|} \\
-\langle H_p - H_q, \nabla G(H_q)\rangle &= -\langle {\eta_1}_p - {\eta_1}_q, -({\eta_1}_q{}^t{\eta_1}_q + {\eta_2}_q)^{-1}{\eta_1}_q\rangle \\
&- tr\left({}^t({\eta_2}_p - {\eta_2}_q)\left(-\frac{1}{2}({\eta_1}_q{}^t{\eta_1}_q + {\eta_2}_q)^{-1}\right)\right) \\
&= {}^t{\eta_1}_p({\eta_1}_q{}^t{\eta_1}_q + {\eta_2}_q)^{-1}{\eta_1}_q - {}^t{\eta_1}_q({\eta_1}_q{}^t{\eta_1}_q + {\eta_2}_q)^{-1}{\eta_1}_q \\
&- \frac{1}{2}tr({}^t{\eta_2}_p(-({\eta_1}_q{}^t{\eta_1}_q + {\eta_2}_q)^{-1})) + \frac{1}{2}tr({}^t{\eta_2}_q(-({\eta_1}_q{}^t{\eta_1}_q + {\eta_2}_q)^{-1}))
\end{aligned}$$

In order to go further, we can express these two formulas using μ and $\Sigma^{-1} = (-\eta_1{}^t\eta_1 - \eta_2)^{-1} = -(\eta_1{}^t\eta_1 + \eta_2)^{-1}$ (cf. Eq. 86):

$$\frac{1}{2}\log\frac{|-({\eta_1}_q{}^t{\eta_1}_q + {\eta_2}_q)|}{|-({\eta_1}_q{}^t{\eta_1}_q + {\eta_2}_q)|} = \frac{1}{2}\log\frac{|\Sigma_q|}{|\Sigma_p|}$$

$$\begin{aligned}
{}^t{\eta_1}_p({\eta_1}_q{}^t{\eta_1}_q + {\eta_2}_q)^{-1}{\eta_1}_q &= -{}^t\mu_p\Sigma_q^{-1}\mu_q \\
-{}^t{\eta_1}_q({\eta_1}_q{}^t{\eta_1}_q + {\eta_2}_q)^{-1}{\eta_1}_q &= {}^t\mu_q\Sigma_q^{-1}\mu_q
\end{aligned}$$

$$\begin{aligned}
-\frac{1}{2}tr({}^t{\eta_2}_p(-({\eta_1}_q{}^t{\eta_1}_q + {\eta_2}_q)^{-1})) &= \frac{1}{2}tr((\mu_p{}^t\mu_p + \Sigma_p)\Sigma_q^{-1}) \\
&= \frac{1}{2}tr(\mu_p{}^t\mu_p\Sigma_q^{-1}) + \frac{1}{2}tr(\Sigma_p\Sigma_q^{-1}) \\
&= \frac{1}{2}{}^t\mu_p\Sigma_q^{-1}\mu_p + \frac{1}{2}tr(\Sigma_q^{-1}\Sigma_p) \\
+\frac{1}{2}tr({}^t{\eta_2}_q(-({\eta_1}_q{}^t{\eta_1}_q + {\eta_2}_q)^{-1})) &= -\frac{1}{2}tr((\mu_q{}^t\mu_q + \Sigma_q)\Sigma_q^{-1}) \\
&= -\frac{1}{2}tr(\mu_q{}^t\mu_q\Sigma_q^{-1}) - \frac{1}{2}tr(\Sigma_q\Sigma_q^{-1}) \\
&= -\frac{1}{2}{}^t\mu_q\Sigma_q^{-1}\mu_q - \frac{1}{2}d
\end{aligned}$$

By summing up of these terms, the standard formula for KL divergence is recovered:

$$
\begin{aligned}
KL(\mathcal{N}(\mu_p, \Sigma_p)||\mathcal{N}(\mu_q, \Sigma_q)) = & \frac{1}{2}\log\frac{|\Sigma_q|}{|\Sigma_p|} - {}^t\mu_p\Sigma_q^{-1}\mu_q + {}^t\mu_q\Sigma_q^{-1}\mu_q + \\
& \frac{1}{2}{}^t\mu_p\Sigma_q^{-1}\mu_p + \frac{1}{2}tr(\Sigma_q^{-1}\Sigma_p) - \frac{1}{2}{}^t\mu_q\Sigma_q^{-1}\mu_q - \frac{1}{2}d \\
= & \frac{1}{2}\left(\log\frac{|\Sigma_q|}{|\Sigma_p|} + tr(\Sigma_q^{-1}\Sigma_p) - d - \right. \\
& \left. \{2^t\mu_p\Sigma_q^{-1}\mu_q - 2^t\mu_q\Sigma_q^{-1}\mu_q -^t \mu_p\Sigma_q^{-1}\mu_p + {}^t\mu_q\Sigma_q^{-1}\mu_q\}\right) \\
= & \frac{1}{2}\left(\log\frac{|\Sigma_q|}{|\Sigma_p|} + tr(\Sigma_q^{-1}\Sigma_p) - {}^t(\mu_p - \mu_q)\Sigma_q^{-1}(\mu_p - \mu_q) - d\right)
\end{aligned}
$$

References

Amari, S. (1997). Neural learning in structured parameter spaces — Natural Riemannian gradient. *Neural Information Processing Society (NIPS)*, *9*, 127–133.

Amari, S. (1998). Natural gradient works efficiently in learning. *Neural Computation*, *10*(2), 251–276.

Amari, S. (2016). *Information geometry and its applications*. Applied Mathematical Sciences. Japan: Springer.

Banerjee, A., Merugu, S., Dhillon, I. S., & Ghosh, J. (2005). Clustering with Bregman divergences. *Journal of Machine Learning Research*, *6*, 1705–1749.

Bogdan, K., & Bogdan, M. (2000). On existence of maximum likelihood estimators in exponential families. *Statistics*, *34*(2), 137–149.

Bottou, L. (1998). Online algorithms and stochastic approximations. In S. David (Ed.), *Online learning and neural networks*. Cambridge: Cambridge University Press.

Bottou, L., & Bousquet, O. (2011). In S. Sra, S. Nowozin, & S. J. Wright (Eds.), *The tradeoffs of large scale learning* (pp. 351–368). Cambridge: MIT Press.

Boyd, S., & Vandenberghe, L. (2004). *Convex optimization*. Cambridge: Cambridge University Press.

Cappé, O., & Moulines, E. (2009). On-line expectation-maximization algorithm for latent data models. *Journal of the Royal Statistical Society. Series B (Methodological)*, *71*(3), 593–613.

Dempster, A. P., Laird, N. M., & Rubin, D. B. (1977). Maximum likelihood from incomplete data via the EM algorithm. *Journal of the Royal Statistical Society. Series B (Methodological)*, *39*, 1–38.

Liu, Q., & Ihler, A. T. (2014). Distributed estimation, information loss and exponential families. *Advances in Neural Information Processing Systems*, *27*, 1098–1106.

Miura, K. (2011). An introduction to maximum likelihood estimation in information geometry. *Interdisciplinary Information Sciences*, *17*(3), 155–174.

Neal, R. M., & Hinton, G. E. (1999). A view of the EM algorithm that justifies incremental, sparse, and other variants. In M. I. Jordan (Ed.), *Learning in graphical models* (pp. 355–368). Cambridge: MIT Press.

Nielsen, F., & Garcia, V. (2009). Statistical exponential families: A digest with flash cards. arXiv:0911.4863.

Petersen, K. B., & Pedersen, M. S. (2012). The matrix cookbook. http://www2.imm.dtu.dk/pubdb/p.php?3274.

Polyak, B. T., & Juditsky, A. B. (1992). Acceleration of stochastic approximation by averaging. *SIAM Journal on Control and Optimization*, *30*(4), 838–855.
Robbins, H., & Monro, S. (1951). A stochastic approximation method. *The Annals of Mathematical Statistics*, *22*(3), 400–407.
Saint-Jean, C., & Nielsen, F. (2014). Hartigan's method for k-MLE: Mixture modeling with Wishart distributions and its application to motion retrieval. *Geometric theory of information* (pp. 301–330). New York: Springer.
Sculley, D. (2010). Web-scale k-means clustering. In *Proceedings of the 19th International Conference on World Wide Web* (pp. 1177–1178).
Shalev-Shwartz, S. (2011). Online learning and online convex optimization. *Foundations and Trends Machine Learning*, *4*(2), 107–194.
Titterington, D. M. (1984). Recursive parameter estimation using incomplete data. *Journal of the Royal Statistical Society. Series B (Methodological)*, *46*(2), 257–267.

Erratum to: Computational Information Geometry

Frank Nielsen, Frank Critchley and Christopher T.J. Dodson

Erratum to:
F. Nielsen et al. (eds.), *Computational Information Geometry*, Signals and Communication Technology, DOI 10.1007/978-3-319-47058-0

The book was inadvertently published with abstracts only in the online version and also without processing the color figures in the print version. Abstracts are added in erratum chapter and the book has been updated with the color figures.

Chapter 1
We give a personal view of what Information Geometry is, and what it is becoming, by exploring a number of key topics: dual affine families, boundaries, divergences, tensorial structures, and dimensionality. For each, we start with a graphical illustrative example (Sect. 1.1), give an overview of the relevant theory and key references (Sect. 1.2), and finish with a number of applications of the theory (Sect. 1.3).

The updated original online version for this book can be found at 10.1007/978-3-319-47058-0

F. Nielsen (✉)
Laboratoire d'Informatique (LIX), Ecole Polytechnique, Palaiseau, France
e-mail: nielsen@lix.polytechnique.fr

F. Nielsen
Sony Computer Science Laboratories, Inc., Tokyo, Japan

F. Critchley
School of Mathematics and Statistics, The Open University, Milton Keynes, UK
e-mail: f.critchley@open.ac.uk

C.T.J. Dodson
Department of Mathematics, University of Manchester, Manchester, UK
e-mail: ctdodson@manchester.ac.uk

F. Nielsen et al. (eds.), *Computational Information Geometry*,
Signals and Communication Technology, DOI 10.1007/978-3-319-47058-0_12

We treat 'Information Geometry' as an evolutionary term, deliberately not attempting a comprehensive definition. Rather, we illustrate how both the geometries used and application areas are rapidly developing.

Chapter 2

In statistical practice model building, sensitivity and uncertainty are major concerns of the analyst. This paper looks at these issues from an information geometric point of view. Here, we define sensitivity to mean understanding how inference about a problem of interest changes with perturbations of the model. In particular it is an example of what we call computational information geometry. The embedding of simple models in much larger information geometric spaces is shown to illuminate these critically important issues.

Chapter 3

We show how information geometry throws new light on the interplay between goodness-of-fit and estimation, a fundamental issue in statistical inference. A geometric analysis of simple, yet representative, models involving the same population parameter compellingly establishes the main theme of the paper: namely, that goodness-of-fit is necessary but not sufficient for model selection. Visual examples vividly communicate this. Specifically, for a given estimation problem, we define a class of least-informative models, linking these to both nonparametric and maximum entropy methods. Any other model is then seen to involve an informative rotation, often embodying extra-data considerations. We also look at the way that translation of models generates a form of bias-variance trade-off. Overall, our approach is a global extension of pioneering local work by Copas and Eguchi which, we note, was also geometrically inspired.

Chapter 4

We discuss an approach called spontaneous data learning (SDL) to open novel explanatory paradigm connecting parametrics with nonparametrics. The statistical performance for SDL is explored from information geometric viewpoint, so that SDL gives a new perspective beyond the discussion for robustness or misspecification of parametric model. If the true distribution is exactly in the parametric model, the theory of statistical estimation has been well established, in which any minimum divergence estimator satisfies parametric consistency. We focus on a collapse of the parametric theory perturbing toward a nonparametric setting, where the true distribution may range from unimodality to multimodality; various estimators are targeted and investigated in a class of minimum divergence. In this context a selection of estimators is explored rather than model selection. Specifically we choose the power divergence class under a normal mean model, where the true distribution is, for example, a mixture of K distributions. Then we observe that the local minima of the empirical loss function for the power divergence properly suggest the K means if they are mutually separated in the mixture distribution, and the order of power is appropriated selected. The resulting method for clustering analysis is shown to spontaneously detects the number K of clusters. Further, we observe that the normalized empirical loss function converges to the true density function if the power parameter goes to infinity. As a result the power parameter combines between the parametric and nonparametric consistency.

Chapter 5
We define the notion of the extrinsic Itô projection of a stochastic differential equation (SDE) on a submanifold. This allows one to systematically develop low dimensional approximations to high dimensional SDEs in a differential geometric setting. We consider the example of approximating the non-linear filtering problem with a Gaussian distribution and show how the Itô projection leads to improved approximations in the Gaussian family. We briefly discuss the approximations for more general families of distribution. We perform a numerical comparison of our projection filters with the classical Extended Kalman Filter to demonstrate the efficacy of the approach.

Chapter 6
Matrix data sets are common nowadays like in biomedical imaging where the Diffusion Tensor Magnetic Resonance Imaging (DT-MRI) modality produces data sets of 3D symmetric positive definite matrices anchored at voxel positions capturing the anisotropic diffusion properties of water molecules in biological tissues. The space of symmetric matrices can be partially ordered using the Löwner ordering, and computing extremal matrices dominating a given set of matrices is a basic primitive used in matrix-valued signal processing. In this letter, we design a fast and easy-to-implement iterative algorithm to approximate arbitrarily finely these extremal matrices. Finally, we discuss on extensions to matrix clustering.

Chapter 7
Stochastic textures with features spanning many length scales arise in a range of contexts in physical and natural sciences, from nanostructures like synthetic bone to ocean wave height distributions and cosmic phenomena like inter-galactic cluster void distributions. Here we used a data set of 35 surface topographies, each of 2400×2400 pixels with spatial resolution between 4 and 7 μm per pixel, and fitted trivariate Gaussian distributions to represent their spatial structures. For these we computed pairwise information metric distances using the Fisher-Rao metric. Then dimensionality reduction was used to reveal the groupings among subsets of samples in an easily comprehended graphic in 3-space. The samples here came from the papermaking industry but such a reduction of large frequently noisy spatial data sets is useful in a range of materials and contexts at all scales.

Chapter 8
This artilce summarizes our work on the clustering of financial time series. It was written for a workshop on information geometry and its application for image and signal processing. This workshop brought several experts in pure and applied mathematics together with applied researchers from medical imaging, radar signal processing and finance. The authors belong to the latter group. This document was written as a long introduction to further development of geometric tools in financial applications such as risk or portfolio analysis. Indeed, risk and portfolio analysis

essentially rely on covariance matrices. Besides that the Gaussian assumption is known to be inaccurate, covariance matrices are difficult to estimate from empirical data. To filter noise from the empirical estimate, Mantegna proposed using hierarchical clustering. In this work, we first show that this procedure is statistically consistent. Then, we propose to use clustering with a much broader application than the filtering of empirical covariance matrices from the estimated correlation coefficients. To be able to do that, we need to obtain distances between the financial time series that incorporate all the available information in these cross-dependent random processes.

Chapter 9

We consider the geometry and model order specification of a class of density models where the square-root of the distribution is expanded in an orthogonal series. The simplicity of the resulting spherical geometry makes this framework ideal for many applications that rely on information geometric concepts like distances and manifold statistics. Specifically, we demonstrate applications of these models in the computer vision field of object recognition and retrieval. We illustrate how invariant shape representations can be used in conjunction with these probabilistic models to yield state-of-the-art classifiers. Moreover, the viability of formulating classification models that take into account shape deformation in an optimal transport context are investigated, yielding insight into the practicalities of working with the parameter space of the densities versus the Wasserstein measure space approach. The free parameters associated with these square-root estimators can be rigorously selected using the Minimum Description Length (MDL) criterion for model selection. Under these models, it is shown that the MDL has a closed-form representation, atypical for most applications of MDL in density estimation. Experimental evaluation of our techniques are conducted on one, two, and three dimensional density estimation problems in shape analysis, with comparative analysis demonstrating our approach to be state-of-the-art in object recognition and model selection.

Chapter 10

We propose a dimensionality reduction method for infinite—dimensional measure—valued evolution equations such as the Fokker–Planck partial differential equation or the Kushner–Stratonovich resp. Duncan–Mortensen–Zakai stochastic partial differential equations of nonlinear filtering, with potential applications to signal processing, quantitative finance, heat flows and quantum theory among many other areas. Our method is based on the projection coming from a duality argument built in the exponential statistical manifold structure developed by G. Pistone and co-authors. The choice of the finite dimensional manifold on which one should project the infinite dimensional equation is crucial, and we propose finite dimensional exponential and mixture families. This same problem had been studied, especially in the context of nonlinear filtering, by D. Brigo and co-authors but the L^2 structure on the space of square roots of densities or of densities themselves was used, without taking an infinite dimensional manifold environment space for the equation to be projected.

Here we re-examine such works from the exponential statistical manifold point of view, which allows for a deeper geometric understanding of the manifold structures at play. We also show that the projection in the exponential manifold structure is consistent with the Fisher Rao metric and, in case of finite dimensional exponential families, with the assumed density approximation. Further, we show that if the sufficient statistics of the finite dimensional exponential family are chosen among the eigenfunctions of the backward diffusion operator then the statistical-manifold or Fisher–Rao projection provides the maximum likelihood estimator for the Fokker Planck equation solution. We finally try to clarify how the finite dimensional and infinite dimensional terminology for exponential and mixture spaces are related.

Chapter 11

This paper addresses the problem of learning online finite statistical mixtures of regular exponential families. We first start by reviewing concisely the gradient-based and stochastic gradient-based optimization methods and their generalizations. We then focuses on two stochastic versions of the celebrated Expectation-Maximization (EM) algorithm: Titterington's second-order stochastic gradient EM and Cappé and Moulines' online EM. Depending on which step of EM is approximated, the possible constraints on the mixture parameters may be violated. A justification of these approaches as well as ready-to-use formulas for mixtures of regular exponential families are provided. Finally, to illustrate our study, some experimental comparisons on univariate normal mixtures are provided.

MIX
Papier aus verantwortungsvollen Quellen
Paper from responsible sources
FSC® C105338

If you have any concerns about our products,
you can contact us on
ProductSafety@springernature.com

In case Publisher is established outside the EU,
the EU authorized representative is:
Springer Nature Customer Service Center GmbH
Europaplatz 3, 69115 Heidelberg, Germany

Printed by Libri Plureos GmbH
in Hamburg, Germany